The Most Secret Quintessence of Life

The Most Secret Quintessence of Life

Sex, Glands, and Hormones,
1850–1950

Chandak Sengoopta

The University of Chicago Press
Chicago and London

Chandak Sengoopta is a senior lecturer in the history of medicine and science at Birkbeck College, University of London. He is the author of two previous books, including *Otto Weininger: Sex, Science, and Self in Imperial Vienna,* also published by the University of Chicago Press.

The University of Chicago Press, Chicago 60637
The University of Chicago Press, Ltd., London
© 2006 by The University of Chicago
All rights reserved. Published 2006
Printed in the United States of America

15 14 13 12 11 10 09 08 07 06   5 4 3 2 1

ISBN (cloth): 0-226-74863-4

This book was published with the help of a grant from the Isobel Thornley Bequest to the University of London.

Library of Congress Cataloging-in-Publication Data

Sengoopta, Chandak.
    The most secret quintessence of life : sex, glands, and hormones, 1850–1950 / Chandak Sengoopta.
        p.   cm.
    Includes bibliographical references and index.
    ISBN 0-226-74863-4 (cloth : alk. paper)
    1. Hormones, Sex—Research—History—19th century.   2. Hormones, Sex—Research—History—20th century.   3. Hormones, Sex—Physiological effect.
    4. Reproduction—Endocrine aspects.   5. Aging—Endocrine aspects.
    6. Gonads—Physiology.   7. Sex (Biology)   I. Title.
    [DNLM: 1. Gonadal Hormones—history: 2. Gonads—physiology.
    3. Aging—physiology.   4. Endocrinology—history.   5. History, 19th Century.
    6. History, 20th Century.   7. Sex. WK 11.1 S476m 2006]
    QP572.S4S46   2006
    612.6–dc22
                                                                                    2005027321

♾ The paper used in this publication meets the minimum requirements of the American National Standard for Information Sciences—Permanence of Paper for Printed Library Materials, ANSI Z39.48-1992.

Jane Henderson, her book

Contents

Acknowledgments

I have worked on this book for more than a decade and at four institutions. Begun at the Johns Hopkins University, the project was continued at what used to be the Wellcome Institute for the History of Medicine and then at the University of Manchester, finally being completed at Birkbeck College. Although I doubt if this prolonged and peripatetic gestation improved the quality of the work, the number of people who have nurtured it over the years is very large indeed. In its earliest form, the project was warmly encouraged by Dan Todes and he has continued to do so with undiminished warmth over all these years. Also at Johns Hopkins, Gert Brieger, Harry Marks, Ed Morman, Susan Abrams, and Vera Cecilia Machline provided enormous help and support, as did Dolores Sawicki, Linda Bright, and Coraleeze Thompson. Chris Lawrence sponsored me for a Wellcome Trust postdoctoral fellowship on the history of endocrinology, and, since then, the Wellcome Trust has generously supported my academic career and research with a succession of grants. My thanks to David Allen, John Malin, Tony Woods, Helen Hawkins, and Henriette Bruun of the Trust's History of Medicine Grants Office for all their cheerful help.

At the old Wellcome Institute, Janet Browne, Bill Bynum, Rhodri Hayward, Stephen Jacyna, Andrea Meyer-Ludowisy, Michele Minto, Michael Neve, and Andrew Wear all helped in ways that I could not possibly list. Vivian Nutton was an admirable chief, always supportive and collegial, but never too pointedly eager for progress reports. Lesley Hall's erudition and interest in the project were invaluable; and Natsu Hattori and Sally Bragg supplied endless encouragement spiced with ruthless wit and not a little wisdom. Roy Porter not only asked some of the most astute questions about the project that I have ever faced (while claiming not to know anything about the subject) but also published in *History of Science* an article that was to become the nucleus of the earlier sections of the book. He also cheerfully wrote references for grants, gave wise advice (usually in the form of jokes), and became a veritable godfather to this project as well as to its author. His untimely death has left a gaping void in the field of medical history as well as in my own life.

At Manchester, John Pickstone prodded me with characteristic energy to finish the manuscript, but the rest of the Manchester crowd—especially Jon Agar, Julie Anderson, Roberta Bivins, Ian Burney, Peter Coventry, Jeff Hughes, Vladimir Jankovic, Gillian Mawson, Helen Valier, and Michael Worboys—never uttered the word "glands" and thus earned my undying gratitude. Sam Alberti supplied an important reference but was otherwise delightfully uninterested in glandular matters. At Birkbeck, Fred Anscombe, John Arnold, Sean Brady, Harry Cocks, Emma Dench, Filippo de Vivo, Leoncia Flynn, Ian Haynes, Matthew Innes, Dave Jones, Mark Mazower, Julian Swann, and Jan Rüger have been wonderful colleagues and friends, while Joanna Bourke and Michael Hunter have gone beyond the call of duty to help out with crucial prepublication tasks and negotiations. My thanks to them all, to the School of History, Classics, and Archaeology, and to the College for providing the most exciting intellectual environment I have experienced in a long and checkered career. My warm thanks to Roger Cooter for facilitating the publication of this book (and my return to London) in every conceivable way and for never asking me the "Cooter Question" at any of the talks I have had to give in his presence. Heiko Stoff provided much intellectual stimulation; his own research was an inspiration to me in the later stages of this project. Adele Clarke, Anne Fausto-Sterling, David Hamilton, Diana E. Long, Ornella Moscucci, and Saffron Whitehead were similarly helpful and their own research as inspirational.

But for a question Punam Zutshi once asked me, I may never have become a historian of medicine; over the years and across the seas, she has always been ready with help, advice, and conversation, whether about nonsense poetry, academic politics, or, indeed, about the "Great Gland Book." At the University of Chicago Press, Doug Mitchell showed his usual warm interest in the project, even when he was coping with a tragic personal loss, and his assistant Tim McGovern was unfailingly helpful. My warm thanks, too, to the helpful comments and questions of the Press's two anonymous referees.

Claudia Rex and Richard Allen shepherded my first book through press some years ago, and it has been an unmitigated pleasure to work with them again. Richard's editorial suggestions and queries did more to improve this book than any author has the right to expect.

I have benefited from the resources of countless libraries; the following, however, are the ones I drew upon most extensively: the John M. Olin Graduate Research Library (Cornell University), the William H. Welch Library (the Johns Hopkins University School of Medicine), the Milton S. Eisenhower Library (the Johns Hopkins University), the New York Public Library, the Wellcome Institute for the History of Medicine Library (now the Wellcome Library for the History and Understanding of Medicine), and the Contemporary Medical Archives Centre, the John Rylands University Library (University of Manchester), the Manchester Central Reference Library, the Radcliffe Science Library (University of Oxford), the British Library, the British Newspaper Library (Colindale), and the University of London Library. A special word of thanks is due to Joanne Crane, the Store Supervisor of the John Rylands University Library, Manchester, who retrieved volume after dusty volume of old medical journals from the basement of the library without the slightest irritation. The Historical Collections of the New York Academy of Medicine provided much material; I thank the academy for funding my visit to New York with a Paul Klemperer Research Fellowship, and Bhavani Balasubramanian, Ed Morman, Lois Fischer Black, Lilli Sentz, Leonore Tiefer, and Caroline Duroselle-Melish for their help and friendship during an exciting month in Manhattan. I am indebted to Eva-Maria Hesche of the Jüdisches Museum Hohenems and to Dr. Hubert Steiner and Dr. John Adler of the Österreichische Staatsarchiv for their generous help in locating unpublished biographical information on Eugen Steinach. My thanks also to the Schering Museum, the Nobel Committee, and the Rare Book Collection of Columbia University. My sincere thanks to the Isobel Thornley Bequest Fund for granting me a subsidy toward the production costs of this volume and especially to its Secretary, Ms. P. K. Crimmin for her patience and collegiality. James Lees of the Institute of Historical Research, University of London, was also of great assistance in this regard. Earlier versions of some sections of this book were published as articles in *History of Science, Isis,* and *Medicina nei Secoli*—my thanks to their editors for allowing me to incorporate revised versions of those articles in this book.

Across the planet, my sister Anuradha continued to be a source of hope and courage, as did old friends such as Aniruddha and Zena Deb, Bappaditya Deb, Kanika and Paritosh Mitra. I cannot say whether this project, which began long before I met Jane Henderson, has been touched by some of her spirit, wit, and vitality but those qualities have certainly revolutionized my own existence, and that revolution has been assisted immensely by Harry and Sally. I dedicate this book to her with admiration, gratitude, and all my love.

This book incorporates material previously published in the following articles. The author is grateful to the journals and their publishers for permitting the use of this material.

"Transforming the Testicle: Science, Medicine and Masculinity, 1800–1950," *Medicina nei Secoli Arte e Scienza* 13, no. 3 (2001): 637–55.

"The Modern Ovary: Constructions, Meanings, Uses," *History of Science* 38 (2000): 425–88.

"Glandular Politics: Experimental Biology, Clinical Medicine, and Homosexual Emancipation in Fin-de-Siècle Central Europe," *Isis* 89 (1998): 445–73. ©1998 by the History of Science Society. All rights reserved.

How does the organism work as a whole? What organizes and coordinates the functions of the diverse parts of the body? The ancient answer was sympathy, especially for situations where organs with no obvious connection affected one another, such as the characteristic flush of the cheeks in tuberculosis of the lungs. Sympathy could, obviously, be mediated through the nerves or the blood. Galen allowed for both, but by the seventeenth century the nervous connections came to be accorded paramount importance.[1] The concept of sympathy remained of importance until the nineteenth century, and the nervous system—the brain, of course, but especially the spinal cord, the ganglia, and the innumerable nervous fibers traversing every tissue and organ—represented the communicative and integrative network of the body.[2]

By 1900, however, this solidistic conception had given way to something rather more complex: the body was now seen to be governed partly by nerves but partly, and in some respects more importantly, by blood-borne substances secreted by certain organs called the ductless glands or endocrine glands or glands of internal secretion.[3] By the early twentieth century, research on endocrine glands had progressed too far for any serious physiologist to deny the importance of the internal secretions (named hormones in 1905 by the British physiologist Ernest Starling), even though their nature and precise mechanisms of action were to remain unknown for many years. As the doyen of British physiologists and pioneering endocrine researcher Edward Schäfer (1850–1935) declared in 1931, "the Old Physiology" had been based on the concept that bodily functions were regulated by the nervous system while the "New Physiology" emphasized the chemical regulation of the organism.[4] The discovery of the internal secretions displaced solidist theories of bodily regulation, bringing about a humoralist orientation, although, as we shall see, it was not a case of one Kuhnian paradigm being succeeded, in one Gestalt switch, by another.[5] Nevertheless, the concept of internal secretions pushed medical thinking in bold new directions. The body was no longer a solid, relatively unchanging entity—the stability of the organism was real, but it depended on a balance of secretions. Change that balance and the characteristics of the organism could change. The humoral body was always potentially malleable, offering unlimited possibilities for what one may call physiological engineering. Few areas of medicine were left untouched by the therapeutic optimism encouraged by such a view, but the greatest attainments of humoral medicine, its champions predicted, would be in the arena of sex.

This book explores the origins, contexts, and consequences of the humoral approach to sex and related areas. One of its primary aims is to chart some of the complex ways in which scientists and the general public came to regard the testicles and the ovaries not merely as sources of gametes but also as secretors of potent chemicals determining not only the narrowly sexual aspects of our existence but also our very beings and lives.[6] "Just as the sex glands are situated in the middle of the body," the physiologist Eugen Steinach once declaimed, "so too are they at the hub of life itself."[7] The American physician and popularizer Henry Smith Williams paraphrased George Du Maurier's novel, *Trilby*, to illustrate the preeminence of the gonads within the endocrine system: "'In ze one class, Trilby, in ze ozzer, all ze ozzer singers'... Well, the glands are like that. In the one class, the gonads... in the other, all the other organs of the body." The gonad was "the star performer."[8] *The Most Secret Quintessence of Life* shows how the gonadal secretions became fundamental to sex and how sex itself became a virtual synonym for life itself.

The word "sex," of course, is merely a convenient piece of shorthand. For the discourses I shall be examining, "sex" generally meant masculinity or femininity as manifested in nongenital characteristics—the attributes called secondary sexual characters by biologists since the time of Darwin—but also in some other, more ambiguous phenomena not patently connected with the body or even necessarily with sex. Take the supposed dullness of the aging mind, for instance—this, according to many experts of the early twentieth century, resulted from a decline in the secretory functions of the gonads. The link between aging and desexualization is, of course, so ancient as to be an eternal verity. In the early twentieth century, however, the aging organism was redefined as one suffering from a deficiency of certain glandular secretions. If the sex glands of the senile could be revitalized, then the aging individual would regain some, perhaps most, of the energy, muscularity, creativity, and enthusiasm of youth.

In spite of such ambiguous elements, my story, for the most part, is about a transformation in concepts of the body. But not just a story of "the" body or even the "humoral" body—it is, rather, about those elements of the body that manifest masculinity or femininity, which I collectively call "the sexual body."[9] It includes everything embraced by the notion of the secondary sexual characters: sexual appearance, sexual attractiveness, sexual function, and all that might be related to those attributes. It goes beyond the standard notion of the secondary sexual characters to include other phenomena and processes not traditionally thought of as sexual but defined as sexual in the glandular discourse at certain times, such as the amalgam of energy, muscularity, and virility we have just discussed. This "sexual" body, of course, is also the *gendered* body, and although I do not use that phrase, gender plays a major role in my analysis.

The following pages chart some of the complicated ways in which this body came to be seen as produced and maintained by the internal secretions.[10] I begin around the middle of the nineteenth century, when, despite occasional prescient speculations and experiments, the reigning model of biological regulation was solidist and neural. It would, however, be a mistake to imagine that the gonads were of secondary importance in this solidist concept of the body. It had, of course, been known since antiquity that they were involved in reproduction, and this involvement was further clarified in the nineteenth century. Also in the nineteenth century, and long before the emergence of the humoral model of gonadal function, the ovaries were used to evolve new theories of the female organism and an interventionist therapeutics of "female" disorders. All of this was sustained entirely by the neural model: the ovary was seen as a neural node affecting innumerable aspects of female physiology. Issues of gender are of fundamental importance here—for, in this period, one looks in vain for even remotely equal interest in the testicles. This apparently barren period of research led, in intriguing and unexpected ways, to the demise of the solidist model on which it was based and was directly responsible, in several important ways, in generating the endocrine hypotheses of ovarian function.

I turn then to the dawn of the humoral age. By the end of the nineteenth century, the endocrine functions of the gonads, other glands (such as the thyroid or the adrenals) and secretory tissues (such as the duodenal mucosa) were established or, at least, strongly suspected. In 1905, all of these secretions—about the chemical nature of which nobody knew anything—were christened "hormones" (from the Greek for "excite" or "arouse") by the British physiologist Ernest Starling, on the advice of a Cambridge classicist. The last years of the nineteenth century had also seen the coming of the renowned organotherapy of Charles-Edouard Brown-Séquard, an immensely popular approach to the treatment of virtually every kind of hitherto incurable disease with extracts from glands or tissues supposed to be causing them. Though it can be (and was) commonly derided, organotherapy did inspire much pathbreaking research into glandular physiology. Brown-Séquard's own organotherapeutic system was quickly shown to have been simplistic and inaccurate, but its underlying vision of a chemically governed body was endorsed so powerfully by experimental research that to this day professional endocrinologists regard the elderly French scientist's dubious self-experimentation to mark the inauguration of modern endocrine science.

The golden years of sex gland research were the 1920s. This was when the testicles (and, to a lesser extent, the ovaries) moved rapidly to the forefront of science as well as popular culture. By the 1920s, the glands were seen to possess virtually miraculous powers, not simply over the narrowly sexual aspects of life or behavior but over the entire body and mind. Hardly anybody interested in explaining sex or gonadal function referred to the nervous system any more. This glandular boom was largely due to worldwide interest in and debates over glandular rejuvenation. I show how rejuvenation was much more than a passing frenzy over "monkey glands": it revolutionized the image of endocrine science, and studying its history challenges some of our conventional ideas about the relationship between laboratory, clinic, and larger society.

It was, however, also during these years of their greatest fame that the sex glands, as I show in the following chapter, began to lose their hegemony over sex and life. Up to the early 1920s, most investigators operated on the presumption that males were males because they possessed testicles which produced the male sex hormone which generated and sustained masculine sexual attributes or, to quote a famous aphorism we shall encounter later, "woman is woman because of her ovaries." From the later years of the twenties, however, such views became harder to sustain. For physiologists and biochemists at any rate, the humoral economy had become definitively ambisexual by the end of the 1930s. The secretions of the sex glands were no longer sex-specific. Males, it was revealed, produced female hormones in significant amounts and females were equally rich producers of male hormones.[11] Just as nineteenth-century medical scientists had come to terms with the universal absence of absolute morphological distinctions between males and females, endocrinologists, too, were compelled to put up with hormonal "ambisexuality."[12] It was the balance of sex hormones that was now held to "make" an individual male or female; to make matters even more complicated, that hormonal balance was produced by a range of glands rather than by the testicles or the ovaries alone. Although details remained obscure, most authorities had, by the end of the thirties, come to regard the gonads as the solo instruments in the concerto of sex, essential but not sufficient in themselves to endow the body with virility or femininity. Other instruments had to work in harmony with them under the baton of a conductor: the enigmatic pituitary gland.

The genesis of the sexual body, then, looked very different in the 1930s from what scientists in, say, 1910 had considered it to be. But this is only half the story and the more striking half, already dealt with by scholars such as Nelly Oudshoorn. There is a more mundane, but perhaps equally instructive part, however, that is almost always overlooked. During the same period, hormones ceased to be the playthings of laboratory physiologists. Thanks to the activities of laboratory-bound biochemists, the hormones of the testicles and ovaries became available as pure, potent extracts or in synthetic form. Clinicians jumped into this exciting new field and there was an explosion in therapeutic experimentation with sex hormones. How important was the new, complex model of the hormonally ambisexual body to these therapeutic schemes? As we shall see, the clinical use of sex hormones did not necessarily follow the reformulation of sexual physiology that had come about during the 1930s. There *were* a few conditions and circumstances when men were treated with female hormones and women with male hormones. On the whole, however, theories of hormonal bisexuality did not really revolutionize the clinic, and I fully agree with Julia Rechter's conclusion that "laboratory breakthroughs did not necessarily dictate the course of clinical work."[13] Sex hormones continued to be used purely empirically or in the light of the older, sexually polarized models of gonadal function.

Whatever the complexities of the new humoral biology of sex, it was, at least, humoral through and through. Solidism was dead. Or so it seemed for some years. In less than a decade, however, the physiology of sex was complicated even further by the return of the nervous system—but in an intriguing new incarnation. The whole endocrine apparatus of sex was found to be ruled not, as it had so far been imagined, by the pituitary gland, but by the hypothalamus, a set of neural centers in the brain, which governed the pituitary and thereby governed all the endocrine glands ruled by the pituitary. The hypothalamus was a component of the central nervous system, but its hegemony over the endocrine system was not a neural, solidist one. Its commands were issued to the pituitary not as electrical impulses but as chemical messages. It was, in short, a part-time gland: a Janus-faced mediator between the nervous and the endocrine systems. After nearly a century of research, then, the sexual body came to be seen as humoral as well as solidist. Neither circulating fluids nor solid structures ran the body by themselves but both were essential. The concept of the neurohumoral body certainly revolutionized the understanding of physiology, but what, if anything, changed in the clinic? And here, as we shall see, there was again no clear impact. Humoral medicine continued to thrive in the field of practical therapeutics and, indeed, received fresh impetus from advances (such as the contraceptive pill) which fall outside the boundaries of my story. The hypothalamus played no role in such clinical innovations.[14]

The Most Secret Quintessence of Life, then, is not simply concerned with sexual *endocrinology*. It deals with the history of endocrine hypotheses of sex gland function but not exclusively. It seeks to analyze how the testicles and the ovaries (or their secretions and their synthetic substitutes) were harnessed into diverse and often divergent *kinds* of scientific and clinical service in the hundred years between 1850 and 1950. There are, however, other themes that I have tried to integrate into this basic narrative. Perhaps the most important of those is what one may call the plasticization of the body: much of the messianism associated with glandular science stemmed from its message that the body was modifiable—significantly, if not infinitely, and in relatively painless ways. The new body, at the physiological level, was far more complex than the older, pre-glandular one, but it was also far more plastic, far more modifiable. The grounds for such modification were, of course, shaped in large measure by cultural, moral, and social imperatives—and I devote considerable attention to those forces—but the means were dazzlingly modern and scientific.

Another theme that I have sought to highlight is the question of the laboratory's relation to the clinic. Until the 1920s, much laboratory work on the endocrine glands was conducted by clinicians, and their experimental research, expectedly enough, was driven by a fundamentally clinical agenda. In 1916, the Viennese physician Wilhelm Falta found it lamentable that experimental investigations of the ductless glands had not kept pace with the clinical use of their extracts. He was not unappreciative of the research the laboratory scientists had conducted but emphasized that "the deficient knowledge of the chemical nature of the active substances given off by the ductless glands constitutes the weak point in the knowledge of the internal secretions." The physiological study of the endocrine glands was "still," he concluded, "in its childhood." Falta defined the tasks of physiology in patently clinical terms. The physiology of the future, he hoped, would provide "a sharp definition of the disease picture to be expected," measure the levels of circulating secretions, determine whether their levels were normal, diminished, or elevated, and, finally, isolate the endocrine substances in forms pure enough to be used for treatment.[15]

What kind of influence did this early clinical orientation of glandular science have on its doctrines and their development?[16] Neither primarily about the laboratory nor exclusively about the clinic, *The Most Secret Quintessence of Life* tries to illuminate the passage between the two without claiming to have explored this interface exhaustively. I offer a series of often contrasting case studies rather than the kind of focused, archive-driven study of a few sharply localized themes that has become the virtual norm for scholarly monographs in the history of science. If I have deviated so remarkably from that norm, it is not because of any lack of respect for it. A broader, more synoptic and largely publication-based approach was compelled by my overall intent to show the *diversity* of uses—intellectual, therapeutic, social, cultural—to which the gonads were put in earlier periods. I have also intentionally avoided confining myself to any one national tradition. This is not in quest of some ideal of total comprehensiveness. The aim, again, is to bring out the intellectual, cultural, and professional diversity of research and application. Certain themes, nonetheless, receive rather more extended consideration than others, such as German and Austrian research or the work of clinical gynecologists. These asymmetries are not unintentional. It is generally recognized that Central European research on the sex glands was crucial to the elaboration of endocrine hypotheses of sex everywhere, as were the contributions of clinical gynecologists. And yet, the Anglophone historiography of endocrinology or sexual biology does not really explore these themes in sufficient detail. I have tried, no doubt imperfectly and incompletely, to recover some of that history.

Much that could be of relevance to my story, however, is not in fact covered in this book. Obviously, I disregard those themes that other scholars have dealt with in depth. There is, for instance, nothing about the institutional history of reproductive biology and its multiple roots in the policy of funding bodies, research institutes, and ideological concerns—Adele Clarke's brilliant study of those issues made them less of a priority for my investigations.[17] Perhaps the most striking omission in this book is that of the industrial connection. Much of the laboratory-clinic interaction in the history of glandular science was regulated by the demands, resources, and priorities of the pharmaceutical industry, and although I hope that I have recognized the importance of this theme, I must leave it to other scholars to address it in requisite depth—as

Nelly Oudshoorn, for example, has done with regard to Organon and Ernst Laqueur's group of endocrine scientists.[18]

I suppose I need not apologize to historians for having no interest in evaluating past theories and practices in the light of current medical knowledge, but a word of explanation may be in order for any clinicians or endocrinologists reading this book. Many of them—and indeed, all readers with a basic knowledge of current biological science—would find the results of some of the experimental studies discussed in this book to be impossible and perhaps even fraudulent. A scientist could well write a book assessing the credibility of those experiments, but I am not a scientist and *Quintessence* is not such a book. All I would like to point out is that although many of those studies were deeply controversial at the time, none of their critics, to my knowledge, charged their authors with deliberate deception. It would be absurd, of course, to deny that the magical aura surrounding early glandular research was generated at least in part by the suggestibility of scientists, patients, doctors, and readers; I am more concerned, however, with investigating what such beliefs led to than in trying to explain why people believed such things. I am less interested in evaluating the truth of past scientific experiments than in exploring the ways in which the results of those experiments (no matter how disputable by today's criteria) came to be endowed with cultural meanings, and the ways in which those interpretations influenced concepts of the body and the medical practices performed on it. My emphasis here is not on how today's science of endocrinology was born. My primary aim is to explore *why*, *how*, and *where* theories of gonadal function were elaborated and used between 1850 and 1950, not whether those theories were right or wrong by criteria that have been evolved since then. It may be worth noting, however, that although I have no interest in evaluating the glandular hypotheses of the past, I do imply—and sometimes emphasize—how endocrinologists' histories of endocrinology have tended, because of their present-centered approach, to disregard some of the historical complexities of glandular research. No endocrinologist today, for instance, has much interest in what an earlier age called rejuvenation. I argue, however, that the development of the discipline cannot be understood if, animated by today's scientific preferences, we ignore the themes that animated the scientists and clinicians of the past.

My historiographic debts to recent monographs and scholarly papers should be evident from the notes. As far as general works are concerned, two have been of especial assistance. One is Victor Cornelius Medvei's *History of Clinical Endocrinology*, a massive and encyclopedic volume covering every aspect of historical knowledge of the endocrine glands, beginning in prehistoric times and concluding with the present.[19] It covers all the endocrine glands, not simply the sex glands; and Medvei, a clinician writing for a clinical audience, does not necessarily approach the subject with the agenda of a professional historian. But it is *because* of those differences that I found his treatise so helpful and so instructive. It taught me how the history of endocrinology is viewed by the professional endocrinologist of our time. My own view is admittedly rather different, but without Medvei's treatise I may not even have known how to frame some crucial parts of my story.

I owe an even greater debt to Hans Simmer and his pupils for their extraordinarily detailed studies of the history of gynecological endocrinology, especially in Germany. Their work has taught me more than I could ever summarize—the innumerable citations on the following pages merely indicate the extent of my reliance on their work, not its depth. It is true, of course, that I do not write from the perspective of a clinical gynecologist and therefore have less interest than do Simmer and his students in charting the genealogy of current gynecological endocrinology. It is also true that I try to pay more attention to cultural contexts than they consider necessary, and, of course, I address issues such as rejuvenation, which they have not dealt with in their work. These differences are doubtless important, but they do not reduce my profound admiration for their scholarly rigor and matchless knowledge of the discourse of early endocrine research.

One pioneering study that has inspired and enlightened me throughout this project is Diana Long Hall's 1976 paper, "Biology, Sex Hormones and Sexism in the 1920s."[20] Its astute analyses of the conceptual underpinnings and cultural contours of early endocrinology have been of far greater assistance to the formulation of my book than the footnotes can reflect. Nelly Oudshoorn's *Beyond the Natural Body* I have already mentioned with regard to the pharmaceutical industry's influence on endocrine research, an important area of investigation where Oudshoorn remains the best-informed guide. Oudshoorn's work, of course, is not simply about the pharmaceutical industry. Its detailed analysis of the deconstruction of masculine and feminine polarity under the impetus of biochemical endocrinology has been of pathbreaking importance. I have tried to build upon her work and to extend it into clinical domains, showing how the deconstruction of gender proceeded at a different pace and to a different depth beyond the biochemical laboratory.

Another recent book from which I have learned much is Anne Fausto-Sterling's *Sexing the Body: Gender Politics and the Construction of Sexuality.*[21] Fausto-Sterling shows us how different systems, different substances (including but not limited to hormones), and different ways of interpretation have produced our notions of the sexual body. Her comprehensive approach to the question is more than I can aspire to; I have only told the story of the *glandular* sexual body and that only up to about 1950.

Finally, a note on nomenclature and related issues. The nomenclature of the sex hormones changed continuously over the period covered by this book and differed greatly across national boundaries. The concept of "estrogen," in the beginning, was a biological one: it was an agent that produced estrus in castrated animals. It was known by 1950 that there were three natural estrogens—estradiol, estrone, and estriol—and that these were produced by all the histological components of the ovary, including by the corpus luteum. Estradiol was the one used most often in treatment and estriol the one found in the highest quantity in the blood of the pregnant female. Names such as "follicular hormone" continued to be used in some contexts but were frowned upon as soon as it was appreciated that the ovarian follicle secreted a whole range of different hormones, including androgens, and that the human testicle, adrenals, and placenta produced ample quantities of the so-called follicular hormone.[22] To make matters worse, the primary sources I am concerned with often identified a hormone by a trade name—e.g., Progynon or Menformon. After the three "classical" estrogens had been identified, the chemical definition of estrogen became the more important one. Even that, however, became inadequate as various nonsteroidal natural substances were discovered with estrogenic actions, and a number of synthetic estrogens were produced in the laboratory. Of the later synthetic estrogens, stilboestrol was the most popular and the cheapest.[23]

Since the chemical identities of the hormones are not my concern, I adopt the crude but convenient practice of following the terminology used by participants, with occasional parenthetical explanations where required. Generally, readers should not have a problem in following the narrative if they can tell when a hormone being referred to was considered to be "female" (such as folliculin, estrogen, Progynon, Theelin) or "male" (as in testosterone, androsterone, testosterone propionate or, more generically, androgen). I use the older term "internal secretions" rather than "hormones" for the period of research before the chemical isolation of pure hormones, and intentionally imprecise phrases such as "glandular science" or "endocrine science" in preference to "endocrinology" for periods before endocrinology became a distinct discipline in its own right.

The Gonads before the Endocrine Era

The early-nineteenth-century body was governed by the nervous system. The nerves carried messages, connected different viscera, and modulated their functions. The gonads, too, were ruled by the nerves and influenced the body through them. Many of these nervous links remained assumptions rather than verifiable anatomical features—the supposed connections between the ovaries and the brain, of which we shall hear much in this chapter, were never really identified in clear, structural terms. This solidistic, neural paradigm, however, generated research and therapeutic interventions at a cracking pace for virtually the entire nineteenth century. Those interventions were diverse and wide ranging and frequently involved the ovaries and (to a much lesser extent) the testicles. The vast majority of clinicians found the neural model of gonadal function justified all clinical procedures quite satisfactorily.

To be sure, a few experimental indications of the endocrine function of the gonads *were* obtained around the mid-nineteenth century, but the experiments proved hard to replicate and their intriguing findings were forgotten for decades. In the late nineteenth century, the clinical vogue for organotherapy led to a resurgence of interest in the chemical secretions of glands, but even then orthodox physiologists and doctors remained chary of extending the idea of internal secretions to *gonadal* function. Physiologist Edward Schäfer, a pioneer in endocrine research, asserted as late as in 1895 that the gonads exerted their influence on the body "without doubt... through the nervous system." Only in 1907 did he acknowledge that it was "highly probable that it is to internal secretions containing special hormones that the essential organs of reproduction—the testicles and the ovaries—owe the influence that they exert on the development of the secondary sexual characters."[1]

Schäfer's personal trajectory was fairly representative of his times. Even those who had few doubts on the reality of internal secretions did not always regard the gonads as endocrine organs. The neural model of gonadal function seemed to answer most biological questions plausibly as well as generate therapeutic innovations that were considered efficacious. Even the turn-of-the-century experimental research of Emil Knauer on ovarian transplantation, which many historians as well as scientists have hailed for setting the seal of science on ovarian endocrinology, developed directly out of the neural paradigm. Even if we are interested *only* in the history of sexual endocrinology, we need to begin with the nineteenth-century concept of the gonads as crucial nodes in the immense network of nerves governing the organism.

"Woman is Woman Because of Her Ovaries"

"During the last twenty years perhaps no organ in the body has been so much written about as the ovary," remarked the gynecological surgeon Lawson Tait in 1883. "Yet," he continued, "much remains to be discovered. To the naked eye nothing could look more uninteresting and unimportant than a human ovary; and yet upon it the whole affairs of the world depend. As far as the individual owner of the gland is concerned—certainly for her comfort, and, if we take with it its appendages, for her life as well, it is the most important organ in her body."[2] Nobody in the nineteenth century knew more about the ovaries or was keener to find out more about them than gynecologists. Nelly Oudshoorn has rightly observed that it was in "the gynaecological clinic" that the ovary was established as "the seat of femininity."[3] Although gynecologists enshrined the ovary as the seat of femininity, they did not, however, conceive of the ovary's feminizing role as a chemical (i.e., hormonal) one.

One of the best illustrations of this nonendocrine concept of ovarian function is found in a midcentury gynecological essay by the young Rudolf Virchow (1821–1902). In January 1848, the twenty-six-year-old Virchow gave a lecture at a conference of the Berlin Society of Obstetrics.[4] The immediate occasion for this lecture was an epidemic of puerperal fever during the previous winter at the Charité hospital, where Virchow had been appointed prosector in 1847. Ostensibly, the lecture tried to outline a comprehensive "developmental history" of the puerperal state, which Virchow defined as lasting from menstruation to childbirth.[5] The lecture remains one of the most articulate expressions of the mid-nineteenth-century model of femininity, in which physiology and cultural imperatives were inseparably entwined.

Virchow spent considerable time discussing the cardinal function of the ovary, which, for him, was the production of the ova. He also believed, like many other medical scientists of the time, that menstruation was *caused* by ovulation.[6] What explained ovulation, therefore, would also explain menstruation. The ova, of course, ripened and were expelled only periodically from the ovary—but what determined that cycle? Was the ovulatory stimulus communicated through the nerves or the blood? This was the crucial question, and in order to answer it, he emphasized, one had to choose between humoralism and solidism. Virchow was renowned for his solidistic doctrine that disease originated in cellular structures.[7] Solidism, however, was more than a matter of attributing the origin of *disease* to solid structures. Virchow's views on the nature of the body and its functions, too, were solidist to the core. Just as he opposed theories of disease based on the humoral concept of dyscrasias, so also did he heap scorn on humoral concepts in physiology.[8] Unsurprisingly, therefore, he argued that if it was indeed the blood that carried the ovulatory impulse to the ovary, then one would have to decide whether the periodicity of ovulation (and menstruation) was caused by periodic changes in the composition of the blood itself or by some mysterious substance produced by an unknown organ and carried to the ovaries by the blood. Nothing, he asserted, was known so far about the first alternative and the second was implausible.

The most persuasive evidence against humoral theories of ovarian function, however, was a case of female Siamese twins. These twins had lived for twenty-two years with what was essentially a single circulatory system but nevertheless menstruated at different times. In the light of this case, it was far more likely, suggested Virchow, that ovulation was caused by periodic changes intrinsic to the ovary itself. What could be more natural than to assume that those periodic, rhythmic changes were mediated by the nervous system? The rhythmic processes of respiration and heartbeat and the contractions of the uterus during labor were all instances of neurologically regulated periodicity—it was simply logical to assume that ovulation and menstruation, too, were similarly regulated.[9]

The most intriguing aspect of Virchow's theory of ovarian function, however, had nothing to do with ovulation or menstruation. Although he did not spell out how the ovaries could affect the whole organism, he was in no doubt that they generated the female sexual body *and* many of the supposed mental qualities of femininity:

> The female is female because of her reproductive glands. All her characteristics of body and mind, of nutrition and nervous activity, the sweet delicacy and rounded-ness of limbs . . . the development of the breasts and non-development of the vocal organ, the beauties of her hair and the soft down on her body, those depths of feeling, that unerring intuition, that gentleness, devotion, and loyalty—in short, all that we respect and admire as truly feminine, are dependent on the ovaries. Take the ovaries away and we get the repulsive, coarsely formed, large-boned, mous-tached, deep-voiced, flat-breasted, resentful, and egoistic virago (*Mannweib*).[10]

Virchow's formulation would be quoted and paraphrased down the years, but, as he had been quick to acknowledge, he was not alone in seeing the ovary as the fountainhead of femininity.

Following the discovery of the ovum by Karl Ernst von Baer in 1827, the uterus had begun to lose its hitherto unassailable position as the defining organ of femininity.[11] Virchow himself quoted the Paris physician Achille Chéreau (1817–85), who, in his 1844 treatise on the maladies of the ovary, had already dismissed Jan Baptista Van Helmont's influential dictum that woman was woman because of her uterus (*propter solum uterum mulier est quod est*).[12] Chéreau had proclaimed, instead, that it was the ovary that de-termined woman's nature and her body—*propter ovarium solum mulier est quod est*—while the uterus was only an organ of secondary, purely reproduc-tive importance. The ovaries were responsible for menstruation, and it was ovarian dysfunction that caused such typically female disorders as hysteria and chlorosis. The Heidelberg anatomist Theodor Ludwig Wilhelm Bischoff (1807–82) had also declared in 1844 that every anatomist, physiologist, and clinician knew that *the* cause of menstruation, sex drive, and all feminine characteristics was to be found in the ovary; the uterus had only a secondary importance.[13] Chéreau and Bischoff had desisted, however, from speculat-ing on the mechanism of ovarian function, and even Virchow, in spite of his unequivocally neurological explanation of ovarian function, had nothing to say about the exact nature of the nervous links between the ovaries and the rest of the organism. Nor could he delineate the physiological processes by which nervous impulses could bring about and sustain "the sweet delicacy" of woman's form and spirit. Despite these problems, he did not consider it jus-tifiable to contemplate any nonneurological mechanism of ovarian function.

One should emphasize that Virchow's invocation of the eternal femi-nine was not simply a cloying and medically unimportant appendix to his neural theory of ovarian physiology. The neuro-gonadal genesis of femininity explained as well as supported the nineteenth-century conviction that sensibility was fundamental to the very definition of femininity. To use the words of Carroll Smith-Rosenberg and Charles Rosenberg, "few if any questioned the assumption that in males the intellectual propensities of the brain dominated, while the female's nervous system and emotions prevailed over her conscious and rational faculties."[14] The Rosenbergs refer to the American context, but, as Joachim Radkau has shown, the situation was not significantly different in German-speaking Europe.[15] Virchow's theory of ovarian function did not simply explain reproductive periodicity by a historically specific model of physiological regulation but placed the ovary and its putative physiology at the heart of a historically specific model of ideal femininity.

The Gland under the Knife: The Ovary as Nerve-Center

Virchow's views on the ovary and its role in generating and maintaining femininity were widely cited in future years—but not always in complete agreement. Later in the nineteenth century, in fact, the gland came to be regarded by many as the likeliest suspect for such "female" disorders as hysteria, nymphomania, ovarian neuralgia, or the imaginatively named oophoromania. True, the ovaries usually seemed to be "normal" in such cases, but outward normality, nineteenth-century physicians warned, was no assurance of inward health and physiological equilibrium; Virchow, some added, had overestimated the importance of the ovary to the female organism.[16] The removal of the ovaries—regardless of apparent normality—became a popular treatment for "functional" conditions as well as in demonstrably organic ailments. The latter ranged from osteomalacia to severe uterine bleeding and even late, inoperable breast cancer.[17]

Although the British surgeon Percival Pott had removed both ovaries of a patient suffering from a tumor involving both glands in 1775 and although the surgeon James Blundell had speculated about the usefulness of extirpating the ovaries in cases of dysmenorrhea in 1825, the actual operation of removing apparently healthy ovaries (as opposed to grossly diseased ones) was independently introduced only in 1872 by American surgeon Robert Battey, the German gynecologist Alfred Hegar (1830–1914), and the British gynecologist Lawson Tait.[18] The rationale for the operation was that many "functional female disorders" were accompanied by amenorrhea.[19] The psychological, somatic, and behavioral symptoms of those disorders, therefore, were interpreted as manifestations of the organism's frustrated effort to menstruate. It followed that they could be resolved if the source of the menstrual impulse—the ovaries—were to be removed.[20]

Although the oophorectomists did not usually clarify how the ovary caused menstruation, it is evident from their practice that they thought of menstruation as the outcome of a neurological process. In the "functional" disorders, the body's nerve force was supposed to be deranged due to amenorrhea; the removal of the ovaries would restore the balance by precipitating menopause. The female organism, historians have repeatedly demonstrated, was seen by nineteenth-century physicians as being particularly finely balanced.[21] Centered on reproductive functions, mature female existence—commencing at puberty and ending in menopause—was a succession of cyclically recurring biological processes and events: menstruation, pregnancy, parturition. This cyclicity was normal, but it was also the fundamental cause of female instability. Virtually all the energy of the female organism—and every individual possessed only a fixed quantum of vital energy in nineteenth-century physiology—was needed to sustain this cyclical rhythm. The balance, even in normal conditions, was a fine one and the slightest alteration in the natural rhythms could jeopardize the entire organic economy of the female. Matters became particularly critical when menstrual flow ceased but the body continued trying to produce it, resulting in a diverse range of symptoms ranging from the hysterical to the neuralgic.[22] The only cure for such symptoms, according to the practitioners of oophorectomy, was to remove the menstrual impetus itself by getting rid of the ovaries.

Not every clinician believed in the operation's utility. but nevertheless by 1906 about 150,000 women, according to one estimate, had their "normal" ovaries removed for a diverse range of illnesses.[23] One American surgeon who detested the operation remarked, "the woman who cannot show an abdominotomy line is looked upon as not in style, nor belonging to the correct set."[24] He was obviously exaggerating or perhaps even concocting an explanation that would condemn the operation while saving his professional brethren from explicit criticism. There is some anecdotal evidence, however, that occasionally, middle-class female patients, in the United States at least, demanded the operation from their gynecologists even when the latter were relatively reluctant to recommend it.[25] It is surely unlikely that oophorectomy was a consumer craze, but it is not prima facie impossible that some patients may have thought of it as a reliable cure for agonies that were only too real for them (even if we regard them as imaginary or socially determined today) and who may have requested it even if their physician had no particular faith in it.

There is much evidence showing that even surgeons championing the operation warned their colleagues against its indiscriminate use. The German gynecologist Alfred Hegar, a founder of the operation and a passionate believer in its efficacy, emphasized that it should be resorted to only after establishing unequivocally that the patient's physical symptoms were *caused* by ovarian pathology. The simple presence of pathological ovaries was not enough: the patient's symptoms had to be clearly linked with the ovarian abnormalities. "All operations which are undertaken without the presence of a disease or anomaly in the sexual system are, according to the present standpoint of our knowledge, unjustifiable," he declared. "The mere presence, however, of a pathological change in the genital system, as has commonly been held, is not sufficient, and a strict proof of the causative connection between that change and the nervous disorder has to be demanded."[26] The operation was not indicated in neurotic symptoms unaccompanied by any "morbid changes in the sexual canal."[27] Most of Hegar's indications for oophorectomy, indeed, were unrelated to nervous or mental symptoms and concerned only with such conditions as uterine fibromyomas, profuse uterine bleeding, or chronic, severe inflammation of the uterine tubes.[28] He operated in cases of nervous disorders only if they had not responded to a long-enough course of conservative treatment and only when accompanied by severe inflammatory conditions in the reproductive tract or serious menstrual irregularities, which, in the light of contemporary medical beliefs in the interaction of (female) mind and (female) pelvis, could legitimately be seen as having caused the nervous condition.[29]

Those who followed him, although far less cautious, still stressed that the removal of the ovaries should be considered in nervous disorders only if a local abnormality had been clearly identified in the reproductive system and unambiguously linked to the nervous or mental symptoms.[30] As an admirer of Hegar clarified, one could expect relief of hysteria after oophorectomy only if the hysterical symptoms were linked to menstruation and were caused by obvious abnormalities in the reproductive system.[31] There was, in fact, wide consensus that, while some cases of hysteria were due to lesions in the reproductive system, many were not and oophorectomy was not indicated in the latter group.[32]

The category of nervous disorder or "neurosis" was defined very broadly in the nineteenth century, a point that Thomas Laqueur overlooked in his remark that Hegar believed that one could exorcise "the organic demons of unladylike behavior" by removing the ovaries.[33] Neuralgias, cardialgia, cramps, vomiting, epilepsy or epileptiform convulsions, severe premenstrual discomfort of diverse kinds from coughs to back pain, and the kind of hysterical paralyses seen in the Salpêtrière clinic of Jean-Martin Charcot were *all* classified as neuroses, and in a significant proportion of cases were attributed to problems in the reproductive tract. The leading practitioners of oophorectomy would have considered only some of these cases to be suitable for the operation and only if other kinds of treatment had failed. The operation would not be indicated in instances of "unladylike behaviour" or even grand Charcotian hysteria unless there was a causal link to reproductive pathology.[34]

To say this is not, of course, to argue that ideas and ideals of femininity played no part in the conceptualization and use of the operation. Reproductive pathology and nervous disorders were linked by the neural model of the female organism, which was a cultural as well as a medical construct. The ideological power of the neural model stemmed not merely from patriarchal attitudes but from its flawless combination of cultural stereotypes with what was perceived as solid biological "evidence."

Pelvis, Brain and Nerves: The Ovaries and Female Neuroses

Hegar and his contemporaries regarded the ovary as a crucial node in the neural matrix of femininity. It seemed self-evident to them that when the ovary developed an irritative focus, the whole network, unstable even under normal conditions, was seriously jeopardized. Conversely, an irritation developing in some other part of the neural network might implicate the ovary, causing the physician to misidentify that gland as the origin of the irritation. "The nervous system," explained Hegar, "is so coherent a whole, that a transfer of the irritation, especially in the morbid state now under consideration, is very easily possible in all directions. . . . Further sympathetic and reflex influences soon come into play, so that the first starting-point is, it must be confessed, often not easily found out."[35] This implied, of course, that many nervous disorders, apparently unrelated to the sexual tract, might actually be linked causally with the latter, especially if the nervous symptoms first appeared "in the nerves starting from the lumbar cord."[36]

This idea was based on the theory of spinal irritation, which held that impulses from an irritative focus in the reproductive system could radiate to the spinal cord via the nerves, from where they might be disseminated all over the body, affecting diverse, apparently unrelated organs.[37] Despite his insistence on the need to exclude non-ovarian causes of female neurotic symptoms, Hegar and many gynecologists of the time subscribed, inconsistently, to the allied notion of reflex neuroses introduced by neurologist

Moritz Romberg in the 1850s, according to which the female reproductive system was the ultimate source of many "neurotic" symptoms, especially those labeled hysterical.[38] "Nervous" and mental disorders of women did not originate in the brain, let alone in the mind, but in the pelvis. Not every physician subscribed so wholeheartedly to Romberg's notion, but even the skeptics conceded that the majority of female nervous disorders was brought about by reproductive problems.[39]

Psychiatrists did not dissent too radically from the essential elements of the neural model. In his monograph-length treatise on hysteria in Ziemssen's authoritative handbook, Friedrich Jolly (1844–1904), Professor of Psychiatry at Strassburg and later successor to Carl Westphal at the University of Berlin, reminded his readers that hysteria was not an exclusively female disorder and hence could not be caused exclusively by female reproductive pathology. But although not the sole cause of hysteria, pelvic abnormality was undoubtedly an important one. Even in healthy women, menstruation and pregnancy were often accompanied by nervous phenomena, demonstrating the specific links between the female reproductive organs and the nervous system. In some hysterics, symptoms were significantly worsened by menstruation or pregnancy, and in many cases, the correction of genital disorders relieved the hysterical symptoms. It was clear that in all such cases, the pressure, inflammation, and irritation caused by local pelvic problems were transmitted to the central nervous system by the centripetal nerves of the sexual organs.[40]

Despite such affirmations of the essential validity of the link between the nervous system and the female reproductive tract, oophorectomies in "functional" cases fell into disrepute even among its enthusiasts towards the end of the nineteenth century. In 1880, the gynecologist August Rheinstaedter, while acknowledging that pelvic causes could indeed give rise to nervous and mental disorders in women, nevertheless complained that the rapid development of gynecology had led to a "hysteromania"—every woman suffering from a migraine or palpitations now believed that she had a uterine malady and could easily find doctors who would agree with her.[41] In 1905, the classic textbook of operative gynecology by Döderlein and Krönig dismissed the removal of normal or slightly pathological ovaries in cases of hysteria as a practice of the past.[42] Even they added, however, that although it was no longer tenable to assume a direct etiological relationship between genital pathology and hysteria, it would be wrong to dismiss any and every connection between the two. Chronic pelvic disorders of no great import in normal women could cause such severe, distressing symptoms in hysterics that surgery was often unavoidable. This surgery, however, should be conservative, corrective surgery, such as the correction of a uterine malposition.[43]

The passions of the female mind were *still* generated in the pelvis: all that had changed was that the therapeutic interventions involving the ovary were becoming less radical.

In support of their position, Döderlein and Krönig quoted the neurologist and psychiatrist Otto Binswanger, Professor of Psychiatry at the University of Jena, who, in a comprehensive 1904 monograph on hysteria, had emphasized that the prolonged treatment of gynecological lesions could actually worsen hysteria. Like Jolly twenty years earlier, Binswanger doubted whether local genital abnormalities could precipitate full-blown hysteria in the absence of a preexisting neuropathic disposition, which he conceptualized in the classic terms of degeneration theory, one of the master concepts of fin-de-siècle psychiatry.[44] He considered it indisputable, however, that the central nervous system could, on occasion, be affected adversely by inflammatory or degenerative conditions of the female pelvic organs. Brain, spinal cord, and pelvis were linked by the cerebrospinal and sympathetic nerves. Despite the new trend toward conservative surgery in gynecology and notwithstanding the obsolescence of the reflex neurosis model of hysteria, the essential link between brain and pelvis had yet to be severed.

Nerves and the Consequences of Castration

The pelvic organs were, of course, considered to be essential not merely for proper mental functioning but also, as Virchow had argued, for the maintenance of physical femininity. The oophorectomists, however, either ignored the possibility of generalized defeminization or denied that the removal of ovaries in *adult* life could bring that about. Hegar, for instance, dismissed Virchow's dictum of woman being woman because of her ovaries as too dogmatic and unsupported by empirical evidence. Refuting Virchow almost point by point, he expressed "very serious doubts whether the female type of configuration, the development of the breasts and of the external genitals, the tone of the voice, the peculiar mental tendency and mode of thought of the female are closely connected with the presence of the ovaries."[45] Elsewhere, he declared that although babies born without ovaries or with damaged ones occasionally developed into unfeminine women, this was far from the norm: masculinization was an exceptional occurrence.[46] Congenital absence of the ovaries, in any case, was frequently accompanied by such unrelated features as idiocy, cretinism, and global deficits in the skeleton—the rare instances of masculinization might, therefore, not even have anything to do with the absence of ovaries.[47] Conversely, striking degrees of masculinization could occur in the presence of apparently healthy ovaries.

Castration of adult animals (such as cows and bitches) stopped menstruation and brought about uterine atrophy but did not usually result in any typical changes in the sexual characters, Hegar pointed out.[48] The consequences of human oophorectomy were identical. The sex glands, Hegar speculated, did not engender the secondary sexual characters; instead, the gonads themselves and the sexual characters were *both* probably engendered by a "sex-determining force" (*geschlechtsbedingende Moment*), the origin and nature of which was completely unknown but which was likely to act through the sex glands in most cases. The gonads, therefore, were important as mediators of this sexualizing force, but once the full adult form had been attained by the organism, the removal of the gonads could not lead to any substantial alterations of the sex characters.[49]

Although candid enough to admit that science knew "little of the influence which castration exerts on the body as a whole, and on the nervous system," Hegar was confident that the removal of the ovaries had no serious effect except uterine atrophy and the cessation of menstruation and ovulation, which, of course, were the very goals of the operation. The possibility of pregnancy, too, was ruled out by the operation, although Hegar was strangely unwilling to do so in absolute terms. "Conception and even completion of pregnancy are possible," he asserted. "But the condition of the patient is such, that this eventuality is not even desirable." Desire for coitus and the ability to have intercourse, however, were retained after the operation, which, Hegar emphasized, was one of the most important differences between male and female castration. In castrated males, "the ability to perform sexual intercourse is almost always lost. . . . The retention of the ability to perform sexual intercourse [in castrated women] also explains the fact that the mental depression, melancholia, and even suicidal tendency, which occur in men after castration, are never observed in females."[50] One far more fervent proponent of the virtues of the operation even declared that for a man who did not want children, marriage to an oophorectomized woman would be the ideal Malthusian marriage, free from the impediments associated with artificial birth control.[51]

The ovary, then, was important to the oophorectomists as an important node in the neural web governing the female organism but not as the exclusive seat of femininity. When they did record their views on the nature of the ovarian influence, they invariably implicated the nerves. Take, for instance, the question of uterine atrophy after the operation. Some of them speculated that the removal of the ovaries terminated the regular nervous impulses they sent to the central nervous system, interfering thereby with reflexes originating in the latter, which normally ensured the proper nutrition of the uterus.

Others suggested that the atrophy could be due to the cessation of trophic nervous impulses sent directly to the uterus by the ovaries. The widely noted tendency to fat-deposition after oophorectomy was explained, among other hypotheses, by nutritional imbalances within the body due to vasomotor disturbances consequent upon removal of the ovaries.[52] It was nerves, nerves all the way—the notion of the ovaries producing *chemical* regulators of femininity was not even a theoretical option.

Not Just for Nerves: The Other Uses of Oophorectomy

Oophorectomy was not used merely for the treatment of intractable nervous afflictions but also in such life-threatening conditions as osteomalacia and breast cancer.[53] The theoretical foundations for these uses were weak but the operation was found empirically beneficial, sometimes much more so than in the functional nervous disorders.

Osteomalacia: The Ovary and the Bones

Osteomalacia is a serious and progressive weakening of bones in adult life—the adult equivalent of rickets.[54] It was far commoner in the nineteenth century than it is today and encountered almost exclusively in women in their thirties.[55] The etiology was unknown and there was no effective treatment.[56] Although it progressed slowly and its course was punctuated by remissions, the patient eventually became bedridden and died of pneumonia or inanition. The condition was especially serious for women of childbearing age. It caused pelvic deformities that made childbirth difficult, and pregnancy itself exacerbated osteomalacia.[57] Pregnant sufferers of osteomalacia were often delivered by a modified caesarean section introduced by the Italian surgeon Edoardo Porro (1842–1902), which combined the caesarean delivery with the removal of the uterus and ovaries. The aim was to prevent future pregnancies. (One gynecologist of the time wondered why future pregnancies could not be prevented by simple tubal ligations, but his seems to have been a voice in the wilderness.)[58] Porro had not introduced his operation for osteomalacia alone but for all cases where it was necessary to prevent future pregnancies. Cases of osteomalacia, however, were often seen to improve after Porro's operation, which encouraged some surgeons to speculate that the removal of the ovaries was of direct benefit in the condition and should be performed even in nonpregnant patients.[59]

One of these surgeons was the gynecologist Hermann Fehling (1847–1925), an admirer and associate of Alfred Hegar, co-founder of the

Zentralblatt für Gynäkologie, and holder of successive chairs at Basel, Halle, and Strassburg.[60] Having obtained impressive results with Porro's operation in osteomalacia, Fehling removed the ovaries of a thirty-six-year-old patient who had been bedridden since the birth of her fourth child. By the time he reported on the case, she was fully mobile again and worked all day as a washerwoman.[61] Understandably, the report had much impact and soon other surgeons began to remove the ovaries in osteomalacia. By 1889, Fehling himself had performed seven bilateral oophorectomies for osteomalacia, and between 1887 and 1906 more than two hundred patients with osteomalacia were treated similarly by others.[62] The number of reported cures was high. Jutta Blönnigen has estimated that of all the reported cases, as many as 63 percent were "cured" and 22 percent were "improved." Moreover, recovery commenced rapidly after the operation and seemed to be sustained over long periods. Only in 6 percent of cases was there no effect at all.

On pathological examination, however, the majority of the removed ovaries did not present any conspicuous morbid features, and the proponents of the operation showed little interest in investigating how the ovaries brought about the bone deformities.[63] Fehling himself ignored the issue until the 1890s and then merely suggested that osteomalacia was probably caused by a morbid hyperfunction of the ovary, which, by stimulating the sympathetic nervous network, increased the flow of blood through the bones. (Only in 1900 did Fehling suggest that the ovarian hyperfunction might be a *secretory* rather than a neurological one.)[64] The hyperemia brought about demineralization of bones and oophorectomy exerted its therapeutic effect by removing the hyperactive ovaries. The notion of a hyperactive ovary was supported by Fehling's observation that patients with osteomalacia were remarkably fertile.[65]

Fehling's neural explanation of osteomalacia as a "trophoneurosis" of the bones was accepted by many practitioners. Even the few critics did not aim at the neural hypothesis before the very last years of the 1890s.[66] In 1893, when the endocrine era had already dawned in some sections of the European medical world, Franz von Winckel criticized Fehling's operation from a perspective that was anything but endocrine. Von Winckel argued that although osteomalacia was often worsened during menstruation, it also flared up in its absence, as during pregnancy and lactation. While it was true that the ovaries were not completely inactive during pregnancy, it seemed logical to him to assume that they were not as active as during menstruation—if, then, osteomalacia, was caused by the ovaries, then one would expect it to *improve* during pregnancy and lactation, which was the very reverse of what actually happened. Oophorectomy, therefore, probably worked in osteomalacia simply by reducing the sensitivity of the muscles and the membrane covering the bones, rather than by removing the fundamental cause of the disease.[67] Others suggested that osteomalacia was caused by bacteria—there was even a serious suggestion that the curative effects of oophorectomy were due simply to the chloroform used for anesthesia, which killed the nitrifying bacteria responsible for the condition.[68]

When endocrine notions did appear in explanations for osteomalacia, they did not enter from the direction one might expect. In 1897, ovarian extracts were introduced for the *treatment* of the condition! This, of course, was diametrically opposed to Fehling's rationale for oophorectomy but was justified with the argument that it was *abnormal* ovarian secretions that caused the disease (hence the benefits from the removal of the ovaries) and the only logical way to treat it was with extracts from *normal* ovaries.[69] The other endocrine glands also came to be implicated and, finally, the disease was attributed to pluriglandular deficiencies.[70] Nevertheless, reports of bilateral oophorectomy in osteomalacia continued to be published almost until 1920 and the procedure continued to be recommended in textbooks as a last resort.[71] As an early exponent had put it pithily, however, oophorectomy in osteomalacia— in spite of all hypotheses, endocrine or neural—remained a purely empirical treatment and the mystery of the disease, in some ways, had actually been deepened by the success of Fehling's procedure.[72]

Cancer: The Ovary, the Breast, and the Surgeon

In 1896, George Thomas Beatson (1848–1933), Visiting Surgeon to the Glasgow Cancer Hospital and a former house surgeon to Lord Lister at the Edinburgh Royal Infirmary, published an account of his treatment of three patients suffering from inoperable breast cancer.[73] Cancer of the breast, Beatson argued, was remarkably similar to lactation:

> We have, under both these conditions, the same proliferation of generations of epithelial cells which block the ducts and fill the acini of the gland; but in the case of lactation they rapidly vacuolate, undergo fatty degeneration, and form milk, while in the carcinoma they stop short of that process, and, to make room for themselves, they penetrate the walls of the ducts and the acini and invade the surrounding tissues. In short, lactation is at one point perilously near becoming a cancerous process if it is at all arrested.[74]

He had also been struck by reports that "it is the custom in certain countries to remove the ovaries of the cow after calving if it is wished to keep up the supply of milk, and that if this is done the cow will go on giving milk indefinitely." British farmers did not spay their cows, but they too negated the action of the ovary by encouraging regular pregnancies, during which, Beatson surmised (as had von Winckel in the context of osteomalacia), the ovaries were inactive or, as he put it cautiously, "we have not the indications of its activity in the shape of the menses." Menstruation, then, opposed lactation, but the secretion of milk, Beatson pointed out, was not controlled by any "special nerve-supply of its own" and could not be affected by interference with the sympathetic or the spinal nerves.

In the 1870s, Beatson decided to test his hypothesis on oophorectomized rabbits. He did not describe his experiments in much detail but summarized his results thus:

> As long as the young ones were at the breast [of the oophorectomized mother rabbit] the milk-supply continued, and when eventually they were taken away the milk-supply ceased: but the creatures increased very much in size, and post-mortem examination revealed that this was due to large deposits of fat around the various organs, and above all, in the lumbar region, where there were masses of pure adipose tissue, showing that the secretion of milk was still going on, but, not being discharged by the usual channels, was deposited in the various tissues of the body as fat.[75]

The only significant difference between lactating breast tissue and breast cancer was that the former continued to a stage "where the cells became fatty and and passed out of the system not only in an innocuous but nourishing fluid—milk." If, now, the suspension of ovarian influence (as during pregnancy) induced the secretion of milk, then would surgical removal of the ovaries stop the neoplastic process in breast cancer by compelling it to proceed to the fatty degeneration seen in lactation?[76] Beatson did not in fact test this hypothesis immediately, waiting instead for almost two decades. Not only was he reluctant to experiment on human beings, but the dramatic new science of bacteriology, he thought, would soon explain cancer as a result of infection.[77]

Finally, however, he relented upon receiving a patient who had already had a mastectomy but whose tumor had recurred, rapidly reaching inoperable status. At first, he treated the patient with thyroid extract in the hope that it might hasten the mucoid degeneration of the cells and thus affect the tumor. That did not occur and Beatson finally returned, with his patient's consent, to his old idea of oophorectomy. Thyroid tablets were continued after the operation but not because of strictly endocrinological reasons. Since the tumor was very large, Beatson feared that the lymphatic system might be overburdened after the rapid dissolution of the cancer cells—the thyroid would act as a "powerful lymphatic stimulant." The patient improved quite significantly and four months after the operation, Beatson noted, "the cancerous tissue has been reduced to a very thin layer" and eight months later, "all vestiges of her previous cancerous disease had disappeared."[78]

In his paper, Beatson described two other patients, both with tumors impossible to resect. In one of them, after oophorectomy and thyroid administration, microscopic examination of the tumor tissue showed extensive fatty degeneration of the epithelial cells.[79] There was no dramatic cure, however: "the disease," Beatson merely suggested, "is in a more quiescent stage." Five years and some other cases later, he reasserted his faith in the value of oophorectomy and thyroid extract for inoperable cancer: the results were sometimes so dramatic that once a patient's physician had exclaimed: "This seems like a romance!" Other surgeons had also begun to report similar successes with the procedure. Sometimes, Beatson suggested, supposedly "spontaneous" cures of cancer were actually attributable to menopause.[80]

What did he conclude from all this? First, that the ovaries were linked to breast cancer and perhaps to all malignancies in the female. More broadly, he began to doubt whether it was correct to attribute "the entire regulation of the metabolic changes in the tissues of the body" to the nervous system. "I am satisfied," he declared, "that in the ovary of the female and the testicle of the male we have organs that send out influences more subtle it may be and more mysterious than those emanating from the nervous system, but possibly much more potent than the latter for good or ill as regards the nutrition of the body."[81] These inscrutable "influences" of the gonads, he hypothesized, normally prevented the somatic cells from turning into undifferentiated, "primitive" germinal cells, capable of unchecked proliferation. (He added parenthetically that he "had never been sure" about August Weismann's absolute separation of the soma from the germ-plasm.) In conditions of "altered secretion" or "any morbid condition" of the ovaries, their inhibitory influence on the somatic cells were removed, enabling the latter to turn, in effect, into primitive germ cells. Cancer cells, Beatson suggested, "will eventually be shown to be special germinal cells corresponding to the ovum-cells elaborated by the ovary."

So, was the ovarian stimulus an *endocrine* secretion? Beatson, of course, lived through the heroic age of organotherapy. Organotherapy, as we shall see in the next chapter, was not necessarily founded in any sophisticated theoretical understanding of the endocrine system, but it patently regarded glands and tissues as producing secretions that had powerful effects on the rest of the organism. As we know, Beatson had himself used hyroid extracts in treating his inoperable patients. Nevertheless, Beatson's ideas on the nature of the ovarian contribution to cancer are difficult to interpret. "It may be an altered secretion," he surmised, "or it may be the migration of cells—it might even be a parasite in the ovarian cells, for it should be borne in mind in regard to the secretions of the reproductive glands 'that unlike other secretions, their essential constituents are living cells' (Stewart)."[82] Had Beatson intuited the endocrine functions of the ovary before the supposed "discovery" at the turn of the century? Although his hypothesis referred to ovarian "secretions" and argued against the exclusiveness of nervous regulation of the body and its metabolism, the secretions he referred to in his ambiguous sentences were the germinal secretions, as Hans Simmer has shown by tracing the original physiological text by George Stewart, from which Beatson had quoted.[83] Organotherapy notwithstanding, the ovary, for Beatson at least, remained a reproductive organ. Indeed, Beatson was even more old fashioned than Hegar, who considered the ovary, because of its neural links, as far more than a purely germinal organ.

Although one can, if one wishes, see premonitions of endocrinology in Beatson's comment about the ovary and the testicle exerting "influences more subtle it may be and more mysterious than those emanating from the nervous system," it is a matter of record that Beatson himself did not. It was only in 1905 that he referred to internal secretions and then only very casually, when he reported that in the light of recent reports about the corpus luteum producing "an internal secretion which causes the fixation of the embryo to the uterine wall and thus controls cell proliferation," he had unsuccessfully experimented with luteal extract in breast cancer.[84] No claim was made about the prescience of his earlier, speculative observations.

The next episode in the ovary-cancer saga was also British, but here, at last, we find a conception of ovarian physiology that even the least whiggish historian would have to acknowledge as essentially endocrine. The protagonist here is the surgeon James Stanley Newton Boyd (1856–1916), senior surgeon to the Charing Cross Hospital, who, ironically, seems to have been completely overlooked by the recent medical champions of Beatson's supposedly "endocrine" treatment of breast cancer. Building on Beatson's reports and on other instances of a "spontaneous" remission of breast cancer after menopause, Boyd presented five cases of his own in 1897.[85] Not all his cases were successful, but nevertheless Boyd confidently postulated a causal link between oophorectomy and the diminution of tumor size. More importantly for us, he hypothesized that "the internal secretion of the ovaries in some cases favours the growth of the cancer, acting either upon the epithelial cells or upon the surrounding tissues."[86] He wondered, incidentally, whether castration would have similar effects on cancers in men but thought it unlikely. Although the gonads developed from the same embryonic rudiments in the two sexes, "nothing," asserted Boyd, "can be clearer than that the ovary and testis differ in most respects."[87] (The nineteenth-century medical man treated ovaries as important but not *so* important as to be wholly immune from therapeutic manipulation or removal. The testicles, on the other hand, were treated as practically sacrosanct.)[88]

In 1900, Boyd presented an analysis of fifty-four cases of oophorectomy in breast cancer, compiled from his own work as well as from that of others.[89] Nineteen of these cases (34 percent) had improved after the operation, thirty four were unaffected or benefited only marginally, and one died of a cause unrelated to the cancer. Depressingly, however, it seemed that even in the cases responding to the operation, the cancer tended to recur within a year in the majority. Nevertheless, "a year or more of useful life seems often to have been gained." It was difficult to explain the wholly unresponsive cases and Boyd speculated that "certain ovaries, probably by pathological variation in their internal secretion, favour the growth of cancer . . . the removal of such ovaries alone will be of benefit." He had not, however, discovered any signs that might indicate this crucial difference, whether clinically or microscopically.[90]

Even though Boyd was undoubtedly thinking in endocrine terms, his hypotheses did not assess the importance of the ovarian secretions to the maintenance of the female sexual body. He mentioned very briefly that after oophorectomy, amenorrhea and atrophy of the uterus and the breasts were constant. Sexual feeling remained unaffected but occasionally, premature aging, growth of facial hair, or flushes were reported.[91] Boyd remained virtually oblivious to the great early-twentieth-century project of explaining the development and maintenance of sexual characters in endocrine terms. His interest in internal secretions of the ovaries was purely clinical, even surgical, and belongs to the history of surgery and cancer therapeutics rather than to the history of endocrinology. Knowledge of (or perhaps more precisely, speculations on) ovarian function followed no disciplinary line as yet and was used in ad hoc ways and to very different purposes across a broad range of medical thought and practice.

The Testicles in the Organic Economy

While the ovary, even when apparently normal, was harnessed to diverse clinical theories and procedures long before any clear knowledge of its endocrine importance had been acquired, the testicle, the symbol of virility for aeons, remained largely free from medical interference except in clear cases of intrinsic abnormality.[92] In the age of solidistic speculation about ovarian functions and invasive surgery based on such theories, the testicles and their possible role in generating, maintaining or vitiating virility received infinitely less attention. It is a well-known fact that as opposed to gynecology, a specific medical discipline devoted to males never developed in the nineteenth century. Even today's andrology is not exactly a prominent member of the medical constellation, and the very name of this new discipline was coined, according to some authorities, by a gynecologist![93] The sexual problems of the nineteenth-century male were taken care of by a variety of practitioners, ranging from elite general consultants to dermatologists. With childbearing ruled out by anatomy and organismal instability ruled out by ideological and physiological convictions, it is hardly surprising that there was no andrology in the nineteenth century.[94] Masculinity, most nineteenth-century thinkers and medical experts would have agreed, was only partly biological, femininity almost wholly so. Woman's body and woman's reproductive organs, consequently, belonged at the heart of medicine while the reproductive functions and appendages of man remained very much on the periphery.

Masculinity remained on the periphery but not, of course, entirely outside the domain of scientific thought. One issue that had long been of concern to investigators, for instance, was the result of castration. Aristotle had observed that male animals, if deprived of their testicles, "really change over into the female state."[95] This remained conventional wisdom for centuries. The British naturalist William Yarrell (1784–1856) provided this classic account of the feminization of the castrated cock:

> The capon ceases to crow; the comb and gills do not attain the size of those parts in the perfect male; the spurs appear, but remain short and blunt; and the hackle feathers of the neck and saddle, instead of being long and narrow, are short and broadly webbed. The capon will take to a clutch of chickens, attend them in their search for food, and brood them under his wings when they are tired.[96]

Charles Darwin, citing Yarrell, asserted that the capon takes on "characters properly confined to the female" and "takes to sitting on eggs, and will bring up chickens."[97] This intriguing phenomenon revealed that "the secondary characters of each sex lie dormant or latent in the opposite sex, ready to be evolved under peculiar circumstances."[98] August Weismann accepted Darwin's hypothesis, adding that "men who have been castrated in their youth retain a high voice like that of the other sex, and the beard does not become developed."[99] The castrated male was a virtual female.[100]

This consensus was based almost entirely on natural-historical observations and on tradition. For the first *experimental* investigations of the issue, we have to look to John Hunter (1728–1793).[101] Only rather incomplete accounts remain of Hunter's experiments on the gonads. He seems to have relied greatly on transplantation experiments to determine their functions. That method would remain central to research on the gonads well into the modern era, and some of Hunter's eighteenth-century experiments seem uncannily premonitory of experimental procedures common in the early twentieth century. In the most celebrated of these (one on which he did not, unfortunately, report in much detail), he transplanted the testis of a cock "into the belly of a hen," possibly expecting, the historian Barker Jørgensen speculates, "a more or less complete sex reversal." That did not occur, leading Hunter to regard the experiment as a failure.

It was Hunter's rather dramatic expectation—if indeed he had harbored that expectation—that may have blinded him to the possibility that the transplanted hens, although they had not miraculously changed into cocks, had developed a mixture of male and female secondary sex characters such as masculinized combs and wattles, results commonly reported after similar experiments during the early twentieth century.[102] Even if Hunter had noted these changes, he is unlikely, of course, to have pondered whether the influence of the testicles on the masculine characters was mediated by the nerves or blood-borne secretions. The polarities Edward Schäfer's generation had to wrestle with—nervous impulses vs chemical ones—were not the dilemmas of Hunter's time. Distant action within the body was brought about by sympathy—whether it was channeled through the nerves or the blood was not an important issue. At a more mundane level, Hunter was less of a natural philosopher than a surgeon. Like many of his contemporary surgeons, he was primarily interested in exploring the "possibilities of severing and reuniting parts of the body."[103] He did not ignore the effects of transplantation, but he was primarily interested in the *feasibility* of transplantation, not in its consequences.

Things changed in the nineteenth century. As with the ovary, the influence of the testicle on the sexual characters and the organism was now seen to be mediated by the nervous system. The reports of localized consequences of unilateral castration—such as the loss of, say, the left antler in reindeer who had lost the left testicle—or the fact that only certain regions of the body sprouted hair at puberty seemed to exclude even the possibility of a blood-borne agent. A sexualizing element circulating in the blood would, presumably, exert identical effects all over the body, which was hardly the case with the sexual characters. Unless one was willing to assume that different parts of the body were chemically so different that some were wholly resistant to the action of the hypothetical chemicals from the testes, there was only one credible explanation. The influence of the testicles was communicated by its nerves to the central nervous system, in the mysterious depths of which those impulses were channeled into other nerves serving the regions of the body affected by testicular influence. The absence of the left testicle in reindeer led, for instance, to the cessation of those trophic impulses communicated to the left antler via the central nervous system. Deprived of those trophic impulses, the antler dropped off. It was an elegant and plausible theory that, declared one of its exponents, explained all available observations and contradicted none.[104]

Nevertheless, the connections between the testicles and the central nervous system were never used to justify testicular interventions in nervous or mental disorders. As noted above in the discussion of osteomalacia, Fehling asserted that rickets was different in males, without explaining why. Even the hormonally inclined Boyd dismissed the possibility of affecting male cancers through the testicles because the ovaries and the testicles were so different. These differences were never listed, never even speculated on—just assumed to be real, ineluctable, universally agreed. The female norm was virtually pathological and the instability of feminine biology was a sexual phenomenon—indeed, almost a sexual character. While male sexuality had its own problems (venereal disease, to take the most obvious example), sex was not seen as central to the male organism. Consequently, the nineteenth-century ovary was subject to constant suspicion, indictment, manipulation, and mutilation while the testicle, whose functions were defined according to a similar neural model, remained largely immune from similar therapeutic interventions.

Chapter 2

The Age of the Internal Secretions

harles-Edouard Brown-Séquard's announcement of his "rejuvenation" marks the official commencement of the age of internal secretions, but many endocrinologists find that origin acutely embarrassing. Even in 1933, when the field of endocrinology had moved to the forefront of medical science, the leading endocrine researcher Herbert Evans waxed indignant about his beloved discipline having "suffered obstetric deformation in its very birth."[1] In 1957, looking back on forty years of the journal *Endocrinology*, San Francisco physician and pioneering clinical endocrinologist Hans Lisser (1888–1964) regarded Brown-Séquard's announcement as "calamitous."[2] It could, however, have been very different indeed—endocrinology *could* have been born in the impeccably scientific laboratories of at least two German professors. Before we explore the allegedly deformed birth and the infancy of endocrinology, let us, then, look briefly at what they might have been.

Berthold's Cocks and Goltz's Bitches: The Origins that Weren't

In 1849, the Göttingen physiologist Arnold Adolf Berthold (1803–61) published a brief paper (preceded by a preliminary communication on the same topic) describing the results of his experiments on six young cocks.[3] Two of the cocks were castrated and allowed to develop into capons. "They were not aggressive," observed Berthold, "they fought with other cockerels rarely and

33

then in a half-hearted manner, and they developed the monotone voice of the capon. Comb and wattles became pale and developed but slightly; the head remained small." In the second experiment, one testis had been removed, with the other, separated from its attachments, left loose in the abdominal cavity. In the third and final group, two cocks were castrated completely but one of their excised testes was then re-implanted on their intestine. The four animals in the second and third groups developed into normal, sexually differentiated cocks: "They crowed lustily, often engaged in battle with each other and with other cockerels, and showed the usual reactions to hens. Their combs and wattles developed as in normal fowls." When the testicular graft was removed from a member of the third group, however, "the fowl began to emit the typical voice of the capon, it no longer showed interest in hens, nor would it battle with other cockerels; in fact it remained at a distance from them and showed the nature of a true capon."[4] Masculine attributes, then, seemed to depend on the simple presence of a testicle and not on their location or number.

Autopsies revealed that the transplanted testes in the experimental animals had lost their original nervous connections—and yet, those grafts had still kept the cocks fully masculine. It followed, therefore, that the masculinizing effects of the testes must be mediated through the blood, rather than the nerves. Berthold did not, however, speculate on the nature of this blood-mediated influence. It is, indeed, not unlikely that his chief interest lay not so much in elucidating the functions of the testicle but, somewhat like John Hunter, in investigating whether testicles could be successfully transplanted. Listing his inferences, Berthold began not with the question of masculinity but by asserting that "the testes belong to the transplantable organs; they may become reattached after removal from the body; actually one may transplant them from one animal to another, and the attachment may occur at the original site of removal or in an entirely foreign location."[5] He also announced that he was planning to replicate Hunter's experimental transplantation of testicles from a cock to a hen, although there is no record that he ever performed this experiment.[6]

Whatever Berthold's own interests, however, his experiments were hailed by later scientists as revolutionary.[7] In his own time, however, they did not have much impact. As the American physician Herman Rubin quipped in 1935, "While Berthold was the Christopher Columbus of the internal secretions, no one even knew he had left home."[8] That was an exaggeration. Berthold's colleague, the well-known physiologist Rudolph Wagner, tried to replicate the experiment without any success. Towards the end of the century, several other scientists also tried their hand at testicular transplantation but they too failed. Despite this lack of corroboration, Berthold's experiments later came to be enshrined as the first scientific demonstrations of the endocrine actions of the sex glands. In 1936, the physiologist and innovative investigator of sexual physiology Eugen Steinach justly wondered why, if Berthold's experiments had indeed been as important as was being claimed, had it taken some eighty years since those experiments to obtain pure sex hormones?[9]

Although Berthold's experiments did not lead to any immediate conceptual or therapeutic innovation, at least one of the attempts at their replication produced important, if serendipitous consequences. In 1897, the young Viennese physiologist Arthur Foges, operating on three-to-five-week-old cocks, attempted to transplant testes from one castrated cock into another. The grafts almost always failed, but in the course of the experiments Foges observed that castration did not lead to total feminization but only to caponization. The caponized cocks could still crow, although they did not usually do so. Far from being timid, they fought constantly among themselves and also with normal cocks and hens. Their combs differed from those of normal cocks but also from those of hens. These observations were soon endorsed by many other authorities (including Artur Biedl, the author of the first and most influential German textbook on the internal secretions) and led to the revision of long-established beliefs.[10] Ever since the pioneering reports of the British naturalist William Yarrell (1784–1856), it had been accepted by most major scientists, including Charles Darwin and August Weismann, that castrated male animals were feminized in appearance and behavior. Each sex, they theorized, possessed the sexual characters of the other in latent forms and these became explicit "under peculiar circumstances."[11] Foges, however, showed that these views were inaccurate.[12] Much, of course, remained to be clarified about the differences between the consequences of prepubertal and postpubertal castration but one point, at least, was reasonably clear. The testicles were essential to the development of full masculinity and a male deprived of them was no longer very masculine, but he did not turn into a virtual female.

Even less influential than Berthold's work was the 1874 report of the Strassburg physiologist Friedrich Goltz (1834–1902) that a bitch with her spinal cord transected at the level of the first cervical vertebra had gone into estrus, mated (with a male toward whom she had previously been antagonistic), and became pregnant with triplets.[13] Goltz's experiment was not concerned with the physical sexual characters but with sexual function and its regulation by the gonads. In an intact animal, Goltz argued, one could assume that estrus was produced by nervous impulses traveling from the sex glands to the brain; that hypothesis, however, was obviously inapplicable when the spinal cord had been completely separated from the brain, since no nervous impulse could reach the brain without going through the spinal cord. Goltz, therefore, concluded that the sex glands exerted their effect on the brain by releasing specific chemical substances into the blood. Despite its acuity, Goltz's hypothesis was virtually ignored. This, Hans Simmer has observed, may have been because of the absence of a theoretical framework that would permit a proper evaluation of Goltz's finding. Only after the conceptualization of a chemical system of regulation akin to the nervous system could one perceive the true significance of Goltz's experiment.[14]

Simmer's comment may well apply to the 1870s, but why, one wonders, was Goltz's work not resurrected and celebrated in the 1890s, when such a model was available? Final answers must await further research, but one important reason for Goltz's obscurity must be the overwhelmingly gynecological character of endocrine research at the turn of the century. As we shall see, the vast majority of gland-researchers then were clinical gynecologists, and the bulk of their research was conducted on the ovary and with explicit clinical motivations. In such a milieu, one could not expect any familiarity with Goltz's decades-old report, which had been published in a physiological journal and which does not even seem to have lingered in the memory of physiologists themselves: when Arthur Foges launched yet another attempt to replicate Berthold's experiments in 1897, he did not reveal any awareness of Goltz's experiment.[15] Moreover, Goltz himself did not add to his work on the gonads and moved on to brain physiology, where he was to become a noted critic of emerging theories of cerebral localization.[16] Small wonder, then, that his report on the bitch was forgotten almost entirely by posterity, so that endocrinology came to be seen as Brown-Séquard's child even by those who were dismayed by that parentage.

The Fluid of Youth: The Rejuvenation of Charles-Edouard Brown-Séquard

The official birth of endocrinology had little to do with laboratory research, even though the father was an eminent experimental scientist, the French-American-Mauritian physiologist Charles-Edouard Brown-Séquard (1817–94). Brown-Séquard possessed sound credentials as a scientist as well as an endocrine researcher. To take only one example, he had shown through a series of experiments involving the removal of the adrenal glands that they were essential to life.[17] In 1889, however, the elderly Brown-Séquard nearly ruined his hard-won reputation by declaring before an assembly of august scientists in Paris that he had "rejuvenated' himself by injections of testicular extracts of dogs and guinea pigs. He was derided at the time and later incurred the contempt of endocrinologists, many of whom thought that Brown-Séquard had almost killed the budding science. "His claims were not confirmed, ridicule and abuse were heaped upon him, and a drought descended upon the field of clinical endocrinology which persisted, with but a few scattered refreshing contributions, for almost 30 years," lamented Hans Lisser in 1957. "The repercussions from this fiasco caused a cynical collapse and darkness followed."[18] None of this, we shall see, was even remotely accurate, but such opinions were common amongst the self-consciously scientific endocrinologists of the twentieth century.

What drove Brown-Séquard to his embarrassing but prescient self-experimentation? Merriley Borell has emphasized that Brown-Séquard's obsession with the testicles was based less on endocrine notions of testicular function than on a more traditional faith in the energizing qualities of semen. The nineteenth century, it is well known, was the century of the spermatic economy, when any inordinate loss of semen, whether through masturbation or excessive venery, was supposed to cause immense damage to health and morals.[19] Since the loss of semen was harmful, the absorption of semen was likely to be of some benefit. Hence, testicular extract could be seen as a likely source of health and renewed vitality in old age, when the production of semen was on the wane. Brown-Séquard may have been mocked by his fellow-scientists, but he inaugurated a clinical revolution. Following his announcement and his subsequent reports, the medical marketplace was flooded with extracts and various preparations of virtually every organ—from the brain to the ovaries, from the testicles to the spleen—which were prescribed for virtually every conceivable disorder that did not respond to orthodox treatment. "Organotherapy, opotherapy, or the Method of Brown-Séquard as it was often called, came," says Merriley Borell, "to be the therapeutic hope of physicians from Cleveland to Bucharest."[20]

In America, Benjamin Harrow of Columbia University, an eminent researcher on vitamins and hormones and somewhat exceptional in his regard for Brown-Séquard's memory, observed in the 1920s that Brown-Séquard had revived the old humoralism "in quite a modern form, and with reasons for its revival drawn from the knowledge of the nineteenth and not the ninth century. One may truly say of him that he is the founder of the conception of ductless glandular function as we understand it today."[21] Although it was and is easy to jeer at Brown-Séquard's claim of rejuvenation—and his own death not long after that announcement certainly did not help his cause—it stimulated a great deal of original laboratory research on the nature and functions of all internal secretory glands, not simply the testes. Borell has emphasized that it was no coincidence that effective thyroid and adrenal extracts were obtained as early as the 1890s. Benjamin Harrow was convinced that the spectacular effects of thyroid extract had forced "those who had scoffed at Brown-Séquard [to] revise their opinion of that illustrious Frenchman . . . glandular treatment became the hope of a world full of maladies."[22] But the hopes were not to be realized as swiftly as people had imagined. Although a dizzying array of tissue and gland extract was tried out on patients, none could replicate the success of thyroid extract, and the outlandish, unsubstantiated claims of magical cures discredited the entire concept of the internal secretions in the eyes of orthodox practitioners. The glandular products, of course, were obtained from animals and must have varied quite astronomically in quality and purity. Questions on the composition and purity of gland extracts plagued clinicians until standardized and synthetic preparations became available in the 1930s. To make matters worse, even the purest glandular extract from an animal was not necessarily of any use in humans. More than three decades after Brown-Séquard, an American physician bemoaned the lack of information on what exactly was contained in the gland extracts that medical practitioners were being encouraged to prescribe.[23] For many later endocrinologists, Brown-Séquard became the symbol of a new kind of quackery.[24] Nevertheless, innumerable physicians and not a few patients *perceived* them to be useful, and such perceptions helped establish endocrinology as a new field in the popular and medical consciousness.

Paradoxically, however, it was not the testicles immortalized by Brown-Séquard that were the first to reap the benefits of this boom. "The thyroids, the adrenals, the pituitary, and the pancreas became the glands of the hour" remarked Paul de Kruif in a popular account of "the male hormone." As far as the gonads were concerned, the ovary received far more attention than the testicles. "Even the ovaries, though sexual, were respectable scientifically," wrote de Kruif. "But the testicles? Shades of Brown-Séquard, they were a bit too hot to handle. The hormone hunters proceeded to begin to unravel the riddle of the human body's internal chemical control while they pretty much left these disreputable organs out of the picture."[25]

Whatever one feels about de Kruif's explanation, he was entirely accurate that the ovaries were to be far more prominent in the annals of early laboratory research on the endocrine glands than the testicles. Those "disreputable organs" would eventually find their place in the sun, but not for at least a couple of decades. The intervening period was to be the age of the ovary.

Vienna, Fin de Siècle: The Case of the Missing Ovaries

In 1922, William Seaman Bainbridge, a New York gynecologist, declared in a lecture: "There was a period when excision of the ovary was the rule rather than the exception and a cystic ovary was a surgically doomed ovary, but the pendulum has swung in the opposite direction.... Happily, the ovary instead of having to prove its right to remain seems now in a position to demand of the surgeon the reasons for its removal."[26] Bainbridge is not an important figure in this story—he appears here only as the representative of a new, conservative gynecology that regarded ovaries with a respect that would have puzzled Battey and Hegar.[27] How did this consensus arise and what was its intellectual basis?

To put it simply, it was a conjoint product of the laboratory and the clinic— with the clinic being the more important partner. It originated not simply from newer organotherapeutic concepts but, perhaps surprisingly, also from the neural model of ovarian function that had justified oophorectomy.[28] Brown-Séquard himself had reported the beneficial effects of ovarian extracts in prematurely senile women and in hysterics. Those early recommendations had led to their use on patients whom endocrinologist George Corner later dismissed as "hysterical girls and cachectic women."[29] As Hans Simmer has shown, however, interest in ovarian functions and the organotherapeutic supplementation or substitution thereof was not found solely among general practitioners trying out ovarian extracts on patients with ill-defined mental or "functional" symptoms.[30] Gynecological specialists were preeminent in this research, and even in 1931 gynaecologist and endocrine researcher Bernhard Zondek declared that hardly any medical field had as intimate a relationship with endocrinology as gynecology.[31] The pioneering experiments on ovarian transplantation conducted in Vienna during the last decade of the nineteenth century by the gynecologists Emil Knauer and Josef Halban—and accorded foundational status by all subsequent commentators—provide excellent illustrations of this special relationship between the science of woman and the science of glands.

The received version of their work, expressed most influentially in George Corner's previously cited lecture, "The Early History of the Oestrogenic Hormones," draws a sharp distinction between earlier efforts at organotherapy with ovarian substance and the "more solid kind of investigation" attempted by Knauer and Halban.[32] Corner's lecture justly continues to be a popular source for the history of research on the ovaries, but his views do not always stand up to historical scrutiny. In the late nineteenth century, as we have seen, numerous women had their ovaries removed for a variety of disorders. Any gynecologist working at a large teaching hospital necessarily encountered oophorectomized patients returning to the doctor with symptoms of premature menopause. Central European practitioners approached these symptoms seriously and tried to find ways of countering them.

Our story begins with the Viennese gynecologist Rudolf Chrobak (1843–1910), who was unique in having received his training in gynecology in a Department of Medicine, and fortunate in having studied experimental physiology with the charismatic Ernst von Brücke (1819–92), whose laboratory, according to Erna Lesky, was an "almost inexhaustible center of stimulation of Vienna medicine in the second half of the century." Chrobak's medical and physiological sophistication alerted him to issues that were rarely addressed by average gynecologists of the time, whom Lesky dismisses, perhaps unfairly, as mere "uterine engineers."[33] In the 1890s, Chrobak found that the satisfaction of conducting technically flawless oophorectomies was often soured (*vergällt*) by the large number of patients who complained that the operation had worsened their condition. Such postoperative symptoms, Chrobak was convinced, largely stemmed from the loss of the ovaries, and, although they resembled those of natural menopause (dizziness, headaches, hot flashes, sweats), they were far more intense. He was so impressed by the seriousness of these symptoms that he began to spare the ovaries in operations where they were routinely removed with the uterus. While he could offer no statistical evidence, he was confident that such conservative operations produced fewer menopausal symptoms.[34]

Chrobak then approached the issue from another angle. He was well aware of recent successes in treating symptoms of thyroid deficiency with extracts of the thyroid gland and reasoned that the administration of ovarian substance to oophorectomized patients might well be useful.[35] He does not seem to have employed it too extensively, however, and reported only moderate success.[36] Neither Chrobak's analogy between the thyroid and the ovaries nor his therapeutic procedure was a radical innovation, although he himself was apparently unfamiliar with the contemporary popularity of identical concepts. The year 1895, in fact, seems to have been particularly crowded with independent and virtually simultaneous "discoveries" of the ovary-thyroid analogy.[37] Although several clinicians of the time reported beneficial effects from the use of ovarian preparations in oophorectomized patients, modern endocrinologists regard them as placebo effects, because the ovarian hormones are not stored in substantial amounts in ovarian tissue and what little is stored cannot be extracted simply by treatment with water.[38] Truly spectacular results, such as the resumption of regular menstrual cycles, were not reported in any case.[39] The historian's point, however, is merely that by the end of the nineteenth century ovarian organotherapy was no longer the preserve of the general practitioner treating "cachectic girls." It was fast becoming the object of specialist gynecological interest.

Chrobak's thyroid-ovary analogy may not have been particularly innovative, but his organotherapeutic orientation led him to take a momentous next step, which, everybody agrees, led to the birth of the modern concept of the ovary.[40] Unsatisfied with the reliability of ovarian extracts, Chrobak wondered whether it would not be better if, instead of administering ovarian substance to the patients, one could actually replace their missing ovaries by transplantation. But was ovarian transplantation even possible? In 1895, Chrobak asked his assistant Emil Knauer (1867–1935) to conduct experiments on rabbits to test its practicability and utility.[41] Transplantation was not, therefore, conceived of as a physiological experiment but as a more effective form of organotherapy.[42] The physiological actions of the ovaries were not directly under investigation and Chrobak did not speculate on the possible mechanism of the ovary's action upon the female organism. Nor did Knauer until later.[43] This reserve is all the more remarkable because a transplanted gland obviously would not sprout nervous links to the rest of the organism and one could not imagine that ovarian transplantation would work unless one also accepted that the actions of the ovary were mediated by chemicals. Nonetheless, Chrobak and Knauer said nothing on the matter.

Knauer began by removing the ovaries from four rabbits and then transplanting them elsewhere in the abdominal cavity of the *same* animal. These autografts "took" well, and in one case the internal reproductive organs (such as the uterus and the uterine tubes) did not show signs of atrophy when examined at autopsy six months after the transplantation. Microscopic examination showed that the grafted ovaries were histologically normal and contained the usual numbers of follicles, some with clearly visible ova in different stages of development. The experiments, therefore, suggested that (a) the ovary could be successfully transplanted in locations different from its anatomical site and (b) in such successful transplants, the ovary remained functional.[44] But exactly how functional and for how long? Was pregnancy possible in an animal with transplanted ovaries? Knauer soon claimed to have succeeded in inducing pregnancy in a rabbit, the ovaries of which had been removed and reimplanted within its own abdominal cavity sixteen months before mating. The pregnancy continued to term and ended in a natural birth.[45] Three years after the transplantation, the rabbit became seriously ill and feeble and Knauer decided to kill it. On autopsy, the findings exceeded Knauer's expectations: the internal reproductive organs were normal and so was the microscopic appearance of the transplanted ovaries. Transplanted ovaries, therefore, could remain fully functional for a significantly long period.[46]

More importantly, given the therapeutic aims of the experiment, would it be possible to *replace* the ovaries of one animal with those of another?[47] Here Knauer fared less impressively. In the first series, all the grafts except two underwent immediate necrosis, disappearing completely at the end of a year. In failed transplants, the uterus, expectedly enough, atrophied and so did the external genitalia. Of the two successful cases in the series, one died after three weeks of the operation and at the time of death, one of its ovarian grafts still appeared to be functional. The second rabbit lived with the grafts for a year and a half but its reproductive organs underwent atrophy and on microscopic examination, it was clear that the ovarian tissue was inactive.[48]

In the discussion following Knauer's lecture on these experiments, another young Viennese gynecologist, Josef Halban, described his own, independent experiments on the subject of ovarian transplantation.[49] Halban's orientation was very different from Knauer's. Although he would later become one of Vienna's best-known clinical gynecologists, Halban's early experimental research had no discernible clinical motivation. It was Halban's aim to study the influence of the ovaries on the development of the female reproductive system and to speculate on the nature, extent, and mechanism of that influence. Here—and perhaps only here—do we have a major, influential instance of late-nineteenth-century research on ovarian physiology that was not directly clinical in its motivation.

In 1897, Halban had removed the ovaries, the uterine tubes, and a piece of the uterus of some newborn guinea pigs. In one group of these animals, he had then reimplanted the removed organs under the skin. The control group did not receive any transplants. The uteri of the latter group remained undeveloped but did not atrophy. The experimental animals, on the other hand, developed normally and even after more than a year, no problems were encountered in the development of the reproductive organs and the breasts—even the uterine tubes transplanted under the skin with the ovaries remained functional. The ovaries, therefore, did not just sustain the functions of the adult uterus but directly governed the maturation of the infantile reproductive organs into their adult form. It did not matter *where* the ovary was—its presence anywhere in the body ensured the proper development of the reproductive system.

Unlike Knauer, Halban was quick to appreciate that his results challenged the neural model of ovarian physiology. No nervous link could conceivably be reestablished after the transplantation of the ovaries under the skin and the ovarian influence on the reproductive system must be mediated, Halban firmly declared, "by the internal secretions, the nature and attributes of which, admittedly, are still wholly mysterious. These experiments compel us to assume that a substance is secreted from the ovaries into the circulating blood, which is capable of exerting a specific influence upon the rest of the genital system."[50] Although Halban's experiments were illuminating in themselves, his greatest contribution was to apply this new and explicitly endocrinological model to Knauer's experimental data. It was with Halban that the ovary began to be transformed from a mysterious, neurologically controlled (and controlling) gland into a source of internal secretion or possibly secretions, the chemical constituents of which were mysterious but whose functions were definable with some clarity.

And yet, this same Josef Halban spent the rest of his illustrious career in arguing that the sex glands merely exerted a protective influence on the sexual attributes: the sexual characters were generated independently *ab ovo*. Much of this conviction was supported by circumstantial evidence: Halban offered no genetic argument at all. He did offer some suggestive experimental evidence from his early experiments, however: for example, the uteri of oophorectomized infant guinea pigs did not actually atrophy; they simply did not develop further. Halban's mature position on the limited role of the sex glands was opposed to that of many other scientists, including Eugen Steinach of Halban's own Vienna. By the end of the second decade of the twentieth century, the latter were arguing for the preeminence of the sex glands in determining and maintaining the characteristics of sex, and it was Halban, but for whose work the history of endocrinology might have been very different indeed, who was their greatest, albeit not very effective critic.[51]

What we need to appreciate from the Knauer-Halban experiments of the turn of the century, however, is that an "empirical," clinically oriented organotherapy was not *displaced* by a "scientific," laboratory-bound endocrinology in one big swoop, as implied by George Corner. Rather, a more endocrinologic orientation toward the ovary grew out of the organotherapeutic perspective. Laboratory research could still be directed with purely clinical aims, as with Knauer. At the same time, one must acknowledge that the experiments of Knauer and Halban, no matter how significant they might seem to us, did not immediately convince every scientist and clinician of the truth of the endocrinological theory of ovarian function. The basic problem seems to have been that transplantation experiments could not *absolutely* rule out a certain "preformation' of the sexual characters. In those of Halban's experiments where the ovarian graft did not survive, the uterus may simply have failed to develop because of unknown intrinsic factors. The failure of the uterus to develop could not, in other words, be attributed solely and definitively to the absence of the ovaries. Secondly, it was highly improbable but not *absolutely* impossible that the results of successful glandular transplantation might be due to the reestablishment of nervous connections between the graft and the rest of the organism.[52] Even a champion of the internal secretory paradigm was compelled to admit, as late as in 1924, that "transplantation experiments cannot absolutely decide the question whether the sexual glands act by nervous reflexes or by internal secretions discharged into the circulating blood."[53] The question, I reiterate, was one of *absolute* certainty. Few denied that the endocrine theory had probability on its side. In any case, even those who believed in the sexual omnipotence of the internal secretions had to admit that their exact histological sources, their nature, and their actions all remained entirely unclear.[54]

All uncertainties notwithstanding, Knauer's demonstration of the feasibility and efficacy of ovarian transplantation aroused considerable clinical interest. Within a year of its publication, his long 1900 paper—which had at last incorporated the notion that the transplanted ovary acted on the reproductive organs through its internal secretions—was reviewed in detail even in distant Pennsylvania by a gynecologist. The reviewer was optimistic about the future applicability of the operation, while emphasizing that it was, of course, far easier to spare the ovaries in an operation than to transplant ovaries later.[55] A large number of researchers, in continental Europe as well as the United States, conducted many experimental ovarian transplantations, in animals as well as humans.[56] The transplantation of ovaries from their original site to another in the same individual (autotransplantation) was usually successful in maintaining menstruation and preventing the symptoms of premature menopause. This could obviously be very useful in human cases where a serious pelvic infection (commonly gonorrheal in the early twentieth century) made it unsafe to leave the ovaries in their normal position.[57] Occasionally, successes were reported in the transplantation of ovaries from one individual to another, but generally this proved far more difficult than autotransplantation and, according to some, practically impossible in humans. Despite the occasional positive report, it remained a deeply unsatisfactory procedure.[58]

The removal of the ovaries, however, was now known to cause definite, serious problems, and gynecologists consequently became deeply reluctant to perform the operation. Nor had experimentalists found any easy, reliable way to replace ovaries in oophorectomized patients. Knauer and his successors had merely shown that the patient's *own* ovaries could be moved around in her body; the possibility of *replacing* missing ovaries with grafts from donors remained questionable. The restitution of ovarian function by extracts was too unreliable and even in the best cases incomplete. Even the most enthusiastic advocate of ovarian extracts had never claimed that they could restore functions such as menstruation. It was safer, therefore, to spare the ovaries as far as was possible.

Thus evolved the new emphasis on conservative surgery, with which we began our exploration of this theme. British gynecologist William Blair-Bell emphasized in 1913 that ovarian transplantation from one woman to another was rarely possible and even the simple autografting of the patient's own ovaries was not always satisfactory. "Thus," he concluded, "there is a very restricted field for this method of treatment, especially when we remember how rarely it is necessary entirely to sacrifice the whole of both ovaries in other pathological conditions."[59] This, of course, was not really what Chrobak had aimed for. He had asked Knauer to find a surgical solution to the problem of premature menopause. What had actually happened, however, was that, thanks to the work of Knauer and others, it became clear that the best surgical solution was the *avoidance* of surgery. Gynecological surgery, instead of rejoicing in the introduction of a new procedure, began to eliminate an established operation around 1900, while a new physiological understanding of the ovary began to dawn.

One Gland Indivisible or Many Glands in One?

Even if one accepted the ovary to be an organ of internal secretion, innumerable questions remained to be answered. One major problem haunting ovarian research in the early twentieth century concerned the histological source of its putative secretion—or perhaps secretion*s*. The ovarian research we have discussed so far had little, if anything, to say on exactly which part of the ovary the internal secretions came from, but that question could not be postponed for very long. The ovary was not a simple, homogeneous piece of glandular tissue: it was a complex organ and its structural features changed significantly at different points of the reproductive cycle. Histologically, the ovary at the beginning of the menstrual cycle was different from the ovary after ovulation; the ovaries of a pregnant woman differed histologically from those of a pubescent maiden. Did such changes affect the character or quantity of internal secretions?

The major event in the ovarian cycle was the expulsion of the ovum from the ovarian (or Graafian) follicle. Once ovulation had occurred, the ruptured follicle was converted into a cellular structure called the corpus luteum. Anatomically, it had been a familiar entity since the seventeenth century, but its functions and significance still remained unclear some two centuries later. Perhaps the most well-known late-nineteenth-century hypothesis on luteal function was that associated with the names of John Beard (1858–1924) and Auguste Prenant (1861–1927). Beard lectured on embryology and vertebrate comparative anatomy at the University of Edinburgh from 1890 until 1920, but had undergone his higher training in Germany.[60] He argued that since ovulation did not occur during pregnancy and the corpus luteum was a prominent feature then, the latter probably represented a mechanism to prevent normal ovulation.[61] He was quite undecided, however, on its *means* of action. In 1898, Auguste Prenant, professor of histology at Nancy and, later, Paris, proposed a more detailed hypothesis. Although the publication of Knauer's experiments was still in the future, Prenant was well informed about recent hypotheses on glandular secretions. Emphasizing the histological similarity between the corpus luteum and other glands either suspected or known to be active internal secretors, Prenant argued that the corpus luteum did suppress ovulation during pregnancy as Beard had surmised but did so through its chemical secretions.[62] This concept of the action of the corpus luteum came to be known as the Beard-Prenant Hypothesis.[63]

Virtually contemporaneously and in all probability independently, the German embryologist Gustav Born (1851–1900) had also identified the corpus luteum as a possible source of important internal secretions. After Born's early death, his hypothesis was investigated experimentally on rabbits by his student, the gynecologist Ludwig Fraenkel (1870–1951), who was assisted on many of the experiments by another young gynecologist, Franz Cohn. According to Fraenkel, Born had been struck by the fact that in instances of successful conception, the mucous lining of the uterus—the endometrium—began to proliferate before the fertilized ovum had reached the uterus. Corpora lutea of significant dimensions did not develop in species where embryos were never implanted in the endometrium. It was likely, then, that the corpus luteum was important in preparing the endometrium for pregnancy.[64]

Fraenkel conducted a diverse range of experiments on guinea pigs to test his mentor's hypothesis; the most illuminating of them relied on removing both ovaries or destroying the corpus luteum by electrical cautery after fertilization but before the implantation of the embryo. Pregnancy was invariably terminated and the uterus atrophied, from which Fraenkel concluded that the fertilized egg could not be implanted in the uterine endometrium in the absence of the corpus luteum. The proliferation of the uterine mucosa and its blood vessels was essential for the implantation of the ovum and this proliferation was induced exclusively, he argued, by the internal secretions of the corpora lutea.[65] If conception did not occur, then the corpus luteum ended its work after triggering the next bout of menstrual bleeding.[66] Although Fraenkel did not completely rule out internal secretory contributions from other histological elements of the ovary, the corpora lutea, for him, were the most important sources of the ovarian internal secretions.[67] The undeveloped state of the prepubescent female reproductive apparatus as well as the post-menopausal atrophy of it he attributed solely to the absence of the corpora lutea.[68]

At the very beginning of his comprehensive 1903 report, Fraenkel stated that he was going to present and prove his theory of the physiological functions of the corpus luteum *and* suggest "important practical applications" and the therapeutic relevance of his theory.[69] Later, he asserted that although the human machine was too complicated to guarantee that even an accurate theory would always lead to genuine, lasting improvements in clinical practice, any theory that claimed to explain a range of physiological and pathological processes of the female organism had to be tested in the real world of the clinic. A theory was of importance only if it explained human maladies or improved their treatment. The utility of his own theories on luteal function was clear, he argued, from the explanation they offered for the variability and unpredictability of ovarian organotherapy. The therapeutic effects of commercial extracts of bovine and porcine ovaries (available under the proprietary names of Oophorin and Ovariin) in menopausal symptoms were sometimes impressive but often slight or downright negligible. Now one knew why. Oophorin was made from the entire ovary whereas it was only the corpus luteum that was responsible (according to Fraenkel himself at any rate) for menstruation and for maintaining the nutrition of the uterus.[70] The secret of a more effective ovarian extract, therefore, was to prepare an extract solely from the corpora lutea.[71]

Fraenkel himself had prepared such an extract from nonpregnant cows (naming it Lutein I) and had used it clinically for more than a year. It had proved to be of enormous use in combating such menopausal symptoms as hot flushes, anxiety, palpitations, and tremors. It acted promptly, reliably, and virtually without unpleasant side effects; even the cost of treatment was not too high.[72] Except for relieving menopausal symptoms, however, Lutein had proven unpredictable in its effects. It certainly did not reestablish menstruation in cases of amenorrhea. Fraenkel had also prepared a luteal extract from pregnant cows (named Lutein II), which had not shown any noteworthy effects even on menopausal symptoms.[73] Fraenkel was compelled to admit that while one could induce uterine atrophy and amenorrhea by removing the corpora lutea, one could not reverse those phenomena by the administration of luteal extract.[74]

Fraenkel emphasized, however, that the clinical relevance of his theory did not depend on the success or failure of organotherapy with luteal extracts. It had other implications too. For instance, it suggested that surgeons should be careful not to remove the ovaries of pregnant women, even in cases of serious ovarian tumors. If the operation was unavoidable, then the tumor should be resected in such a way as to leave the corpus luteum intact.[75] More speculatively, he suggested that antibodies could be synthesized against different components of the ovary, which might be used for "total sterilization by biochemical means" of women or for the cure of osteomalacia, which, he emphasized, was invariably alleviated by bilateral oophorectomy.[76]

Fraenkel's work stimulated his peers, but few accepted his claim that the corpus luteum was *the* secretory organ of the ovary. Many medical scientists were beginning to suspect that the ovary was a consortium of at least three important kinds of glandular tissue, each producing internal secretions of importance for the maintenance of the sexual attributes of femininity. The Graafian follicle and the interstitial cells were the two entities most commonly thought to be responsible for the endocrine actions of the gland.[77] In 1910, Artur Biedl concluded from a fresh examination of Fraenkel's evidence that he had not even succeeded in proving beyond all reasonable doubt that the corpus luteum was essential to sustain pregnancy. The very diversity of ovarian structure made it difficult to pinpoint any particular part as being responsible for any particular action. All the processes that Prenant, Born, and Fraenkel had attributed to the corpus luteum could, decided Biedl, "result, with equal probability, from hyperfunction of the interstitial cells and reduction of activity on the part of the other tissues, particularly of the Graafian follicles."[78] In Britain, gynecologist Louise McIlroy remarked in 1913 that although rabbits certainly needed corpora lutea to sustain pregnancy, "cases of oöphorectomy during early pregnancy in women, in which no subsequent interference with the progress of the pregnancy took place, prove that laboratory experience must not necessarily be made to apply clinically."[79] Until the late 1920s, in fact, the role of the corpus luteum would remain controversial and undecided.

But for all its ambiguities, Fraenkel's work presented a novel vision of the ovary: a gland whose secretory activity was periodic—the corpus luteum, of course, was not a constant structure—and located (almost entirely) in certain transient bodies. This sophisticated physiological model had a strong clinical dimension and that should not surprise us in the least. Hans Simmer, perhaps the greatest historian of ovarian endocrinology, has criticized Fraenkel and his interlocutors "for obscuring the issue by clinical observations" and for combining "observations on a menstruating female [presumably a *human* female!] with those obtained from a non-menstruating animal."[80] I would argue, instead, that the interpenetration of laboratory data with clinical observations in Fraenkel's work demonstrates, once again, that in the formative years of endocrine science, the laboratory bench was far closer to the clinic than present-day scientists or clinicians appreciate. Fraenkel was a busy practicing gynecologist, and, although his allegiance to experimental science was unquestioning, he was not cauterizing corpora lutea *solely* to garner physiological data. Many of the gynecologists who investigated ovarian functions in the late nineteenth and early twentieth centuries could stand up and be counted with the most competent physiologists of the time—but they were not *simply* physiologists, with the possible exception of the early Halban. For the vast majority of this generation, physiology, pathology, and gynecological therapeutics were parts of one inseparable whole. Their quest was for usable physiological knowledge: a physiology that would help evolve more efficacious forms of treatment. The clinical observations were not distractions from the issue, as Simmer suggests. They *were* the issue.

Glands, Gender and Profession-Building: The Triangular Quest of William Blair-Bell

The relationship between clinical gynecology and glandular research was intimate, but its form and character varied in different historical contexts. So far, we have heard much about the laboratory-clinic link in the gynecology of the German-speaking lands. In other national contexts, however, the same conceptual connections could serve very different ends. In Edwardian Britain, for example, the separation of gynecology from surgery and midwifery occurred relatively late, and gynecology remained inferior in status to internal medicine and surgery for a relatively long period. Those striving to raise the status of gynecology in Britain fought a complicated battle, and one of their major weapons was the discourse of glands. The way forward and upward for gynecology was to emphasize its alliance with the glittering new science of glandular secretions.[81] This particular use of glandular science was unnecessary as well as unknown in Central Europe.

The Liverpool gynecologist William Blair-Bell (1871–1936) is usually remembered today only as the first president of the Royal (formerly British) College of Obstetricians and Gynaecologists and a leading figure in the struggle to establish the college, the idea of which was opposed by powerful sections of the medical profession. "We his followers," wrote the gynecologist V. B. Green-Armytage, "must never forget our debt to his fanatic zeal and dynamic force of intellect, for to him ... we owe our College. ... He subordinated everything personal and professional to the attainment of his one ambition—a Royal College of Obstetricians and Gynaecologists."[82] As one of his peers put it more pithily, Blair-Bell was "the restless, lovable torch-bearer who never forgot nor allowed anyone else to forget that he was bearing a torch."[83] Blair-Bell was indefatigable in fighting for the recognition of gynecology as a true science and an autonomous medical specialty—and his claims were based, to a significant extent, on the science of sex gland function.[84]

For Chrobak and Knauer, glandular physiology had been subsidiary to glandular therapeutics; for Blair-Bell, on the other hand, the knowledge of sex-gland function was necessary in order to *transform* gynecology and obstetrics into one genuine, unified science.[85] The British medical context was particularly important here. For the medical elite of nineteenth-century Britain, an ideal doctor was one who could treat the whole body; well into the twentieth century, therefore, gynecology was regarded as yet another manifestation of the new and pernicious trend of medical specialization.[86] To assign a sector of the medical profession solely to the treatment of one *kind* of problems was anathema to the medical elite. It was all the more deplorable when that specialization was on one system in one sex.[87] Moreover, women patients were seen as crucial to the success of any medical practice, whether general or special. Physician Thomas Clifford Allbutt declaimed in 1885 that

> provincial medical men know well what, up to the present, they have had to expect when one of their lady-patients migrated to the "London gynaecologist." ... It meant lodgings in town, the doctor's brougham at the door three or four times a week, sixty or seventy guineas to pay at the end of the season, and a deluded and neurotic patient as the end of it all.[88]

Gynecologists themselves were well aware of the importance of women patients. Also in 1885, gynecologist James Hobson Aveling, the co-founder of the British Gynaecological Society, told medical students:

> The successful management of the diseases of women is the key to general practice and forms a large portion of your work. Women, as you know, enjoy, and always find time for gossip with one another. ... Woe to the unhappy practitioner who has failed in his treatment of their troubles; his condemnation will be widely heard.

On the other hand he who has been successful will have the trumpet of fame sounded with extravagant force.[89]

Small wonder, then, that the physicians and surgeons in elite practice opposed the formation of a gynecological specialty.

"Unfortunately," William Blair-Bell wrote toward the end of his life, "for the last hundred years gynaecology has been of little interest to the physiologist, anatomist, and biochemist. So the credit for practically all modern advances have fallen to clinicians. Now, in the immediate present, it appears that . . . the scientific aspects of our subject are attracting the attention of specialists in various branches of biology."[90] At the beginning of his career, he had instead had to lament the failure of past medical scientists "to take the female genital organs within their ken." This, he had appreciated, was not entirely bad news for him and his confréres: "the modern gynaecologist" had thus been bequeathed "a field for research wherein discoveries await him at every turn."[91] From his own earliest days as a gynecologist, Blair-Bell himself was a devoted researcher-clinician, and the endocrine system was one of his primary interests.

Unlike most contemporary glandular researchers, Blair-Bell looked upon every ductless gland in the female body, rather than simply the ovaries, as of sexual importance. His most well-known work, *The Sex Complex* (1916), was based on the conviction that *"femininity itself is dependent on all the internal secretions."*[92] The ungainly title of the book, far from indicating any allegiance to psychoanalysis, was his omnibus term for all the sexually active internal secretory organs. The ovaries, of course, were crucial components of this ensemble, but their endocrine functions were intimately and complexly linked with those of the others.[93]

Blair-Bell conducted much experimental research—removing the ovaries in pregnant cats and rabbits, for example—but as an experimenter, he was, frankly, an amateur compared to the gynecologists of Central Europe. His findings essentially replicated those of other workers, suggesting that pregnancy could not be maintained in the absence of the ovaries. He then transplanted ovaries, again replicating what others had already found: that a successfully transplanted ovary was perfectly adequate in maintaining the secondary sexual characters. His particular interest, however, was in exploring the role of the ovaries in the general metabolism of the female organism. Removing the ovaries of six cats, he collected their urine for varying periods and found that calcium excretion was diminished by one-half after the removal of the ovaries. Relating this experimental finding with the clinical observation that osteomalacia was often cured after bilateral oophorectomy,

Blair-Bell concluded that the ovaries promoted the excretion of calcium.[94] Overall, bilateral oophorectomy lowered metabolism as a whole—and in women, whose metabolism was "much more easily disturbed than that of lower animals," it could also lead to psychoses and neuroses.[95]

Blair-Bell was also greatly interested in the effects of oophorectomy on the other ductless glands, but his data were far from illuminating. The thyroid, for instance, seemed to remain unaffected in cats, but in rabbits oophorectomy led to pronounced thyroid changes of unclear significance.[96] The thymus (in cats) and the adrenals (in cats as well as rabbits) were also enlarged after oophorectomy, but again the physiological significance of this change remained obscure.[97] Perhaps frustrated by the situation, he then took the opposite approach. What were the metabolic and specifically genital consequences of the removal of the major ductless glands? Some of his findings seemed to be significant: removing the thyroids from cats resulted in uterine atrophy, for instance, and histological features suggested an increase in pituitary secretion.[98] Despite some interesting results, however, the data was too fragmentary to support any definitive claims on the polyglandular basis of femininity. Since it was the aim of *The Sex Complex* to establish that femininity was "produced" by the combined action of all the ductless glands—and, indeed, that femininity amounted to no more than that polyglandular symphony—the book can only be described as an interesting failure, although somewhat ahead of its time in its emphasis on the entire endocrine system.

Not all of Blair-Bell's glandular studies, however, were so inconclusive. He had a particular fascination for the pituitary and lamented its relatively low profile in contemporary research.[99] In 1909, he co-authored an important paper claiming that the extract of the posterior part of the pituitary gland produced intense contractions of the uterus.[100] Although not the discoverer of this phenomenon, Blair-Bell played a major role in popularizing pituitrin (as the extract came to be called) to expedite labor. Its real utility, however, was eventually found to be in arresting postpartum hemorrhage. This was an extraordinary achievement for the time—the pituitary was still an enigmatic organ and it was not even known that Blair-Bell's pituitrin was actually a mixture of two separate hormones.[101] Although of little use in supporting grand theories of woman's place in nature, pituitrin was invaluable in reinforcing the scientific claims of the gynecological profession and in expanding its clinical armamentarium.

This was precisely the kind of research Blair-Bell himself considered worthwhile. He had as much contempt for the pure laboratory scientist as for the pure clinician. A genuinely scientific approach, for him, spanned laboratory and clinic. "In the days that have passed—almost to the end of the last century—the clinician," he wrote, "claimed a very large proportion of all the scientific advances made in medicine and surgery. It is only of recent years that others, working as 'professional' scientists—if I may so term them—have invaded the province of gynaecology and obstetrics."[102] At the same time, he urged younger clinicians to appreciate how deeply the "patience and courage, knowledge, and scientific attitude of mind engendered by research" benefited one's clinical practice.[103] "It is not so very long," he warned them, "since gynaecologists were regarded as accoucheurs trying to do surgery, and obstetricians as midwives."[104] The implication was clear: the art (and profession) of gynecology would founder if the science was neglected, but being a laboratory-bound scientist was not going to achieve much either.

Blair-Bell had always been firm in his desire to keep the laboratory in its place. It was imperative, he had asserted in 1913, that gynecologists embrace the laboratory sciences, but "it cannot too strongly be emphasized that laboratory work is carried out in order that the knowledge so gained may be applied to clinical practice, which thus sets the seal on or checks experimental work." Germany and America, he warned, had acquired an enormous advantage over Britain in medical research by instituting "the combined laboratory-ward system."[105] By 1932, he knew, however, that the golden age of the clinician-researcher was over: much of what had constituted the science of gynecology in the earlier decades of the century had now passed into the hands of full-time laboratory scientists. "Clinicians who take their problems to the laboratory and try to solve them there," he admitted, "are now regarded by the professional scientist as amateurs . . . the gynaecologist of the present realizes that there will be little or no scientific recognition for him. . . . Nevertheless, our great and important subject has been placed on its present scientific foundation by the endeavours of scientifically minded clinicians during the last fifty years."[106]

Blair-Bell's scientific activities have not been entirely ignored by historians, although he is unlikely to have been flattered by their opinions. Ornella Moscucci, for instance, has acknowledged him as a pioneer of gynecological endocrinology, while emphasizing that the historian is struck less by the scientific merit of his contributions than by his rearticulation of the "enduring ideology" that the very identity of woman was constituted and dominated by her sexual functions.[107] Questioning whether Blair-Bell's work even justifies the label "scientific," Roy Porter and Lesley Hall have rebuked him for not maintaining an "open mind in the face of contradictory evidence" and for using scientific arguments "to shore up existing prejudices."[108] I think it is more useful to see Blair-Bell's work as an illustration of the interpenetration and mutual reinforcement of science, professional aims, cultural beliefs, and social stereotypes.

Even if one wishes to focus solely on his sexual ideology, it would be worthwhile to note his self-contradictions. He declared, for instance, that it was senseless to consider the male as necessarily superior to the female and waxed eloquent on how her intellect was "a source of personal pleasure and pride" to the human female, while stating in the same breath that "it must surely be recognized by all that the male mind and masculine form are suited to the business of life" while "the central motive of a normal woman's existence is the propagation of the species."[109] Although he subscribed to a polarized view of complete and opposed sexes, he asserted in 1915 that "however much maleness or femaleness there may be in any individual, there is always a certain amount of the opposite sex latent: certain men are somewhat effeminate, and many a woman has but the smallest balance of femininity in her favour."[110] He thought that bilateral oophorectomy for trivial complaints like menstrual pain was "unscientific and reckless": "I know of a sad case of a young girl who had both ovaries removed for dysmenorrhoea," he told a medical audience in 1920. "I can hardly control my words in stating what I think of such an unscientific procedure. . . . There is no reason for removing the whole of the ovarian tissue in any woman, unless the lesion be malignant."[111]

And yet, this same man considered masturbation to be so harmful that he proudly reported: "In one case the patient's distress and remorse at her own evil ways, which she found impossible to check, were such that we excised her clitoris and nymphae. This method of treatment may be adopted with excellent results if the right type of case be selected: the girl who is not suffering with excessive sexuality, but, rather, with the fascination of a bad but pleasant habit, to the detriment of her moral and physical equilibrium."[112] But he added in the same article: "A woman is not judged by the standard of masculine sexuality. The average man is supposed to be immoral, and undoubtedly he is. A woman, if she have the same feelings, as is often the case, either becomes ostracized or may suffer from the restraint imposed. Social exigencies, in fact, establish the relative standards which suit the community best, if not the individual."[113] He berated the "modern woman" for her "rejection of maternal functions," wondering, however, whether the rejection was part of "Nature's plan for securing the disappearance of Man to ensure further evolution."[114] Later, he pleaded with "normal" women to fulfill their reproductive obligations, "for it is the normal woman alone who should perpetuate the race and maintain the dominance of home life, without which men are handicapped both mentally, physically and as citizens."[115]

This is the point where his sexual ideology merged most seamlessly with his professional agenda: "If this be accepted," he went on, "the scope for gynaecology and obstetrics of the highest type is wider than it has ever been. We should aim not only at the best ways of dealing with the abnormalities of the genital system, but also throw ourselves whole-heartedly into the task of encouraging women to maintain a normal psychological outlook in regard to the special attributes and functions of their sex."[116]

"The language and practice of gynaecology," according to Chris Lawrence, "demonstrated to Victorians, on a day-to-day basis, the enormous determining power of the female reproductive parts. From this determinism flowed naturalistic prescriptions which defined the role of middle-class women in Victorian society."[117] The post-Victorian Blair-Bell's ideas on female nature and appropriate feminine conduct differed little in essence from those of his Victorian forebears. The discourse of glands, however, enabled him to frame his "prescriptions" in the language of a new, authoritative science. His views on femininity were integral to his larger professional project. To see women as sexual beings and machines for reproduction was also to see them as the raison d'etre of his own profession. The gynecologist was not just a medical specialist: he served the human species by ensuring the safety of the vessels through which it was perpetuated. And the reproductive functions of women could be understood and kept in good condition only by proper attention to their ductless glands.[118] We must, therefore, see Blair-Bell's glandular research, sexual ideology, and professional vision as different aspects of the same project, the overriding purpose of which was to establish the scientific and professional status of gynecology. Just as there was no specific discipline of endocrinology in the early twentieth century, there was no one approach to the study of ductless glands even among gynecologists.

The Oldest Endocrine Treasure Trove?

Although the testicles have always been associated with virility, scientific interest in its functions, as we have seen, remained low for many years. Brown-Séquard did draw the testicle into the medical mainstream, but the organotherapeutic boom stimulated more research on the thyroid and the adrenal than on the testicles. From around 1900, however, one senses a change.[119] In sharp contrast to the ovarian research of the period, this new trend was almost entirely nonclinical in motivation, and most of its early achievements occurred in the laboratory.

The elaboration of modern concepts of testicular endocrinology was greatly dependent upon the elucidation of testicular histophysiology. Franz

Leydig (1821–1908), in 1850, had identified that the testicles comprised at least two distinct groups of cells: the germinal cells (which produced spermatozoa) and the interstitial cells, which came to known as Leydig cells. There were other cell-groups of histological importance (such as the cells of Sertoli)—physiologically, however, the two former groups seemed to be the most active. Initially, it was widely believed that the function of the Leydig cells was to process nutritional material for the production of spermatozoa. Soon, however, it was noticed that the destruction of the sperm-producing cells—as seen naturally in undescended testes—did not interfere with the development of the secondary sexual characteristics. Boys with undescended testicles usually grew up to be fairly typical men.[120]

Similar findings were obtained after experimental destruction of the germinal tissue, as in the early-twentieth-century work of Julius Tandler and Siegfried Grosz, both of Vienna. Tandler and Grosz had irradiated the testes of roebucks with x-rays. The animals showed none of the typical changes associated with castration, such as the loss of antlers. The internal secretions of the irradiated testicles could, therefore, be assumed to have remained undisturbed. Microscopic examination revealed massive destruction of the generative tissue but almost total conservation of the interstitial cells.[121] The experiments of Pol André Bouin (1870–1962) and Paul Ancel (1873–1961), anatomists at the University of Nancy and then at Strassburg, provided further persuasive evidence: they ligated the vas deferens, thereby causing pressure atrophy of the generative tissue. The interstitial cells survived and the experimental animals showed no loss of masculine sexual characters. In other experiments, they showed that if a rabbit was castrated unilaterally and the vas deferens of the intact testicle resected, there was no diminution in sexual activity. When the remaining testicle was examined, however, the generative tissue was seen to have degenerated completely while the interstitial cells had multiplied luxuriantly. It was clear that the sexual characters and the sexual instinct of the male depended entirely on the interstitial tissue.[122]

A major investigator of the testicles during this period was the physiologist Eugen Steinach (1861–1944).[123] Graduating in medicine from the University of Vienna in 1886, Steinach became a full-time physiologist and spent more than two decades at the German University in Prague, where he was initially the First Assistant of the renowned physiologist Ewald Hering (1834–1918). Steinach established a laboratory for general and comparative physiology at Prague, which is supposed to have been the first of its kind in the German-speaking world, and published a number of papers on topics such as the physiology of blood capillaries and the comparative physiology of the iris, most of which he later dismissed as slight.[124] He did, however, consider one of his Prague publications—on the comparative physiology of secondary sexual characters in frogs and rats—to be of high quality and referred to it until the end of his life.[125] This paper marked Steinach's somewhat hesitant entry into the field that would hold his attention until the end of his life: sexual physiology.

It was written in reply to an 1887 study by the Russian physiologist Iwan Romanowitsch Tarkhanow (1848–1908).[126] Tarkhanow had claimed that in male frogs, sexual desire manifested itself only when the seminal vesicles were full. The pressure exerted by the semen on the walls of the vesicles, he argued, triggered off a nervous reflex that led to mating. Steinach disagreed, asserting that the seminal vesicles had nothing to do with sexual maturation or mating. The removal of seminal vesicles in frogs did not diminish sexual activity in frogs during the mating season; nor was mating hampered by empty seminal vesicles. Nor did the removal of seminal vesicles interfere in any way with the sex lives of rats. On introducing females in heat into the cages of these rats,

> What I actually saw almost bordered on the incredible. After the customary dally-ing the operated male rats repeatedly mounted the females, who strongly defended themselves. This sexual battle subsided to a certain extent after two days, but even in the later hours of the evening it was noticeable that the sexual excitability of the operated males, though abated by repeated union with the females, remained unweakened, just as in the control animals.[127]

What did, however, diminish sexual activity in frogs and rats was castra-tion before puberty, although even after this drastic procedure pubescent animals did reveal some sexual excitability. In rats castrated *after* puberty, however, sexual activity was not even reduced.[128] These experiments con-vinced Steinach that although certain degrees of sexual development and activity were independent of the gonads, the mating urge and the secondary sexual characters matured to their adult male form only under the influence of the testicles. He did not, however, explain how the testes exerted their in-fluence. All he said was that an impulse from the testes affected the cerebral centers regulating sexuality. A nervous link was not ruled out, and there was no allusion to internal secretions.[129]

Steinach's experimental research entered a new, remarkable phase after he returned to Vienna in 1912. In the city of Krafft-Ebing, Freud, and Halban (not to mention Otto Weininger and Egon Schiele), Steinach dedicated himself to the study of sex. All of this work was done at the Institute for Experimental Biology, established in 1903 by three independently wealthy biologists and housed in a former animal house called the Vivarium in

Vienna's amusement park, the Prater. The British biologist D'Arcy Thompson commented that it "was known to every naturalist who came to Vienna." Biologists, according to Arthur Koestler, referred to it as the "Sorcerers' Institute."[130] The Institute supported research on every aspect of biology, but the traditional descriptive and comparative modes of biological research were rejected in favor of invasive, innovative experimentation. This was in response to the embryologist Wilhelm Roux's insistence that the purpose of all biological research was to obtain causal explanations of biological processes and phenomena, not simply detailed descriptions. Causal explanations could be constructed only on the basis of experiments that changed and distorted biological processes.[131] The analytical biologist deduced the character of normality from the intentional, planned creation of abnormalities.[132] Planned distortions enabled the investigator to determine how the processes concerned worked under normal conditions. Roux once compared his own experiments to "the insertion of a bomb into a newly established factory . . . with the purpose of drawing an inference about its inner organization."[133] This, essentially, was the approach of the scientists at the Vivarium and, particularly, of Steinach himself.[134] Transplantation was a particularly favored technique of distortion, and Steinach made good use of this procedure.[135]

When Steinach returned to the investigation of sexual physiology around 1910, he built upon the arguments of his 1894 paper against Tarkhanow. Now, however, the concept of internal secretion was emphasized from the start.[136] Admitting the "preexistence" of a degree of sexuality and of some secondary sexual characters, Steinach nonetheless insisted that the attainment of sexual maturity depended entirely on the internal secretions of the gonads. So far as somatic maturity was concerned, the basic importance of the internal secretions was by now widely admitted; what remained to be established was the extent of sex gland influence on the behavioral and mental aspects of sexual maturity, such as sexual interest in the other sex during the breeding season. Were these induced by the purely chemical influence of the sex glands or by nervous impulses, perhaps even nervous impulses from the gonads themselves?[137]

Steinach began by establishing that the central nervous system, outside the breeding season, *inhibited* sexual activity. The elimination of this inhibitory influence by transection of the brain stem elicited the clasp reflex in frogs during any season. Castrated frogs regained their sexual interest on injection of testicular extracts from sexually active males into their spinal lymph sacs: within a few hours after the injection, traces of the clasp reflex began to be elicited and within forty-eight hours, the reflex was fully developed in the vast majority of cases. No other reflex was affected by the injections, and the action on the somatic sex characters (such as the thumb-swellings of mating frogs) was much slower.[138] Steinach then attacked the problem from the other end. Assuming that the internal secretions of the testes must, in order to negate cerebral inhibition, bind with the brain tissue, he injected brain and spinal cord extracts from actively mating frogs into castrates—at which the latter promptly became sexually active. Brain extracts from castrates and intact females did not have any such effect; nor did extracts from muscles or liver.[139] Ovarian extracts exerted a weak and inconstant positive effect, from which Steinach concluded that the ovaries produced a related substance that was conducive to mating. The conclusion from these experiments was that the frog testes secreted a chemical that acted on specific regions of the central nervous system to negate the continuous inhibitory influence exerted by the center on mating behavior.[140]

Steinach then turned to the "rich and more complex" expressions of the mating urge in mammals. As in his earlier paper against Tarkhanow, he chose rats to represent mammals.[141] Feeding castrated infant rats with testicular tissue from freshly slaughtered mature rats, rabbits, and guinea pigs did not, however, produce any effect on sexual development.[142] Steinach, therefore, returned to the idea of transplantation. Well aware of the difficulties encountered by earlier researchers such as Foges, he emphasized the superiority of his technique. Of his forty-four attempted transplants, at least one graft had survived in twenty-seven cases.[143] In those cases, sexual development had proceeded normally, but the animals with two surviving grafts had often become hypersexual. They forcibly mated with females who were not in rut, a phenomenon Steinach had never seen in intact males. The unsuccessful cases did not develop into completely asexual animals. They did recognize females in heat and followed them around, but erection and mating did not occur. In short, not everything in sexual development depended on the glands, but complete sexual maturity was unattainable without their secretions.[144]

These experiments did not simply establish the role of the testes in mammalian sexual development. They showed, Steinach claimed, that the *degree* of somatic and functional masculinity was related to the degree of secretory activity of the testicles. With partially functioning grafts, one obtained partly masculine subjects, and with nonfunctioning grafts, the animals remained sexually undeveloped. As far as sexual behavior was concerned, the development of an immature male animal into a sexually active male depended on the action of the testicular secretions on the central nervous system. This effect Steinach called "erotization" (*Erotisierung*). Once achieved, erotization lasted for a long time even in the absence of the testes, a claim supported by his own earlier studies of rats castrated *after* puberty as well as by reports on the persistent sexuality of human eunuchs.[145]

Was the internal secretory activity of the testis linked with the sperm-producing function of the gland, or were they two autonomous functions? Although approvingly citing the investigations of Bouin and Ancel, Steinach asserted that neither irradiation nor vasoligation completely destroyed the generative tissue. Some regeneration of the germinal tissue was unavoidable. Only his own experiments, he declared, provided conclusive evidence on the matter.[146] The microscopic examination of successful testicular grafts revealed that not a single generative cell had survived in the grafts.[147] The interstitial cells (or Leydig cells), however, had not only been spared but they had proliferated beyond their usual numbers.[148] Autotransplantation experiments, in which the testicles had been excised and re-transplanted elsewhere in the animal's own body, revealed the same phenomenon. These radical claims were not supported with the detailed histological evidence one would expect. Since the animals themselves were behaving in the boisterously sexual ways that were normal for intact rats of their age, it was now obvious, at least to Steinach himself, that the development of the masculine sexual characters depended *entirely* on the interstitial cells.[149] Always fond of coining colorful but not strictly accurate neologisms, he now christened these cells collectively as the "puberty gland" (*Pubertätsdrüse*).[150]

Even if one accepted all of Steinach's contentions, many aspects of puberty gland function remained obscure. Were the effects of the puberty gland sex-specific? Would the transplantation of an ovary be as effective as a testicular graft in masculinizing a castrated male? If it was, then one would have to conclude that the puberty glands of the male and the female were not sex-specific: either could stimulate the development of male *or* female sex characters, i.e., *homologous* as well as *heterologous* sex characters. Secondly, did the puberty glands merely stimulate the development of preexisting elements, or did they transform wholly undifferentiated tissue into sex-specific attributes? One could answer these questions only through experiments involving the transplantation of heterologous sex glands.

Steinach commenced by grafting ovaries in castrated male guinea pigs and rats.[151] All the animals were young enough not to have developed any marked somatic sexual characters at the time of the transplantation.[152] The transplants succeeded in fewer than half the cases. The grafted ovarian tissue showed well-developed follicles on microscopic examination—unlike testicular grafts, ovarian grafts did not lose all the generative elements—but hyperplasia of the interstitial tissue (the puberty gland) was also apparent.[153] The crucial question, of course, was not a histological one at all. Had the successful ovarian graft feminized the castrated males? Only then could one establish that the action of the puberty gland was specific to each sex. If, on the other hand, the animals had been masculinized, then obviously the puberty gland of any sex could induce the development of male as well as female sexual characters.

The ovarian grafts had not in fact stimulated the development of the penis, seminal vesicles, or the prostate in the male castrates. Usually, castration did not lead to immediate cessation of penile growth, but even this exiguous development was prevented in castrates with ovarian implants. "The penis," observed Steinach, "no longer deserves its name and appears to have been reduced to a clitoris." If the ovarian graft did not "take," however, then the penis did grow slightly as expected in castrates. The skeletal dimensions of successfully "ovarianised" castrates acquired characteristic feminine dimensions, and their fur did not grow into the rough and thick coat typical of the male. "Indifferent" features such as the male nipple, areola, and mammary gland grew to female proportions and did so at an accelerated tempo. Even histologically, the mammary tissue was indistinguishable from actual female breast tissue. In the partly successful cases, where the ovarian grafts had been gradually resorbed, the nipples had enlarged during the period the graft was viable and then regressed a little when the graft had perished, remaining at an indeterminate stage between male and female.[154] The functions of the male and the female puberty glands, therefore, were not identical but sex-specific: each stimulated the development of *homologous* sex characters.[155]

Steinach, however, was not concerned solely with the somatic sexual characters but also with sexual behavior. When his "feminized" animals reached puberty, they did not display characteristically male mating behavior and showed no interest in females in heat. Instead, they behaved exactly like females. The feminized rats held their tails high while being pursued by males, which, according to Steinach, was a typically female trait aiding olfactory identification of sex and the degree of heat. An occasional raising of the tail, he conceded, might occur in males—whether intact or castrated—but so sustained a pattern was found only in females. "Real" males, furthermore, would never let themselves be pursued: they would turn around and fight the pursuer. The feminized animals, on the contrary, showed the characteristic defense-reflex of females: the raising of a hind foot and sharp backward strike to prevent being clasped by an unwelcome male. The most decisive sign, of course, was that the feminized animals were treated exactly as females by males.[156]

Intriguingly, these feminized males were often *more* feminine than average females of their kind. The feminized male, for instance, was even smaller in stature than the usual female, and "his" breasts were comparable to the breasts of a pregnant female rather than a young female of "his" age. The feminized males even lactated; infants of the species recognized them as females and followed them around for milk. The experimental animals obliged by suckling them in what Steinach considered to be the typically feminine way. For Steinach, lactation and maternal care of the young constituted "the highest expression of femininity" (*der höchste Grad weiblicher Eigenart*).[157] This "hyperfeminization" of his experimental subjects, according to Steinach, was due to an extraordinary proliferation of the interstitial cells of the ovarian graft.[158]

Steinach now tried to masculinize females by castration and testicular grafting. This proved to be rather more difficult than the feminization of castrated males. There were, however, a few successful cases. In the "masculinized" females, the nipples, mammary glands, and uterus remained uninfluenced. "Indifferent" characters such as build and fur were transformed into typically masculine forms. The behavioral manifestations followed suit. The masculinized females pursued females in heat, fought with other males over possession of females, and clasped them in the typically male way.[159] What remained to be determined were the possibility and consequences of dual transplantation of ovaries and testes in the same individual. Could such a transplantation succeed? And if it did, would it turn the recipient into a hermaphrodite? In castrated animals implanted with testicles and ovaries, prolonged survival of both grafts was not as frequent as the survival of single grafts in his previous experiments. In successful cases, however, the results fulfilled Steinach's expectations. The grafts lost their germinal elements and turned into typical puberty glands.[160]

The experimental "hermaphrodites," when fully grown, were masculine in build and appearance. The female puberty gland's inhibitory influence on musculoskeletal growth, Steinach suggested, had been nullified by the male puberty gland's stimulatory influence. The fur, too, was masculine, and the penis and the seminal vesicles showed no stunting. The male puberty gland could, therefore, induce the development of homologous characters, even in the presence of a functioning female puberty gland. The nipples, areolae, and the mammary tissue of the animals, however, had taken fully feminized forms. Each puberty gland could, therefore, exert its usual stimulatory influence over the homologous characters but failed to inhibit the development of heterologous characters. As soon as the ovarian graft was removed, the breasts returned to the usual male condition and the animal developed in a generally masculine manner. The removal of the testicular graft, on the other hand, led to sustained feminization.[161]

The sexual behavior of the experimental "hermaphrodites' was alternately masculine and feminine. The same animal was at one time attacked as a rival by other males but was subsequently chased by those same animals as an object of sexual interest. The latter (feminine) phase lasted for about two to four weeks, coincided with lactation, and recurred at intervals of two to three months. The female puberty gland, therefore, was *cyclically* active. Feminization occurred only when the secretory levels were at their highest. When the female hormones ran low, the originally male-erotized brain operated in its normal male mode.[162]

Steinach concluded from these experiments that the male and female puberty glands were antagonistic in their functions. Each stimulated the growth of homologous sex characters and inhibited the development of heterologous sex characters. The male puberty gland, for instance, induced the growth of coarse fur, whereas the female puberty gland inhibited it; the female puberty gland stimulated the growth of the uterus while the male puberty gland inhibited uterine development.[163] The antagonism was discernible even at the histological level. When testicles and ovaries were grafted adjacently, they often grew into one another and formed an "ovotestis." These grafts perished even more rapidly than other unsuccessful grafts; when examined microscopically, there was a strong impression of "a battle between tissues" (*ein Kampf zwischen den Geweben*).[164] The difficulties experienced in transplanting ovarian tissue into uncastrated male rats also testified to the power of sex gland antagonism. Clearly, Steinach was conflating ideas about the antagonistic functions of sex gland *secretions* with notions of a direct, histological antagonism between male and female puberty glands, an issue that would later assume considerable importance.[165]

But these experiments did not merely establish the confused and confusing concept of sexual antagonism. They also established that neither somatic nor psycho-behavioral sexual characters were laid down permanently *ab ovo*. Sexual development was under the control of the internal secretions. This control, to be sure, fluctuated across time: the earlier one transplanted an ovary, the greater the feminization of the recipient. The sexual characters stabilized over the individual's life span, becoming less and less amenable (but probably never *absolutely* unamenable) to transformation by the internal secretions.[166] Steinach was a firm believer in the sexual polarity of the gonads. If puberty glands were naturally hermaphroditic, then, he argued, cases of feminization or masculinization would be far more frequent than they were.[167] In this, he was quite old fashioned—Otto Weininger and Sigmund Freud in his own city and many others in neighboring Germany had argued exactly the opposite. Steinach would later come to align himself with them, and some of his later work would actually strengthen the concept that *all* individuals—all human individuals at any rate—were partly male and partly female in a biological sense.

But even this early phase of Steinach's research on sexual biology repays historical attention. At the conceptual level, it gives us a comprehensive series of illustrations of the application of Wilhelm Roux's precepts on distortive experimentation. Roux's urgings, reinforced by the institutional ethos of the Vivarium, encouraged Steinach to appreciate that the best way to explicate sexual development may be to distort, impede, or modify it and then deduce the causal processes of normal sexual development from the results of those experiments.[168] The creation of masculinized and feminized monsters was indispensable to understanding how normal masculinity and normal femininity were developed. The theoretical consequences of the experiments were hardly less interesting. Unlike Halban's view that the endocrine secretions merely brought preformed sexual characters to fruition and "protected" them, Steinach proffered a bold, almost entirely developmental (and endocrine) theory of sex.[169]

Forgotten by endocrinologists and just beginning to interest historians, these experiments were accorded high importance by some of the most prominent scientific contemporaries of Steinach. Nobody in early-twentieth-century Central Europe exercised greater authority on endocrinology than Artur Biedl, and, in the English-speaking world, Francis Hugh Adam Marshall's *Physiology of Reproduction* was the most authoritative conduit for information on new endocrinological approaches to reproductive biology. Both Biedl and Marshall made considerable space for Steinach's experiments, without necessarily agreeing with all of Steinach's claims.[170] As the American biochemist Benjamin Harrow would later observe, Steinach "commanded the respect and admiration of his colleagues throughout the world, though some of his interpretations of his work have not gone unchallenged."[171] Other scientists of the time attempted to replicate Steinach's results and their experiments too were widely discussed and debated in the contemporary medical literature.[172] As a stimulus to research and discussion, Steinach's work was simply nonpareil. Although he never did receive the Nobel Prize, he was nominated for it six times between 1921 and 1938.[173]

Steinach's science, respectable as it was by the standards of the time, was nonetheless a science shaped fundamentally by contemporary ideas of gender. The naked anthropomorphism of his portrayal of sexual behavior in experimental animals was noted even at the time by other scientists.[174] What was novel about Steinach's research, however, was that in spite of its own roots in a traditional, polarized view of masculinity and femininity, its results served ultimately to undermine some of his and his culture's traditional notions of masculinity and femininity. Steinach's experiments, in spite of his own explicit reservations, suggested that the supposed polarity of the sexes was far from absolute. Masculinity and femininity were not immutable givens. Rather, they were chemical phenomena that were malleable at least up to a point. Gender and its attributes, therefore, were always potentially unstable. The few historians who have examined Steinach's work have all remarked on the combative metaphors of struggle and battle that he used to denote the polarity of the sexes. What they have tended to overlook is that by showing repeatedly how easily those polarities could be reversed, his apparently bizarre experiments also subverted the fundamental concept of sexual polarity, including Steinach's own beloved concept of antagonism.

This subversive dimension was reinforced by Steinach's lack of interest in the genetic determination of sex. Although there was growing interest in the genetic basis of sex in the early twentieth century, Steinach never even mentioned it in any of his published writings. In one private letter, however, he conceded that sex *was* indeed determined by genetic factors but insisted nevertheless that—and this, for him, was crucial—the *characteristics* of sex could always be modified by modulating the functions of the sex glands.[175] What fascinated him was the malleability of sex and the growing power of science to modify the expressions of sex. Compared to this, the *determination* of sex faded into metaphysical insignificance. Neither the physical sexual characters nor the complex psychological and behavioral manifestations of sex were governed by the chromosomes. Without the internal secretions an organism would remain, for all practical purposes, sexless. Indeed, Steinach argued that the embryo, at its inception, was neither one-sexed, as had recently been argued by Josef Halban, nor bisexual, as was widely believed by many physicians and scientists of the time.[176] For him, the embryo was *asexual* until the development of the gonads. If a male gonad developed, then sexual differentiation proceeded in one direction; if a female gonad developed, then it proceeded in the opposite direction. Incomplete differentiation of the gonads led to hermaphroditism.[177] With the inappropriate kind of internal secretion, moreover, an organism's sexual characters and behavior might be the reverse of what was determined by the chromosomes.[178]

Even those who had greater respect for the genetic basis of sex than Steinach could agree with the substance of his assertions: German biologist Richard Goldschmidt, after an admiring discussion of Steinach's earliest experiments, agreed that the genetic factors were determinants, while the internal secretions were the "completers."[179] In 1922, British zoologist Julian Huxley put it even more starkly. Chromosomes, in vertebrate species like man, played

> their chief rôle in early development, ending by building up either a male or a female gonad in the early embryo. This, once produced, takes over what remains of the task of sex-determination. It secretes a specific internal secretion which in a male acts so as to encourage the growth of male organs and instincts, to suppress those of females; and vice versa in a female. . . . In mammals the activation of the sexual instincts of one or the other sex appears to be completely or almost completely under the control of the internal secretions of the reproductive organs.[180]

In America, endocrinologist Roy Hoskins observed in 1933: "From the moment of fertilization some organisms are made of male stuff and others of female stuff. Nevertheless in such creatures as ourselves the sex potentialities are not realized except under the secondary influence of the gonad hormones."[181]

The message, then, was clear and simple: genetic determination of sex was essential but not sufficient. (Nor, Steinach may have added in an undertone, was it very interesting to the physiologist.) Broadly speaking, most clinicians and even biologists seem to have experienced little difficulty in accepting that although the genetic determination of sex was of primary significance, the soma of vertebrates remained sexually "bipotential" to a certain extent, or, in Knud Sand's terminology, "versible." The sexual characters could, therefore, be modified in the direction of the other sex by natural or experimental factors, of which hormonal interventions were, of course, the most obvious.[182] Further research served only to put the chromosomes ever more firmly in their place. Having reported the cases of two happily married "females" who turned out not only to possess testicles but to be chromosomally male (i.e., XY) but wished to continue being "women," the biologist Emil Witschi and gynecologist William Mengert suggested in 1942 that the pervasive feminization of these hermaphrodites may well have been due to modifier genes and hormonal influence from their mother. They admitted in conclusion: "We had become accustomed to look at human sex determination as a solved problem, as a toss-up between X and Y chromosomes. . . . We begin to realize that aberrations due to modifying genes and special hormonal conditions are much more prevalent than ever suspected."[183] In 1944, British surgeon Lennox Broster, an expert on virilism caused by adrenal problems in women, declared that "the sex chromosome mechanism is the sex determining mechanism, but under certain circumstances this may be overridden and the sex determined in other ways."[184] Steinach's views on the genetic determination of sex, in short, were far from maverick.

More broadly, Steinach's research encapsulates the drama and the potential of early research on gonadal functions. Although based in conservative notions of gender polarity, endocrine concepts of the development of the sexual body soon challenged those founding principles, leading in turn to scientific and medical initiatives to reestablish traditional categories and reconceptualize the sexual body. All of this, moreover, occurred against the backdrop of growing popular interest in glandular science and its apparently boundless possibilities. It is to those complicated themes that we must now turn.

Chapter 3

Sexuality, Aging, and the Gonads: New Physiology and Old Values

> Gonadal hormones, transfusion of young blood, magnesium salts....
> All the physiological stigmata of old age have been abolished . . . at sixty
> our powers and tastes are what they were at seventeen. Old men in
> the bad old days used to renounce, retire, take to religion, spend their
> time reading, thinking—thinking! Now—such is progress—the old men
> work, the old men copulate, the old men have no time, no leisure from
> pleasure, not a moment to sit down and think.
> —ALDOUS HUXLEY[1]

The 1920s were the heroic age of the endocrine glands, and specifically of the gonads. The role of the sex glands in forming and maintaining the sexual body was now more or less accepted, and cultural and medical attention turned toward the *correction* of deficiencies and deviations of the sexual body. If the sex glands generated the perfect male or female body, then could imperfect or otherwise aberrant specimens of those bodies be perfected or, at least, normalized by glandular means? The interwar decades were the heyday of such utopian projects, and there was some apparent justification for this faith in the glands. The availability of potent extracts of the thyroid and the adrenals in the waning years of the nineteenth century had endowed organotherapy with new respectability, and at the beginning of the 1920s came insulin, with its miraculous effects on diabetics.[2] Even though "none of the hormones later discovered," as David Hamilton reminds us, "had quite the revolutionary effect on clinical medicine as the thyroid extracts and insulin," it seemed during the twenties that glands and their secretions held the key to vast, hitherto unmanageable areas of life.[3]

The "Chemical Perfectibility of Human Life"

Glands preoccupied everybody after the Great War: scientists as well as journalists, clinicians as well as physiologists, novelists as well as scientific popularizers. Some of the most intractable problems—crime, senility, feeblemindedness—could, it seemed, be resolved by glandular magic.[4] Nowhere was this optimism more intense than in the United States. Some of the greatest popularizers and proselytizers of the new science were American. One, the biochemist Louis Berman, was particularly eloquent. Intelligence and temperament in children were not adequately stimulated by the traditional techniques of suggestion and education, he declared. It would make children far brighter if they were fed with glandular extracts, which acted "upon the roots of both the chemical constituents and the reactions of the cells."[5] "We are the creatures of these glands!," exclaimed another endocrine popularizer. "Why Johnny is slow with his studies, while Willie is at the head of his class, is merely the difference in the amount of chemical fluid produced by a little gland in his throat.... [The internal secretions] are not only the arbiters of reactions and emotions, but... they actually control character and temperaments, whether for good or ill."[6]

"Glands! Potent word in our present state of scientific knowledge!," exclaimed endocrinologist Max Goldzieher in 1935. "In its longing for a universal panacea, it [the public] would make wholesale and unjustified application of each tiny fact in glandular therapy as it is announced." Goldzieher wrote these critical words, however, in an introduction to an educational psychologist's treatise advocating the feeding of thyroid and pituitary substance to "slow" and "defective" children. Such glandular treatment would, the author claimed, boost national efficiency and develop "inadequate human material into efficient adults."[7] In Germany in 1933, psychiatrist Walter Cimbal proclaimed that endocrine research was the key to improving the mental functioning of Germany's undeveloped post-Great War generation.[8] Others used glandular science to explain criminality. An American treatise on "the new criminology" asserted in 1928 that most crimes were caused by "disturbances of the ductless glands in the criminal or through mental defects caused by endocrine troubles in the criminal's mother."[9] "Perverts of the Jack-the-Ripper type, the individual showing definite traits of Sadism, masochism and other aberrations, are not only of distinctly abnormal endocrine types," wrote an American physician in 1928, "but could even be relieved of these abnormalities through proper treatment of their glands."[10] There was no need for society to concern itself with improving people's conditions of life, because the environment did not produce criminals. In every case, the determining factor was "the mental, nervous and glandular soundness of the individual."[11] There seemed to be no limits to what glandular treatment could do to improve society.

This kind of hypothesis spread quickly to the popular sphere. In Dorothy Sayers's 1921 novel, *The Unpleasantness at the Bellona Club*, a society lady whose daughter is engaged to an aspiring gland doctor declared that for thousands of years, society had been wrong about criminals: "Flogging and bread-and-water, you know, and Holy Communion, when what they really needed was a little bit of rabbit-gland or something to make them just as good as gold."[12] In Agatha Christie's *The Murder at the Vicarage*, published in 1930, the village doctor observed: "Too much of one gland, too little of another—and you get your murderer, your thief, your habitual criminal. . . . I believe the time will come when we'll be horrified to think of the long centuries in which we've punished people for disease—which they can't help, poor devils. You don't hang a man for having tuberculosis."[13]

A notion of distinct endocrine temperaments or personalities also emerged. Louis Berman was one of the earliest and certainly one of the most flamboyant exponents of the art—and the acknowledged or unacknowledged source for many a later writer on glands. "A man's nature," he declared, "is essentially his endocrine nature"; extraordinary natures, it followed, must be based in extraordinary "endocrine traits."[14] Broadly speaking, human beings were divisible into "the adrenal centered, the thyroid centered, the thymus centered, the pituitary centered, the gonad centered, and their combinations."[15] These types were the "great kingdoms of personality," but variations and intermixtures, of course, were common. Berman proffered much evidence for his contentions from the lives of the famous or the infamous. Charles Darwin, for example, "was a poor animal, the poorest of animals, because he possessed poor adrenals. What saved him was his congenitally superior pituitary (the nidus of genius) and the overacting thyroid."[16] Napoleon's "sharply outlined features and a powerful lower jaw, combined with oddly small plump hands, long straight black hair, and dark complexion, all point to the pituitary, with a secondary adrenal effect."[17] Studying the photographs of thirty-four winners of the Congressional Medal of Honor for outstanding courage in the Great War, Berman triumphantly noted that "twenty-five exhibited the somatic criteria or hormonic signs of the ante-pituitary type." The anterior pituitary was the source of courage (with some assistance from the adrenals), intelligence, and judgment, and this fountainhead of all that was good in man, he informed his readers, was governed by the gonads.[18] Other writers followed Berman with ever-more hormonally colored explanations of the personalities of historical figures such as Julius Caesar, the Prophet Muhammad, Byron, Flaubert, and Dostoyevsky.[19]

Endocrine personality theory was not the preserve of popularizers alone. In a work on the differential diagnosis of gland disorders, Allan Winter Rowe, Director of Research at the Evans Memorial Hospital, Boston, and a leading endocrine clinician of the time, described those with depressed thyroid function as "self-centered, negative, and truculent" and those with depressed ovarian function as "both voluble and shrill in denunciation of a social environment, unappreciative alike of her many excellencies and profound sufferings." The latter kind of personality, Rowe added, "presents fertile soil for the growth of the major forms of hysteria."[20] Practical implications were easily imaginable. As one physician explained, if a woman, after some years of a happy marriage, turned into a nag and the husband asked for a divorce, one should investigate the wife's ovaries. Perhaps they had gone sluggish and the marriage could be saved by treatment with ovarian substance. Science, the author of this formula happily reported, "can now change love's chemistry to the original condition that gave connubial bliss. There is no longer any necessity for law courts and divorce. Science can do what the law cannot do, restore the original impulse."[21]

Some of these psycho-glandular discussions frequently took a swipe at psychoanalysis.[22] Gynecologist Samuel Bandler, for example, agreed that "we are, in the final analysis, very much the expression of the activities of the endocrines" and that our subconscious mental functions depended on gland-induced instincts and emotions. "How much closer to the true and wholesome are we led by these facts," he exclaimed, "than by the aberrations of the theories of Freud."[23] Louis Berman was only slightly more respectful toward psychoanalysis. The "dark places in human nature seem to have become the sole monopoly of the Freudians and their psychology. But only seemingly," he observed. "For all this time the physiologist has been working. Beginning with a candle and now holding in his hands the most powerful arc-lights, he has explored two regions, the sympathetic nervous system and the glands of internal secretion, and has come upon data which in due course will render a good many of the Freudian dicta obsolete. Not that the Freudian fundamentals will be scrapped completely. But they will have to fit into the great synthesis which must form the basis of any control of the future of human nature. That future belongs to the physiologist."[24]

Others agreed. Gynecologist Bandler prophesied in 1921 that in a few years, there would be very few cases of mental retardation, insanity, cancer, or diabetes, because endocrine treatment would cure them very quickly in their early stages. "When the next war comes, if it does at all," Bandler continued, "soldiers before going over the top will not be given alcohol: they will be given endocrine cocktails and the adrenal cortex will be an important ingredient. And if the world, in the near future administer to its diplomats, to its highest officials, to its legislators, and to its people the proper endocrines, especially anterior pituitary, and inhibits the adrenal cortex a little bit, there may be no more wars."[25] Clearly, much of this speculation expressed not simply the urge to comprehend human individuality in one simple scheme but the hope that deviations from accepted norms could, in the future, be *controlled* by glandular treatment. As the science of the glands progressed, the disturbances of physique, conduct, and character brought about by endocrine imbalance could be corrected.[26]

The internal secretions governed human nature, and whoever controlled the glands would "control human nature." The future, promised Louis Berman, would bring the "chemical perfectibility of human life." Rhapsodizing over the omnipotence of the glands, he prognosticated: "Internal glandular analysis may become legally compulsory for those about to mate before the end of the present century." Endocrine enlightenment would bring a truly positive eugenics and ensure "selective breeding for the production of the best endocrine types. . . . It should become possible to produce new mutations, good and bad. . . . All the physical traits, stature, color, muscle function, and so on, offer themselves for improvement, as well as brain size, and the intellectual and emotional factors which have dominated man's social evolution." Knowledge of the glands offered nothing short of a whole new science of man or, indeed, a new heaven and a new earth: "The chemical conditions of [man's] being, including the internal secretions, are the steps of the ladder by which he will climb to those dizzy heights where he will stretch out his hands and find himself a God."[27]

Even some of the most serious research of the time—and there *was* plenty of such research—might leave a modern endocrinologist nonplused, as it did one of their forebears. In 1923, Swale Vincent, a veteran endocrine researcher and Professor of Physiology at the University of London, warned that the burgeoning science was in danger of turning into "a formidable kind of quackery." "There is no subject upon which so much utter nonsense has been talked as upon internal secretion," he scolded, "and organotherapy, or at any rate a large part of it, may be defined as the application of this nonsense to practical medicine."[28] Vincent's, however, was a minority voice.[29]

His remarks were made at a meeting of the Royal Society of Medicine, which was chaired by Sir Walter Langdon-Brown, an enthusiast for the internal secretions who, although welcoming Vincent's "cold douche of scepticism" as "refreshing," was a great believer in an empirical organotherapy as long as it was conducted in a critical spirit. "The empirical method," Langdon-Brown countered, "has always been suspect by the laboratory . . . unfairly suspect, when one remembers that cod-liver oil was used for many years, to the scorn of the experimentalist, until vitamins were discovered, and then he scoffed no more."[30]

Many champions of the new science worried about quackery, but few shared Vincent's views in full. As eminent a scientist as Ernest H. Starling, Jodrell Professor of Physiology at University College London and the man who, in 1905, had introduced the term "hormone," declared that endocrine science was "almost a fairy tale." Moving to sex and its origin, he was in no doubt that "the whole differentiation of sex, and the formation of secondary sexual characters, are determined by the circulation in the blood of chemical substances."[31] Or take Artur Biedl, Professor of Experimental Pathology at the German University in Prague and the author of the first European textbook on endocrinology. "The endocrines singly and in their totality," he asserted in the mid-1920s, "have an effectual influence on birth, formation, and development, on the somatic and psychic condition of the individual . . . they determine his metabolism and the activity of all his tissues, organs and systems. The endocrine formula of a person is his fate."[32] These words were worthy of Louis Berman himself.

The medical correspondent of *The New York Times*, alluding to the "almost continuous hysterical manifestations of concern for . . . glandular welfare," complained in 1922 that "a war-ridden world has given place to a gland-ridden world . . . until a long-suffering public is really in doubt as to whether world comity or the world's glandular condition should occupy first place in local, State, national and international deliberations."[33] His own newspaper, however, in item after enthusiastic item, disseminated the utopian message that glands would change the world. And it was far from alone. Even Swale Vincent had to acknowledge that "hardly a branch of biological science does not now call in the assistance of endocrine hypotheses. . . . The conception of internal secretion has also invaded the realms of the social, ethical, and moral sciences. Even the doctrine of original sin has been abandoned in some quarters for that of endocrine perversion, and the bad behaviour of children and adults has been attributed to disordered function of the ductless glands."[34]

Vincent's heart was so unreceptive to the magic and romance of the glands that he could not even understand why the world had suddenly gone gland-crazy. American biochemist Benjamin Harrow, although not a contributor to the glandular hysteria of the time, was far more perceptive. "The achievements [of endocrine science], judged by rigid scientific standards, are no more than modest," he remarked, "but the possibilities are limitless. It is because of these vast possibilities that an imagination, not sufficiently tempered by self-criticism, is apt to enlarge a molehill into a mountain."[35] Glandular research had become the symbol of a glorious future when the human organism would be physically and mentally re-engineered into a more perfect form of life.

In this brave new world, sex would play a central role. To quote the ever-eloquent Berman, "over no domain of the body have the endocrines a more absolute mandate than over that of the whole complex of sex. Both as regards the primary reproductive organs, their size and shape . . . as well as the physical and mental traits lumped as the secondary sexual, puberty, maturity, and senility, voice changes and erotic trends, virility and femininity, the internal secretions are dictators at every step."[36] And if the sex glands controlled sexual identity and sexual behavior, then could human sexual anomalies be cured by endocrine means? What glands could bring about and control, glandular treatment could surely modify: *that* was the grand tune to which science as well as popular culture marched in the twenties.[37]

Glands, Sexual Identity, and the Therapy of Desire

Some of those therapeutic dreams had already begun to materialize in Steinach's Vienna. Not content with changing the sex of rats in his laboratory, he had marched into the clinic and larger society in 1918 with his remarkable claim that he had discovered a cure for male homosexuality.[38] He had, as we know, already speculated that human hermaphroditism and homosexuality might be consequences of sexually undifferentiated gonads.[39] At one level, the 1918 announcement merely took this hypothesis to its logical conclusion. It was, however, more than a matter of logic alone.

From the late nineteenth century, the scientific study of sexuality and the treatment of "perversions" was an area where the interaction of "science" and "society" was particularly dramatic, intimate, and explicit. Steinach's involvement in the treatment of homosexuality was no exception. "The nineteenth-century homosexual became," in the words of Michel Foucault, "a personage, a past, a case history, and a childhood, in addition to being a type of life, a life form, and a morphology, with an indiscreet anatomy and possibly a mysterious physiology."[40] Pursued most actively in the German-speaking lands, this medicalization of homosexuality was concerned almost exclusively with males, and biological explanations of sexual orientation were often pushed most energetically by emancipationists demanding the social, moral, and legal destigmatization of homosexuality. If sexual orientation was biologically determined, then an aberration such as homosexuality was neither a vice nor a progressive disease but a simple biological anomaly like a cleft lip.[41]

One of the most prominent upholders of this theory was Magnus Hirschfeld (1868–1935), a practicing physician in Charlottenburg, Berlin, and the founder of the Institute for Sexual Science, a privately run institution dedicated to research on all aspects of human sexuality, and male homosexuality in particular.[42] From 1896 until the outbreak of World War I, Hirschfeld published extensively and almost exclusively on homosexuality and appeared frequently in court as an expert witness. As a clinical researcher, he interviewed and examined numerous homosexuals: his works, crammed with case histories, statistics, and occasionally, photographs, spoke the language of empirical science.[43] Describing his knowledge of homosexuality as "unequalled," British sexologist Havelock Ellis hailed Hirschfeld's 1914 treatise *Die Homosexualität des Mannes und des Weibes* (Homosexuality in Man and Woman) as "not only the largest but the most precise, detailed, and comprehensive . . . work which has yet appeared on the subject."[44]

Not simply an expert, Hirschfeld, from his days in medical school, had aspired to use science to end legal and cultural discrimination against homosexuals. In 1897, he and some of his associates established the Scientific-Humanitarian Committee, which demanded equal rights for homosexuals with an intensity, consistency, and tenacity unmatched at the time by other groups, whether pro-homosexual or anti-homosexual. The Committee, which included representatives of a variety of professions, was guided by a scientistic philosophy reflected in its motto: "justice through science." The Committee petitioned the Reichstag repeatedly for a repeal of Paragraph 175, the German statute on sodomy; held lectures and public meetings to disseminate "scientific" information on homosexuality; and published the *Jahrbuch für sexuelle Zwischenstufen* (Yearbook for Sexually Intermediate Forms) from 1899 until inflation killed it in 1923.[45] The central motif of the Committee's argument was explicitly biological. Its very first petition to the Reichstag, which was composed by Hirschfeld and submitted in 1897, asserted that recent scientific research had determined that homosexuality was the consequence of a simple developmental error. No legal or moral guilt could attach to so involuntary a condition.[46]

Late-nineteenth-century physicians had regarded homosexuality as the manifestation of degeneration, a protean and vaguely conceptualized hereditary pathological condition that blighted the entire organism, revealing itself through diverse behavioral as well as physical abnormalities.[47] Alcoholism, tuberculosis, and homosexuality, for example, were all *signs* of underlying degeneration. So were cleft-lips, misshapen ears, and unretractable foreskins. These conditions were not necessarily hereditary, but the degenerate taint that brought them about was. It was not just inherited but amplified over generations—its manifestations assuming greater and greater severity—until the last member of the tainted line was killed off by the sheer burden of accumulated pathology.[48] The degenerationist view of homosexuality was displaced around the turn of the century by a newer developmental theory, which explained homosexuality as a simple anomaly that was, strictly speaking, pathological but not a sign of underlying degeneration.[49]

Embryologists of the time believed that human genitals developed in fetal life from a sexually undifferentiated rudiment into distinct male or female forms. Evolutionary biologists pointed out that such an ontogeny faithfully reflected the phylogenetic descent of the human species from hermaphroditic ancestors.[50] It was well-known, however, that this ontogenetic process did break down occasionally, leading to the birth of a hermaphrodite. Developmental theories of homosexuality claimed that what was possible for the genitalia could well be possible for the brain and the psyche.[51] Due to an error of development, a human embryo might be born with male genitals but, to use the words of American physician James Kiernan, a "femininely functionating brain."[52] The development of the genitals and that of the psyche, in short, could follow different paths, leading to different outcomes. The German classicist, lawyer, and homosexual activist Karl Heinrich Ulrichs (1825–95) had argued similarly long ago, claiming that the male homosexual represented "a female soul in a male body."[53]

Magnus Hirschfeld and the activists of the Scientific-Humanitarian Committee seized upon this developmental hypothesis of homosexuality with great enthusiasm—but with one crucial modification. Hirschfeld denied that the developmental error in homosexuality led to the feminization of the psyche alone. Acknowledging that the male homosexual almost invariably possessed normal genitalia, Hirschfeld pointed out that the rest of his body, however, was neither clearly male nor clearly female. It was, rather, a harmonious fusion of the two. In 1903, he declared: "Of the 1500 homosexuals that I have seen, each was physically and mentally distinct from a complete male."[54] Homosexuality, in short, was a morphological but nongenital form of hermaphroditism.[55] Homosexuals, Hirschfeld asserted, usually presented numerous subtle and multiple signs of morphological sexual ambiguity, which were obvious to a trained observer.[56]

Fundamental to Hirschfeld's approach was what he called the principle of universal sexual intermediacy. This stipulated that there was no absolute, qualitative distinction between the male and the female. All humans were placed on a spectrum stretching between the hypothetical poles of absolute masculinity and absolute femininity. *Every* human being was partly male *and* partly female (i.e., *intermediate* between the absolute male and the absolute female) but the *degrees* of maleness and femaleness varied between individuals.[57] Sexual intermediacy, furthermore, was not confined to the body or the mind. The genital hermaphrodite was, obviously, an intermediate form, but so were the male homosexual, who was more female than the average man, and the adolescent tomboy, who was more male than the average girl of her age. Qualitatively, they all belonged together, representing different degrees of sexual intermediacy. But although much of what Hirschfeld wrote suggests that he thought of the human species as naturally but variably androgynous, he was inconsistent enough to argue that the instances of sexual intermediacy had come into being due to *anomalous* biological processes, which needed to be investigated.[58]

From 1912, Hirschfeld had speculated about possible chemical bases of gender and sexual behavior, and he had, in fact, followed Eugen Steinach's experiments with interest.[59] Shortly before the outbreak of the Great War, Hirschfeld had traveled to Vienna, visited Steinach in his laboratory, examined the masculinized and feminized animals, and published appreciative articles on the experiments. Hirschfeld had even suggested that Steinach transplant ovaries and testes simultaneously in a castrated animal and had been delighted to learn that Steinach had independently commenced such experiments.[60] Their relationship was to reinforce the work of each. Hirschfeld's clinical work, of course, was part of his broader political program for the emancipation of homosexuals, and he used Steinach, in effect, as a political ally. Steinach made good use of Hirschfeld's clinical evidence that male homosexuals were not simply psychosexually drawn to men but were somatically feminized as well.[61]

If homosexuality was due to a glandular anomaly, could it be corrected? Hirschfeld had initially dismissed this as an impossible dream.[62] Not so Steinach. What, he asked, would happen if the testicles of a homosexual were removed and replaced with "normal" glands?[63] Others, too, had asked that same question. In 1916, biologist Richard Goldschmidt, well known for his studies of intersexuality in moths and butterflies and well acquainted with Steinach's laboratory studies, had suggested that homosexual orientation was probably due to deficiencies of gonadal secretion and might, therefore, be cured by gonadal extracts or the implantation of a "normal" gonad.[64] In 1917, the Leipzig physician and specialist in sexual disorders, Hermann Rohleder, a great admirer of Steinach's experimental studies and much influenced by Hirschfeld's clinical reports, had speculated that in the most severe cases of homosexuality, the implantation of a "normal" testicle should masculinize the male homosexual.[65] But it was Steinach who was to act on the hunch.

Transplantation of the testicles was not a novel technique at the time. Reports of ostensibly successful transplants of testicular tissue between individual men had been appearing in the medical literature for some years.[66] In 1916, American urologist G. Frank Lydston had reported four cases of transplantation conducted for various indications, including "defective or aberrant psychic or physical sex development and differentiation—inversions and perversions." Some of his patients had lost their testicles through infection or trauma and then developed what Lydston called effeminate bodies.[67] The results of the transplantation were frequently impressive.[68] In Steinach's own circle, the urologist Robert Lichtenstern, who had once worked in Steinach's laboratory, had also reported remarkable success with the operation. In 1915, Lichtenstern had been consulted by a soldier who had lost both his testes in war and subsequently lost his libido completely. The soldier was also losing his body hair, putting on fat, especially around the neck, and was profoundly depressed about his condition. Around the same time, Lichtenstern had to remove the undescended testicle of another patient under his care and decided to transplant the removed gland into the soldier. Steinach was present at the operation. The patient experienced lasting improvement in his condition after the grafting: within four weeks, his sex life was back to normal, the facial and body hair had started to grow and the deposits of fat that had appeared since the castration had disappeared.[69]

Lichtenstern and Steinach were now convinced that testicular grafting exerted definite masculinizing effects on the recipient. There was, then, no reason not to try it on a homosexual. Their first subject was a thirty-year-old homosexual, whose testes needed to be removed in any case because of tuberculosis. Steinach emphasized that the subject had a rounded, effeminate body; this, of course, suggested a glandular anomaly. The donor was a healthy and sexually "normal" man, who had an undescended testis that needed to be removed. Twelve days after Lichtenstern performed the "surgical exchange," the homosexual subject reported having erections and erotic dreams of a heterosexual nature. He had sex with a female prostitute six weeks after surgery, and many times subsequently. Gradually, his voice deepened and his body became more masculine. In less than a year, he married, and wrote to his physicians: "My wife is very satisfied with me. . . . I am disgusted to think of the time when I felt that other passion." Steinach and Lichtenstern discounted the possibility that this metamorphosis had resulted from suggestive influences, pointing out that the patient had also become somatically masculinized after the transplantation.[70]

The operation also produced expected results in several other cases. Steinach and Lichtenstern did not recommend its unrestricted use in cases of homosexuality, claiming merely to have demonstrated one way of overcoming a condition that was "unpleasant and dangerous" for affected individuals as well as for society in general. Steinach followed up Lichtenstern's cases with histological examinations of the removed testes. There were only five specimens. They contained very few interstitial cells but more intriguingly, showed many large, unusual cells that were not normally present in testicular tissue. Reminiscent of the lutein cells of the ovary, their microscopic appearance was so characteristic that even a physician with little experience of histological work, Steinach claimed, could spot them without difficulty. These were christened F-Cells and Steinach was confident that here was the source of the female hormones that made the male homosexual desire men.[71]

Scientists and clinicians debated Steinach's theory vigorously. Biologist Julian Huxley, in a 1922 lecture to the British Society for Sex Psychology, observed that while much more research was needed to decide the issue, there was "no theoretical objection to the possibility."[72] A Dutch physician suggested that Steinach's findings explained not simply true homosexuality but also the so-called pseudohomosexual activities amongst prisoners and boarding school boys. The pseudohomosexuals were actually bisexuals with incompletely differentiated sex glands. Although the male hormonal element might eventually prove dominant, the female element was strong enough to manifest itself in youth.[73] Other responses to Steinach's homosexuality-cure were more ambivalent. Some clinicians, while not denying that homosexuality might have *a* glandular cause, pointed out that homosexuals and their behavioral patterns were too diverse to be caused by any one factor.[74] To my knowledge, nobody in the course of the debate even alluded to the possible glandular basis of *female* homosexuality, although, before Steinach's work hit the medical headlines, Hermann Rohleder had suggested ovarian irradiation in females with strong homosexual tendencies.[75]

At this point in time, the psychoanalytic explanation of homosexuality offered little opposition to Steinach's glandular hypothesis. Freud himself was impressed enough to declare that psychoanalysis aimed only to identify the "psychical mechanisms" of sexual object choice and their instinctual origins: "There its work ends, and it leaves the rest to biological research, which has recently brought to light, through Steinach's experiments, such very important results."[76] Steinach himself reportedly told his admirer, the German-American journalist George Sylvester Viereck (1884–1962), that in homosexuality "the [testicular transplantation] operation, to be successful, should be followed, as a rule, by psycho-analysis." "The surgeon," explained Viereck, "coöperating with the psycho-analyst, restores endocrine and psychic harmony—a task that would be too difficult for either without the aid of the other."[77]

Among surgeons, Lichtenstern's first transplant led to a brief vogue for such operations.[78] Richard Mühsam reported that he had successfully reversed homosexuality in two cases referred to him by Hirschfeld but added that histological examination had failed to find any F-cells in the removed testicles.[79] There was worse to come. Mühsam found that although testicular transplantations led to quick heterosexualization, evinced by erotic dreams, it was a fleeting phenomenon and dissipated quite rapidly. By 1926, he declared that the clinical results of the operation were so poor that he had stopped performing it.[80] Other surgeons followed a similar path.[81] American surgeon Max Thorek, one of Steinach's transient allies, recalled how Steinach, "over some fine Mocha in his home in Vienna," had explained his theory of homosexuality to Thorek and his wife. Impressed, Thorek had tried the operation in one male homosexual, but it was "entirely unsuccessful" and he never attempted it again.[82] The strongest opposition came from histologists. The so-called F-cells, they declared, were simply atypical forms of usual constituents of testicular tissue.[83]

Only Hirschfeld remained steadfast in his support. He publicized the transplant operation in his journal, referred patients to surgeons for the operation, and himself conducted a histological investigation of testicular tissue from homosexuals in 1920.[84] Even he and his associates, however, failed to find F-cells in those specimens, although they did not let that prevent them from deciding that the specimens did not quite measure up to histological norms.[85] In that same year, Hirschfeld declared that Steinach's claims would undoubtedly be upheld by future research, establishing once and for all that homosexuality was completely biological and congenital. Rather than Ulrichs's "female soul," it was a female *gland* within a male body that generated a homosexual man's desire for other men.[86]

For Hirschfeld, Steinach's theory was not just another hypothesis—he needed it scientifically as well as politically. Steinach's experimental findings, Hirschfeld claimed, had established the validity of his own clinical hypotheses about the constitutional femininity of male homosexuals.[87] In debates with orthodox physicians, some of whom (such as the renowned psychiatrist Emil Kraepelin) continued to believe that homosexuality could be acquired after seduction and should not, therefore, be decriminalized, Hirschfeld responded by citing Steinach's "demonstration" of the innate biological nature of homosexuality. Homosexuality could never be acquired unless one's biology had already made one a homosexual.[88]

The absence of confirmatory evidence for Steinach's theory soon made these claims sound quite weak. When the not overtly unsympathetic psychiatrist Kurt Blum reviewed a large sample of all applicable studies in 1923, he concluded that Steinach's endocrine theory of homosexuality had failed to find any convincing confirmation, which, Blum emphasized, meant that Hirschfeld's congenitalist view of homosexuality, too, remained unproven.[89] Hirschfeld finally lapsed into silence, and in one of his last major works, transcribed without comment the following passage from the letter of a patient who had undergone the Steinach transplantation but remained homosexual in orientation: "The value of Steinach's gland transplantation was greatly overvalued in medical circles of those times. I have examined the literature without finding a single case in which the transplantation produced a lasting effect."[90] Steinach, too, soon fell silent on the homosexuality-cure. At the very end of his life, however, he argued that F-cells could well be found in the testicles of heterosexuals because most human males "contain in their organisms minute vestigia of a female character even though under normal conditions they never come to functional expression."[91] He did not, however, explain why other investigators had never managed to find the F-cells in the testicles of *homosexuals.*

"Dr. Steinach Coming to Make Old Young!"

> That the effective years of life might be materially extended is not at all improbable. Living tissue does not necessarily age. In the culture tube, cells even of the higher animals can be propagated generation after generation without loss of vigor. Had we but the necessary knowledge, the cells of the body, too, might be preserved for longer periods in an efficient state. The control of their metabolism is the core of the problem. The spring of eternal youth will probably never be found, but old age may at least be made more livable if not significantly deferred.
>
> —R. G. HOSKINS[92]

Homosexuality was but one of the riddles that Steinach's glandular research sought to explain. Senility was another. A committed developmental mechanist, Steinach heeded Wilhelm Roux's call to analyze not simply the origin and maintenance of organic forms but also their involution. There was reason to suspect that senility reflected a decline in sex gland function. Common experience suggested that the differences of gender were blurred in the extremes of life:

> Just as it is often difficult to distinguish between the face of a little girl and that of a little boy, so the shaven face of an old man resembles that of an old woman.... Naturally temperament and disposition begin to lose their typical expression for the different sexes, the old man revealing only traces of his former masculine aggressiveness and the old woman but feeble remnants of her modesty and forbearance.[93]

The somatic sexual characters (such as the seminal vesicles in males) were present at all ages, developed fully only after puberty under the influence of the sex-glands, but regressed to near-infantile dimensions in old age. Interestingly, the shriveled seminal vesicles of elderly rats resembled the undeveloped ones of those castrated before puberty. Senility, in short, could be the result of biological desexualization.

Steinach, as we have seen, was convinced that it was the interstitial cells of the testes that produced the internal secretions of the gland—if one could induce the proliferation of those cells, he argued, then the secretory deficits of senility ought to be compensated.[94] Theoretically, one could provide the interstitial cells with the space to proliferate by destroying the germinal cells. Since it was well-known that ligation of the vas deferens caused germinal atrophy, Steinach attempted to rejuvenate senile male rats by bilateral vasectomy.[95] Within a few weeks of the operation, he reported, the previously lethargic, underweight, and almost lifeless rats had become active, gained weight, developed a glossy new fur, and regained sexual interest. Some hitherto decrepit animals were now tireless sexual performers, having intercourse as many as nineteen times in fifteen minutes.[96] (His rats, Steinach claimed with Benthamite fervor, were housed in cages that were under constant observation. "Their whole lives and all their drives are perpetually visible to the experimenter. No change in temperament, no rise or fall in strength and mobility, no indication of lust or absence thereof is hidden from the examining gaze.")[97] The seminal vesicles, too, had regained their former dimensions and microscopically, the testicles showed marked proliferation of the interstitial cells and degeneration of the germinal tissue. Many of these striking findings were reported in a brief communication issued in 1912, which concluded with a promise to explore the applicability of this technique to humans.[98]

Eight years later appeared Steinach's monograph-length paper on rejuvenation, dedicated to Wilhelm Roux on the occasion of his seventy-fifth birthday and published in Roux's own journal. It caused an immediate uproar not because of its exhaustively detailed accounts of experiments on rats but because it incorporated the case-histories of three human subjects of the operation.[99] Impressed by the results of his experiments on rats, Steinach, in 1918, had requested his old associate, the urologist Robert Lichtenstern, to determine whether aging men could be rejuvenated as easily by vasectomy as elderly laboratory rats.[100] Lichtenstern, already struck by the virilizing effects of testicular transplantation on elderly men, was keen to explore whether the same results could be brought about by reactivating an underfunctioning "puberty gland" by Steinach's procedure.[101] He performed his first "Steinachs" on elderly or prematurely aged men during operations for hydrocele, prostatic hypertrophy, or testicular abscess.[102] It was essential, he emphasized, to perform the procedure without informing the patient in order to exclude suggestion.[103]

The very first patient was a coachman, only 44 years of age, but presenting "a typical picture of premature senility without organic disease." He had lately been unable to work for long hours and had lost weight and appetite. His skin was dull, his hair grey and scanty, and his muscles weak. The patient also had a hydrocele, and the vasoligature was performed without his knowledge during an operation for the former, performed on 1 November 1918. There were no dramatic consequences for the first three months. Then gradually, his appetite increased, he gained weight, and his appearance became hale and hearty; a year later, his hair had grown thicker and he reported that he now carried "loads up to 220 pounds with ease." His hair grew so luxuriantly that he had to shave twice as often as earlier and have his hair cut twice as frequently.[104] Eighteen months after the operation, the patient "with his smooth, unwrinkled face, his smart and upright bearing" looked like "a youthful man at the height of his vitality."[105]

The news spread like wildfire with reports of successful rejuvenation by the Steinach operation soon pouring in from all over the world.[106] The *New York Times* reported in 1923 that every major American and European city already had a number of surgeons specializing in the operation.[107] By 1926, New York alone reportedly had more than a hundred surgeons who performed it "especially in cases of acute premature loss of vitality."[108] One of them, Steinach's admirer and tireless publicist Harry Benjamin, made sweeping claims for the operation. "The point that so far has not been brought out with necessary clearness," he observed,

> is, that by the Steinach operation the patient is given a more or less massive and continuous dose of his own gonadal hormone. . . . All symptoms due to senility, including sexual impotence, as a rule improve after the operation. . . . A few other possible indications would be: beginning arteriosclerosis, hypertrophy of the prostate, eunuchoidism, mental depression, and cases of dementia praecox, where we suspect gonadal deficiency as cause.[109]

The gospel found many takers at the time. Writing in 1940, Steinach himself referred to more than two thousand operations performed by surgeons in Vienna, New York, London, St. Petersburg, Copenhagen, Chile, Cuba, and India.[110] There were no accusations of fraud, although there was, as we shall see later, much scientific criticism of Steinach's claims.

Thanks to the Steinach operation and the simultaneous vogue for monkey gland transplants (on which see below), the 1920s turned into the decade of the testicle. One can hardly think of another era when one bodily organ featured so consistently and so prominently in the world's news as well as in the specialist medical press. From *Good Housekeeping* to *Medizinische Klinik*, there was no escape from the latest news and debates on the secretory functions of the testes, even though the newspapers referred "politely but vaguely" to "glands" rather than to testicles or gonads.[111] In 1924, the *New York Times* published a full-page special feature entitled "Science Promises an Amazing Future." In it, a number of leading scientific figures provided brief reports on the recent advances in their fields and likely future discoveries. Glandular research had its own section, contributed by the leading scientist R. G. Hoskins, Professor of Physiology at Ohio State University and editor-in-chief of the best-known professional journal of the still-new science, *Endocrinology*. Hailing the introduction of insulin and the ongoing research on pituitary and ovarian hormones, Hoskins announced that "the problem of rejuvenation will probably be solved by endocrine researches. Old age and death are merely the 'running down' of certain necessary cells. Certain endocrine products are known to have a profound influence in building up body cells."[112]

In the popular domain, reactions, of course, were even more hyperbolic. When the Viennese surgeon Adolph Lorenz, well known in America for his bloodless orthopedic operations, revealed that he had had a "Steinach," an illustrated feature in a popular American magazine began with the assertion,

"Every man more than sixty years of age should have an operation to rejuvenate himself. In twenty-five years every man of sixty will be rejuvenated by surgery." "It is simple—a scratch. It is no worse than having your hair cut," Lorenz told his interviewer:

> There are no bad results and one may resume work in a short time. For me—I was an old man. I was done—a broken man, overworked, and I felt that my work was finished. Everyone who knew me had given me up. . . . There was nothing to lose by submitting to the operation. I wanted to go on with my work—to feel alive again—so I went to Dr. Victor Blum in Vienna and got back my vigor and health. For five years I have felt—oh, how is it you say it?—great—strong![113]

Science had finally found the fountain of youth.

What, however, was the physiological basis of rejuvenation? In old age, Steinach argued, the tissues and organs did not themselves degenerate but functioned poorly because of undernutrition.[114] Regeneration and fresh growth could be induced even in conditions of near-atrophy by improving blood flow.[115] The secretions of the sex glands, Steinach claimed, did just that. The sustained, reactivating effects of the operation were not seen until the interstitial cells had proliferated sufficiently to raise the secretion of the male hormone substantially. This usually took a few months and sometimes as much as a whole year.[116] Steinach's follower, surgeon Peter Schmidt speculated that the sex gland secretions reduced the tension of the blood vessels by reducing the irritability of the sympathetic nervous system, thus leading to improvement in circulation, and Steinach himself endorsed this explanation.[117] Erwin Horner, another surgeon who worked closely with Steinach, suggested that elevation of the male hormone secretion resulted not only in hyperemia of other parts of the body but of the testicle too, thereby stimulating it to secrete ever higher amounts of sex hormone. The testicular hormone, furthermore, probably stimulated the other endocrine glands, bringing about a reactivation of the entire endocrine system and, therefore, the rejuvenation of the organism as a whole.[118]

Many other explanations were also offered.[119] So little was known, however, about the nature and physiological actions of the sex-gland secretions that no detailed explanation of the rejuvenative effects of the Steinach operation was conceivable. An almost mystical faith in the powers of the sex glands hung over all speculations. "Sex," declared Steinach, "is the root of life. Just as it produces physical and psychic maturity, induces and preserves the period of flowering, shorter or longer, here richer, there poorer, so it is also responsible for the withering of the body and gradual loss of vitality.

Sex is therefore the obvious means for natural stimulation or 'activation' in youth, and also the instrument for methodical 'reactivation' in old age. Sex is not only the measure for the rise, peak, and fall of the currents of life, but also, up to a point, for their restitution."[120]

Demand for the operation was considerable. At the age of sixty-nine, the poet William Butler Yeats, who was then passing through a period of bad health and creative barrenness, had a "Steinach." "It revived my creative power," wrote Yeats in 1937, "it revived also sexual desire; and that in all likelihood will last me until I die." It was not just his concupiscence that was stimulated by the operation but also his art, resulting in a crop of late poems ranked with his best work.[121] The sixty-seven-year-old Sigmund Freud opted for the operation, hoping not so much for rejuvenation as for a delay in the recurrence of his cancer. He told journalist George Viereck that the Steinach operation "sometimes arrests untoward biological accidents, like cancer, in their early stages. It makes life more livable."[122] As for his own experience, the reports are contradictory. He confided in Harry Benjamin that "he was very satisfied with the result. His general health and vitality had improved and he also thought that the malignant growth of his jaw had been favorably influenced."[123] Freud's biographer Ernest Jones, however, was certain that Freud had not experienced any benefit from his "Steinach."[124]

Innumerable less famous men also availed themselves of the Steinach operation. An American journalist estimated in 1926 that more than eight thousand men had already undergone the operation in the United States and "perhaps three times that number in Europe."[125] Glandular rejuvenation became a cultural motif. The following exchange is from Noel Coward's 1930 play, *Private Lives:*

> ELYOT: Would you be young always? If you could choose?
> AMANDA: No, I don't think so, not if it meant having awful bull's glands popped into me.
> ELYOT: Cows for you, dear. Bulls for me.
> AMANDA: We certainly live in a marvellous age.
> ELYOT: Too marvellous . . .
> AMANDA: It must be so nasty for the poor animals, being experimented on.
> ELYOT: Not when the experiments are successful. Why in Vienna I believe you can see whole lines of decrepit old rats carrying on like Tiller Girls.
> AMANDA (laughing): Oh, how very, very sweet.[126]

Many case histories were published, demonstrating the wonders of the Steinach operation, most of which emphasized the same features as in

Steinach and Lichtenstern's first patient. Here, for instance, are excerpts from a typical case history reported by the London sex therapist and surgeon Norman Haire:

> Case I. A.B., aged 53 (novelist). *March 14, 1922.* Complains of lack of energy, slowness of sexual act, feeling of tiredness and inability to make any effort. . . . His clothes hang on him—they would go twice round him. . . . He looks like a man of seventy or more. . . . Skin very dry and wrinkled. . . . No evidence of any severe organic lesion. . . . *March 19, 1922*—Double-sided vasectomy and vasoligature. . . . *May 8, 1922*—Writes that he is much better and feels happy. . . . *March 19, 1923*—Patient came to see me, exactly one year after the operation. Looks a new man. Has put on weight . . . his skin looks different. . . . He has written a new book, several short stories, and some one-act plays. . . . Initiative, power of concentration, and memory have improved. He does not remember ever feeling better, and thinks his literary powers are greater than ever before. Renewed self-confidence and joy in life. Sexual intercourse entirely satisfactory to both himself and his wife. He no longer feels exhausted after it. The mental condition of the whole family has improved as a result of the change in him.[127]

Not every instance of rejuvenation was as successful, but the results on the whole seemed to encourage much optimism. Moreover, it was a simple operation, and there was only one reported case in which the operation was implicated in a fatality.[128]

The chief indication for the operation was premature senility, marked by loss of energy, memory (especially for figures), or impotence.[129] Subsidiary indications proliferated steadily, including the stimulation of the "inactive" testicles of young eunuchoid males.[130] Looking back on ten years of experience, Viennese surgeon Erwin Horner warned that it was wrong to expect the revitalization of the organism to proceed steadily. As was generally the case with all living phenomena, reactivation proceeded in fits and starts.[131] Many others emphasized that the operation was *not* a panacea and could not restore an irreparably damaged organism to its original condition. The "Steinach operation is no cure-all, no wonder treatment, that will revolutionize medicine," emphasized Harry Benjamin, "but it has a definite value in balancing endocrine disturbances, relieving symptoms of age and prolonging efficiency—no more and no less."[132] Such modesty, however, did not always mark the views of Steinach's professional allies, some of whom could be as hyperbolic as any newspaper reporter. In 1931, for example, Peter Schmidt expressed the hope that the rejuvenation of older men might soon become as routine and mandatory as the vaccination of children.[133]

Medical and popular interest in rejuvenation was predominantly related to men. The rejuvenation of aging women, however, was not neglected. Obviously, the Steinach operation could not simply be applied to women, since females had no vas deferens to ligate. Steinach had reported, however, that ovarian transplantation gave especially good results in rats, probably by direct endocrine actions of the implant as well as indirect stimulation of the senile ovary of the recipient. The transplanted ovaries reenergized the senile female—appetite was stimulated, the animal gained weight and moved in a lively, energetic manner. Then, the sexual cycle was revived and the animal went on heat as in its youth. The rejuvenated animal's own ovaries literally returned to life and began to ovulate again. The hitherto senile female was now fertile, and one of Steinach's rejuvenated rats actually became pregnant, gave birth, and suckled the young.[134] Whether this is physiologically possible or even credible by *today's* standards is not the issue here; the point is to appreciate how the sex glands were constructed as fountains of youth in the 1920s, and how such constructions could lead to specific clinical interventions. Although Steiach, despite his optimism about the rejuvenative effects of ovarian transplantation, did not himself attempt it in humans, others did try the operation, and the possibility did not go unremarked in the literary world of the time.[135] In Mikhail Bulgakov's satirical novel *The Heart of A Dog* (written in 1925 but suppressed until the 1960s in English and until 1987 in Russian), the eminent surgeon and rejuvenator Professor Philip Philipovich Preobrazhensky tells a female patient, "I am going to implant some monkey's ovaries into you, madam." "Oh, professor—not *monkey's?*," groans the patient.[136]

Although he never used ovarian transplantation for human rejuvenation, Steinach was eager to develop a reliable, relatively simple means of rejuvenating women, and he experimented with different techniques with characteristic enthusiasm.[137] There were two overlapping phases in this research. During the first, he sought to stimulate the woman's own ovaries to secrete more hormone. In the second, he combined those techniques with injections of a hormonal extract named Progynon that he had himself developed in collaboration with the pharmaceutical firm Schering-Kahlbaum. The first phase had begun early with an investigation conducted with radiologist Guido Holzknecht, Vienna's pioneering radiologist, around 1913.[138] The premise of the experiments was identical to that of Steinach's rejuvenation technique for men: if the germinal tissue of the gonads could be selectively destroyed, then the secretory elements would proliferate beyond their usual mass.[139] The easiest way to destroy the germinal cells of the ovary was by x-rays, to which the hormone-secreting cells were less vulnerable.[140] The technique was tried on guinea pigs; results were slow to appear but in about 40 percent of irradiated animals, the breasts and nipples eventually enlarged and the uterus and breasts were found to be hypertrophied at autopsy. Histologically, the ovary was packed with hypertrophied secretory cells.[141]

In his 1920 report on rejuvenation, Steinach alluded only very briefly, however, to the possibility of using this method to rejuvenate women.[142] His friend and disciple Harry Benjamin published more on using x-rays to rejuvenate women and regularly prodded Steinach for information and insights on rejuvenating women, not only from scientific interest (although that was strong enough) but also, as he stressed in a 1922 letter, because the matter was extraordinarily important from the financial point of view.[143] In one early article, Benjamin emphasized that the effect of x-rays on bodily organs (and especially endocrine glands) was not necessarily destructive. In low doses, they could exert a stimulatory effect, especially on the female organism. This was the conviction of some Central European experts at the time, although it was contested by others, including Guido Holzknecht himself.[144] Benjamin, however, had no doubts whatever: "For the retardation of age," he wrote, "I, myself, have treated over one hundred patients showing signs of a congenital or acquired hypofunction of the ovaries."[145]

The majority of Benjamin's patients were around the menopausal age, some considerably older: most of them complained of symptoms typically associated with menopause (such as hot flushes and sweats) and what Steinach had identified as the characteristic symptoms of "premature senility"—lack of physical and mental energy, weak memory, and a general falling-off in skills "most vital to the individual in question."[146] In his first reported case, Benjamin dealt with a sixty-four-year-old woman, who found it hard to concentrate and felt her imaginative ability had declined. No physical disorder could be discovered, and Benjamin diagnosed her as a case of "mental sterility."[147] In June 1922, she received another two treatments. When examined in August, she reported that "nothing could tire her any more."[148]

This patient was the novelist Gertrude Atherton (1857–1948). Although largely forgotten today, she was prominent in her time, not least because of her penchant for themes condemned then as sordid or "daring."[149] After years of churning out novels and journalistic pieces, Atherton, for all her panache, had succumbed to a barren phase around the age of sixty-five. It was, she recalled in her autobiography, as if her "mental dynamo refused to tune up." Salvation came one morning in the shape of a newspaper feature on Steinach's rejuvenation technique. It quoted Harry Benjamin's remark that "women were running to the Steinach clinic from all over Europe, among them Russian princesses who sold their jewels to pay for treatments . . . that might restore their exhausted energies and enable them to make a living after the jewels had given out."[150] Now, here was a brilliant idea for a new novel! Atherton immediately made an appointment with Benjamin, ostensibly to gather material, but found herself telling him instead about her own "mental sterility."[151]

Soon, she found herself to be Benjamin's first known female patient. The treatment was painless but unimpressive. There was no improvement in her condition for a month. And then all of a sudden, she

> had the abrupt sensation of a black cloud lifting from my brain, hovering for a moment, rolling away. Torpor vanished. My brain seemed sparkling with light. . . . I almost flung myself at my desk. I wrote steadily for four hours. . . . It all gushed out like a geyser that had been "capped" down in the cellars of my mind, battling for release. That geyser never paused in its outpourings until the book was finished, five months later.[152]

This was her rejuvenation novel, *Black Oxen*, the story of which (ignoring the subplots) centered on a mysterious European aristocrat, Countess Zattiany, who appears in New York in the 1920s, entrancing the city—especially its males—with her beauty and sexual allure. She bears a strong resemblance to a sometime New Yorker, Mary Ogden, but because Ogden would be in her late fifties by then, people speculate whether the countess was her illegitimate daughter. Actually, the countess is none other than Mary Ogden herself, miraculously rejuvenated by the Steinach treatment in Vienna.

Published in 1923, *Black Oxen* turned out to be the biggest success of Atherton's career, earning her more than thirty thousand dollars in royalties in one year and occupying first place on the bestseller list of *Publishers Weekly*. (The far more lastingly famous *Babbitt* by Sinclair Lewis came in fourth.) It was banned from the Rochester public libraries because of its immoral content, and the *New York Times* voiced concern over the novel in an editorial. H. L. Mencken observed that Atherton seemed to "find the primary springs of human character in the ductless glands."[153] "I am the mother," Atherton herself told Harry Benjamin, "and you the father of this book."[154]

Atherton did Steinach and Benjamin proud: parts of her novel read like publicity tracts written by them. The author was swamped with letters from readers asking for information on the Steinach technique. "Poor Dr. Benjamin!," she recalled, "I nearly ruined him. Women besieged him, imploring him to give them the treatment free of charge or at a minimum price. . . . I met several of those patients with whom the treatment had been as successful as with me. I also had enthusiastic letters from others who, living abroad, had gone to Steinach's clinic or to Dr. Schmidt in Berlin."[155] Although Atherton was being overly optimistic about the number of patients she referred to Benjamin and Steinach—the correspondence of the two men is full of complaints about the scarcity of patients!—she did render them an invaluable service by publicizing their work, especially amongst women. Even the mercurial Steinach—who often craved publicity but reacted furiously at anything that struck him as even remotely *unwissenschaftlich*—was appreciative of her efforts, especially of the occasional patient she sent to him in Vienna.

Techniques for rejuvenation of women changed far more over Steinach's career than they did for men. Radiation soon began to be combined with diathermy (i.e., warming the whole body and the ovarian regions electrically), the rationale being that the function of the ovary was stimulated by the heating in itself and, when x-rays were administered afterwards, the gland was more radiosensitive and responded to a smaller dose of x-rays than would otherwise be the case.[156] Benjamin also began to irradiate other endocrine glands concomitantly, especially the pituitary and the thyroid.[157] As Steinach changed his techniques, he offered them (sometimes *gratis*) to Atherton and the novelist was unfailing in submitting to them and singing their praises.[158] In her 1931 novel *The Sophisticates*, she included a scene of pituitary diathermy, not used for rejuvenation but to help an ambitious journalist cope with his enormous workload.[159] (In the background of the plot, however, hover rumors that various women of the town had also submitted to "the endocrine reactivation treatment to renew their energies, and possibly their looks.")[160] Throughout her life, Atherton remained a staunch believer in Steinach and his theories, although she regretted the public's "false emphasis on the physical aspects" of rejuvenation.[161] Even in 1948, at the age of ninety, ailing and close to death, Atherton was taking hormone tablets prescribed by Benjamin, although she had by now come to accept that "when one has reached the venerable age of ninety, one cannot expect too much."[162]

But Atherton apart, how many women opted for glandular rejuvenation? Numbers are unavailable—largely because of the loss of crucial archival records, especially Steinach's own papers—but impressionistically, it is hard to resist the suspicion that in spite of enthusiastic endorsements such as Atherton's, rejuvenation for women was less popular than the Steinach operation for men. For one thing, the procedures for women were complex and less standardized. "Rejuvenation in women demands the expenditure of much more time than in the case of men," admitted Steinach's disciple Peter Schmidt. "It is also much more complicated, and needs considerable modification to suit each individual case, whereas the rejuvenation of males by the Steinach operation is an almost invariable procedure."[163] That grateful subject of the Steinach operation, surgeon Adolph Lorenz, was more blunt. "The ladies," he admitted, "are unfortunate. The operation, simple and harmless for the male, would be difficult and dangerous for woman. We saw Gertrude Atherton's *Black Oxen* in the movies but thought it far-fetched."[164] Moreover, it was thought that female rejuvenation could not be successful if attempted too late in life. "Eve cannot afford to wait so long for her rejuvenation as Adam," lamented George Sylvester Viereck. "Men of seventy and over have been successfully Steinached. In women of so mature an age, the attempt would be almost hopeless. The most favorable time is the period immediately before, during or shortly after the change of life."[165] One clinician who was not entirely skeptical about the Steinach treatment wrote in 1939 that the results of ovarian irradiation were "most problematical, since by the time a woman applies for rejuvenation therapy most of the ovarian tissue has been atrophied and there is nothing left to reactivate."[166]

Technical difficulties and biological uncertainties, however important, may not have been the sole reason for the lack of popularity of rejuvenative procedures for women. Men, according to most practitioners, hoped (indeed, expected) their efficiency, sexual desire, and vitality to be reawakened after the Steinach operation, but women, apparently, hoped to regain beauty. "For the woman of modern times," Peter Schmidt declared, "the loss of her good looks entails serious forfeits in occupational or social life":

> The reactivation of a climacteric woman may re-establish menstruation, may lower blood pressure, may improve functional capacity in various ways, but she will be inclined to say "thank you for nothing" unless at the same time you have improved her looks by effacing the signs of age in the skin of her face.[167]

Gertrude Atherton's rejuvenated heroine, one recalls, regained not only vitality and sensuousness but also her youthful beauty. While one must not make too much of scattered observations from interested parties, it is surely not implausible that because of their conviction that men and women sought different things from rejuvenation, physicians sought to rekindle not youthful vitality in a global sense but only *those* attributes of vitality that they considered to be culturally appropriate to the client's gender—efficiency and strength for men, beauty and sex appeal for women. The latter task, needless to say, proved a bit beyond their reach, and perhaps, although I cannot do the issue any justice here, many of the potential female clients of gland rejuvenators found the knife of the aesthetic surgeon more appropriate to their needs.[168]

Also, rejuvenation for women necessarily implied the elimination of reproduction. In men, too, the Steinach operation for rejuvenation was designed, of
course, to achieve exactly this. Once liberated from their reproductive obligations, the sex glands would turn into fountains of youth and vitality. "Instead
of giving life to children," Benjamin once wrote regarding the operation for
males, "aging men were to be made to give life to themselves."[169] For females,
too, life and vitality were available only at the cost of forgoing procreation, but
for a female to choose vitality over motherhood during her reproductive years
was obviously no easy choice in cultural terms. Nor, indeed, was society of
the time likely to approve of a post-menopausal woman's desire to reawaken
sensuous femininity (which rejuvenation was widely supposed to induce)
with the aid of science.[170] The Steinach operation for men promised exactly
what was commanded by social imperatives: energy, vitality, and creativity.
Even the sexual prowess that was supposed to come with these, although
not approved of by traditional moral codes, was surely more acceptable for
men in their seventies than for women of the same age. The heroine of
Black Oxen declares proudly to her new thirty-four-year-old lover Clavering,
"I do not merely *look* young again, *I am* young. I am not the years I have
passed in this world, I am the age of the rejuvenated glands of my body." She
immediately adds, however: "Of course I cannot have children. The treatment is identical with that for sterilization. This consideration may influence
you."[171] If, then, Steinach's rejuvenation procedure was more popular with
men, one important reason for that may have been that the treatment for
men challenged fewer social and cultural norms. Needless to say, much work
remains to be done on this issue, but it is possible that the gender differences revealed in the rejuvenation story, when thoroughly explored, may
show that cultural values and expectations were so integral to the science and
practice of rejuvenation that the word meant different things for different
genders.

Monkeying with Glands: Voronoff and the Other Path to Youth

> How did the traffic get so jammed?
> bedad it is the famous doctor who inserts
> monkeyglands in millionaires a cute idea n'est-ce pas?
> —E. E. CUMMINGS[172]

The current fashion among conservative physicians is to deride all reports of
benefit from gonad implantation. . . . Undoubtedly mere suggestion plays a large
role but in fair perspective the total weight of the evidence indicates with a

considerable degree of probability that testicular substance is of some genuine utility in the treatment of those conditions vaguely connoted by the term "lack of virility."

—R. G. HOSKINS[173]

The Steinach operation, Harry Benjamin complained in 1921, was "frequently confused with gland transplantation, for instance of the monkey gland and with all forms of gland treatments."[174] The colorfully named "monkey gland" operation involved the grafting of testicular tissue from chimpanzees or baboons into aging men had nothing to do with Steinach's procedure. Thanks to David Hamilton's comprehensive study of monkey gland transplantation, I shall provide only the briefest outline of its history and contexts.[175]

Although many surgeons experimented with testicular grafting, the use of chimpanzee testicles in man was the innovation of Serge Voronoff (1866– 1951) of Paris.[176] Born in Voronezh, Russia, Voronoff emigrated to France when he was eighteen. He trained in medicine and worked for some years in Egypt as a physician to the court of Khedive Abbas II, where he married the daughter of Ferdinand de Lesseps of Suez Canal fame. In 1910, he returned to Paris and set up a fashionable surgical practice. Although he had not shown any serious interest in any kind of research so far—he was already forty-four—he now developed a keen interest in transplant surgery, in which there was wide international interest at the time. He made the acquaintance of Alexis Carrel, whose innovative work on organ transplantation was already very well known, and was soon writing to him about his not very successful efforts to transplant ovaries in ewes. Not surprisingly, the motivation of Voronoff's research was clinical: although not professionally a gynecologist, he clearly aimed to find a dependable surgical treatment for human infertility. Then came the First World War, when he served in the French army with Carrel himself. After a serious illness, he left the army in 1917 and, having also divorced his wife, joined the prestigious Collège de France—associated with Claude Bernard and Brown-Séquard, to mention only two famous names in the history of endocrinology—where he financed his own research unit from income from the Standard Oil shares that had come with his former laboratory assistant and second wife, the American Evelyn Bostwick, daughter of a business partner of John D. Rockefeller. (She died in 1921 after a year of marriage, leaving Voronoff a wealthy man.)[177]

Now commenced his lifelong interest in aging and its glandular treatment. Senility, for Steinach as well as Voronoff, was nothing other than demasculinization. Steinach's research on rejuvenation had begun with the observation that castrated rats showed changes indistinguishable from the features of senility. Voronoff proceeded on the same analogy but referred it to his observations of human eunuchs while in Egypt. He selected other examples from history: Abelard never wrote after being castrated and Goethe remained sexually active until the end of his long life.[178] The presence of fully functional testicles seemed to confer intellectual and physical exuberance, and in their absence the individual aged, like a eunuch, before his time. "In the manifestation of his physical and intellectual qualities, varying according to the individual, man himself," declared Voronoff, "is worth whatever his sex glands are worth."[179] Could the replacement of exhausted sex glands make an individual virile, creative and youthful again?

Collecting old rams from all over France, Voronoff tried to rejuvenate them with grafts of testicular tissue from young rams. The results he reported were as astonishing as the ones Steinach would report with his rats.[180] Reporting on his ram transplants to the French Surgical Congress, Voronoff remarked that such grafts could work in humans. Since French law prohibited the removal "of any part or parts of the body, even in death," using human testicles was out of the question.[181] In a post-Darwinian age, however, that was not an insuperable obstacle: in the absence of human glands, the anthropoid testicle would serve eminently well. As he insisted later,

> the glands of monkeys ... are the only ones that can furnish grafts which will find among human tissues the same conditions of life that they had originally. To use the glands of other animals is to ignore completely the laws of biology: they could never be, in the human organism, anything but foreign bodies. The anthropoid apes form a race very close to the human race. Their embryology, their dentition, the analogy of the skeleton, the skull and the internal organs, furnish abundant proof of the biological parentage of man and monkey. ... The blood of the chimpanzee differs less from that of man than it does from that of other species of monkeys.[182]

A monkey gland graft, in short, was "really not a hetero-graft. ... It may be termed 'homeo-graft,' the organs of one species finding identical conditions of life in the organism of a nearly allied species."[183] Or, as a reviewer for the *British Medical Journal* put it more succinctly, the use of monkey glands for rejuvenation was "a return, as it were, to our ancestry for refreshment."[184]

Voronoff alone had performed about a thousand monkey gland transplants in men by the end of 1926—he had set up his own monkey farm on the Italian Riviera, employing a former circus animal-keeper to run it—and had been invited to contribute an article on rejuvenation to the *Encyclopaedia*

Britannica.[185] And in America, perhaps the most enthusiastic land for glandular treatments of all kinds, monkey glands soon became all the rage—E. E. Cummings's "famous doctor who inserts monkeyglands in millionaires" was not a figment of the poet's imagination. Although signs of "psychical and sexual excitation" appeared immediately after the operation, these, Voronoff emphasized, were transient. The initial months following the transplantation were not marked by any revitalization: "This period is one of disappointment, disillusion and discouragement," Voronoff admitted. But after this commenced a period of progressive global restitution that was every bit as remarkable as anything the practitioners of the Steinach operation could boast of.[186] In the *New York Times*, Voronoff claimed that monkey gland transplants could "put back human aging by twenty to thirty years" and confided that each of his chimpanzees cost about five hundred dollars.[187] But the wealthy surgeon could well spare the cash, and it was a good investment for the fame he craved. His American acolyte, Chicago surgeon Max Thorek, recalled that soon, "fashionable dinner parties and cracker barrel confabs, as well as sedate gatherings of the medical élite, were alive with the whisper—'Monkey Glands.'"[188] Thorek admitted later that "Voronoff's doctrine proved to be one of the most profitable theories that ever fell into the spacious laps of quacks" but insisted that it had, nevertheless, "acted as a most potent stimulant in the advance of endocrinology. . . . Voronoff's announcement was itself a hormone, arousing an intensified study of the internal secretions."[189]

Monkey gland transplantation, like the Steinach operation, was applicable to men alone. Acknowledging the difficulties in rejuvenating women, Voronoff remarked: "In the meanwhile I can only offer this consolation: the mortality statistics of every land prove that women live much longer than men. Hence they already have the advantage of us and consequently may still wait a few more years before the experiments in course of development bring them the remedy which is to intensify and prolong their existence."[190] There were rumors that Voronoff had transplanted human ovaries into a female chimpanzee called Nora, and the chimpanzee was going to deliver a half-human baby. The rumors provided the plot for Félicien Champsaur's 1929 novel *Nora, la guenon devenue femme* (Nora, The Monkey Turned Woman).[191] Little research has been done on Voronoff's attempts to develop a female form of monkey gland transplantation, but there are intriguing hints that he did try a few transplants of chimpanzee ovaries into women.[192]

Monkey gland treatment, expectedly enough, faced severe criticism from many physicians and surgeons. Eugen Steinach considered the technique to be worthless. Testicular grafts, he claimed, were accepted by the body only if the host was a bilateral castrate. Existing gonads, no matter how weak or how senile, would always resist the implanted gonads and bring about their rejection. This was not some discovery or conviction of Steinach's own but based on the general dictum of the eminent Johns Hopkins surgeon William S. Halsted that glandular transplants "took" only when the original gland had been removed. This principle was endorsed so widely that it came to be known as Halsted's Law.[193] The rejection, it was believed, was all the swifter if the implanted tissue came from another species. "It amounts to self-delusion," Steinach declared, "if people seriously believe that the transplantation of chimpanzee testicles into human beings can produce anything but rapidly passing effects."[194] Steinach did not even consider Voronoff as a legitimate scientist and observed brusquely to his American collaborator that "Voronoff is not to be taken seriously and therefore, not to be cited."[195] Other scientists could be more supportive. Roy Hoskins, for example, asserted that the reported benefits of testicular transplantation were too numerous to be dismissed as the results of suggestibility; even when his own experimental study of testicular transplantation in senile rats did not yield striking results, he blamed his own technique rather than the concept itself. He found "no room for reasonable doubt that testicle grafting is capable of producing definite somatic changes and definite erotization."[196]

An obvious criticism was that the grafts could not possibly do much good, because they would be quickly absorbed by the host's body. Voronoff claimed, however, that a successful graft led to reactivation for a period of three to five years, by when the graft was completely fibrosed and lost its secretory powers. Looking back at the end of the 1920s, he claimed that if one defined a successful case as one where improvements were maintained for more than two years, the success rate with monkey gland transplants was 90 percent in premature senility and 74 percent in the true senility of men between 70 and 85. In cases of simple impotence in patients between 25 and 55 years, 89 percent regained potency after transplantation. "Failure of resection of the vas deferens done outside France"—i.e., the Steinach operation—had provided him with thirty-six patients aged 50 to 65, and his operation had rejuvenated 60 percent of them. As for his own graftees whose grafts, as expected, had ceased to function after about five years, a second graft repeated the miracle for 91 percent of forty-two patients aged from 65 to 70. Without providing any details, he also reported that the monkey gland transplant had also worked in 57 percent of cases of "sexual inversion."[197] Voronoff's case histories, as Swiss surgeon Benno Slotopolsky commented, were so brief and consisted so largely of the patients' own reports that one could not really assess his results by objective clinical criteria.[198]

London consultant surgeon Kenneth Walker, who at least initially was no skeptic about the utility of testicular transplantation and had operated on ten recorded cases, maintained, however, that all scientific experience with transplants indicated that no matter how good the vascularization of the graft, it "undergoes from the very beginning a process of absorption," and the longest he expected a graft to remain viable was two years. However, that was no reason not to graft testicles because the graft, before being absorbed, might revitalize the patient's own testicles.[199] This was not an uncommon conviction. Robert Morris, the New York ovarian transplant pioneer, had reported the case of a eunuchoid man, who had received a testicular graft, which, however, gradually disappeared. The patient's own shriveled testicle, however, now became "a growing testicle proper," accompanied by a growing epididymis and spermatic cord! "Coincidental with the development of the testicle, evidence of its presence became gradually more and more marked, to such a degree that the patient has recently asked if he might marry."[200] The possibility that a graft might itself disappear but give new life to the host's own, hitherto inadequate, gonad opened "a vista which has not been anticipated."[201] Walker speculated that such revivifying effects might also be exerted on other endocrine glands by the monkey gland graft. While the testicular graft itself would perish, it might leave as its legacy a literally rejuvenated endocrine system and, therefore, a revitalized, resexualized body.[202]

By the early 1930s, however, Voronoff's star was on the wane (as was Steinach's). With increasing knowledge of hormones and surgeons becoming more cautious about transplantation, monkey gland operations no longer made the headlines. Even those who believed in the possibility of transplantation itself experienced serious doubts that testicular transplantation was of much worth for reactivation of the senile.[203] Voronoff's great supporter Max Thorek confessed in his autobiography: "I knew that the quest of rejuvenation had long been a jack-o'-lantern, leading other investigators astray. Time proved that I was to be no exception." He continued to believe that monkey gland transplants could in certain selected cases lead to some improvement in senile debility, but these improvements, he had discovered, were

> more transient than permanent, and when the fountain of youth within us runs dry, not all that the manufacturers sell, and nothing that the surgeon can plant, will bring back the magic waters. . . . It is cruel to seek rejuvenescence with the idea that vigor borrowed from another's ovary or testicle will fertilize the barren and make puissant the impotent. This belief made science the ally of crime.[204]

Quackery or the Road to Salvation?

Modern endocrinologists and gerontologists find it embarrassing to refer to Steinach or Voronoff.[205] The author of an otherwise superficial history of gerontological medicine observed that "just as the editors of Soviet encyclopedias correct historical 'errors' by simply scissoring out entire chapters of previous editions, so gerontology has tried to consign its early deviationists to oblivion. The controversial pioneers, the rejuvenation 'bootleggers,' who had the field almost entirely to themselves for the first half century, are only rarely touched upon in the literature."[206]

The rejuvenation craze, of course, had its bizarre aspects. The public response could be particularly hysterical. Peter Schmidt reported that after he had reported in a lecture in New York that there had been some visual and auditory improvement in one of his patients after the Steinach operation, a local newspaper announced that "The Steinach Operation Cures the Blind and the Deaf." Within a day, Schmidt recalled, "dozens of blind and deaf persons had called to consult me; I was continually being summoned to the telephone; and I received hundreds of imploring letters from the blind and the deaf." Journalists added fuel to fire. Hailing Steinach as "the greatest modern biologist," George Sylvester Viereck asked, "Does Professor Eugen Steinach bring us the knowledge that the serpent promised Eve, that shall make us like gods? . . . If we halt the insidious advance of Age, may we not, in time, challenge Death?"[207] Such examples could be multiplied almost without end.

It may be more surprising to note that the medical press of the era did not take rejuvenation lightly. Doubts, to be sure, were voiced often enough, but there was no accusation of quackery. In Britain, for instance, Ernest Starling told the Royal College of Physicians in 1923 that although further observations were necessary before the results of the Steinach operation could be accepted unreservedly, "it must be owned that they are perfectly reasonable and follow, as a logical sequence, many years' observations and experiments in this field." The sex hormones, Starling emphasized, were "chiefly responsible for determining the secondary sexual characters. Among these . . . must be classed the whole of a man's energies. Virility does not mean simply the power of propagation, but connotes the whole part played by man in his work within the community."[208]

The concept of rejuvenation reflected as well as contributed to shifts in the conceptualization of aging. Perhaps the most important of these was what Thomas Cole calls "the medicalization and devaluation of old age from the late eighteenth to the early twentieth centuries."[209] At the end of the eighteenth century, the Viennese physician Gerhard van Swieten had doubted that the weaknesses of old age were pathological and dismissed the need for any medical intervention in that natural process.[210] By the early twentieth century, there appears to have been a sea change in medical attitudes: old age was beginning to be stripped not only of its traditional associations—for example, wisdom—but even of its status as a natural season of life. Many views of the aging process were propounded, many of them attributing senescence to specific degenerations or deficiencies. They ranged from Ilya Ilyich Mechnikov's theory that senility was brought about by the destruction of higher tissue by macrophages driven berserk by large intestinal microbes to more modern hypotheses, such as the exhaustion of some vital ferment, the hardening of arteries or, as with Steinach and many others, endocrine deficiencies.[211]

Such theories were criticized by some medical scientists for confusing consequences with causes and for emphasizing one or another set of features of aging in isolation from the underlying process of involution: "The phenomena to which senescence is ascribed by these authors," argued American physiologist T. Brailsford Robertson, "must arise from some much more general underlying process, and although . . . the factors enumerated by them may actually be responsible for the death of senescent individuals, yet they are not in themselves causes, but on the contrary, consequences of senescence itself."[212] Charles Stockard, Professor of Anatomy at Cornell University Medical College, was far more negative. He charged the rejuvenators with confusing cause and effect. The decline of gonadal function in late life was "one of the many symptoms of the senile complex, and being a symptom it can scarcely also be the causal factor in aging." To revitalize the gonads might lead to some "temporary stimulation, but always without counteraction of the general aging symptoms." Rejuvenation could only be attained if the whole ensemble of influences causing aging could be counteracted: to supplement one aged organ in the hope of influencing *the* cause of aging was to court disappointment.[213] Others questioned why the gonads should be implicated in questions of senility and rejuvenation when castration had never been proven to shorten lives.[214] Other medical critics might accept the endocrine explanation of aging but refuse to locate the endocrine deficiency in the gonads alone. "Just as all bodily conditions and their chemical reactions are under the influence of the entire hormone apparatus," emphasized Artur Biedl, "so is old age." Not just the gonads but the whole endocrine system—from the pituitary to the adrenals, from the pineal to the thyroid, had to be rejuvenated. "Old age and youth are questions of the endocrine harmony."[215]

Such criticisms notwithstanding, theories attributing aging to problems located in particular organs or systems proved popular with clinicians, no doubt because they permitted many of the debilities of age to be seen as curable conditions. "No longer," declared Peter Schmidt, "shall we look upon senility as an unavoidable evil, as a physiological process fatally determined. We are now able to regard old age as an illness, as a result of decay in the reproductive gland."[216] In America, one physician propagandizing for a rival technique of rejuvenation proclaimed: "Old age is now coming under the same control as germ diseases. Yes, verily, old age is, like any disease, on the verge of conquest."[217]

Expectedly, much medical criticism claimed that the Steinach operation produced its effects solely by suggestion. One German physician pointed out that many of the so-called signs of rejuvenation were regularly achieved by mass suggestion at Lourdes or in Indian temples.[218] The champions of the operation replied that identical results had been repeatedly produced in patients "Steinached" without their knowledge in the course of another operation. Even Benno Romeis, for many years a vociferous critic of Steinach, emphasized that although the overblown hopes associated with the procedures introduced by Steinach and Voronoff had not been fulfilled, there was no real proof that they achieved all their supposed results solely through suggestion. It was, he added, not unlikely that in a certain percentage of humans, the ravages of premature senility could be alleviated by hormonal treatment.[219]

More than the matter of suggestion, however, it was the term "rejuvenation," with its dramatic connotations of a literal return to youth, that preoccupied medical critics. Steinach himself and his most enthusiastic followers agreed that the term was less than ideal and tried to replace "rejuvenation" with the more neutral "reactivation" or "restitution," but neither ever really caught on.[220] Although more accurate, these terms lacked the magical ring and cultural resonance of rejuvenation, a word evoking images of purification, immortality, and redemption. The magic of the word guaranteed popular fascination, but it was that very magic which raised the hackles of orthodox physicians and scientists, who argued that no medical procedure could possibly deliver what the word rejuvenation promised.[221] They challenged the rejuvenators to prove that their procedures extended the life span in an absolute sense, which any genuine rejuvenation would surely do.[222] Steinach riposted that since one could never predict exactly how long an individual would live, it was nonsensical to talk about extending the lifespan.[223] "The only symptoms we can expect to relieve by a Steinach operation," Peter Schmidt emphasized, "are those which are the direct outcome of the ageing process."[224] The most that one could claim, according to one writer, was that "in building up general health conditions the operation naturally will increase the power of resistance against the inroads of other diseases, and in that way lengthen the normal span."[225] Harry Benjamin pointed out philosophically that although it was impossible to say whether the operation literally prolonged life, it certainly extended "man's life of activity, that part of man's life where he works and creates."[226]

For some medical men, however, even that was a perilous prospect. Aging was a natural process and it was, they protested, a transgression of natural law to turn the clock back, especially when that involved the restitution of sexual powers.[227] Still other critics focused on the reawakening of the libido after the operation, and the eugenic dangers of enabling the senile to reproduce.[228] Peter Schmidt responded that "sexual activity per se has a rejuvenating influence. There is no remedial measure in the world so effective as the sexual act in invigorating the circulation . . . and in producing a health-giving alternation from extreme tension to complete relaxation."[229] Such exuberant celebrations of the therapeutic effects of coitus were, of course, counterproductive, providing valuable ammunition to critics that the Steinach operation or a monkey gland graft, even if successful, led merely to the brief revival of sexual potency in elderly debauchees, who would come to regret it pretty quickly.

American pathologist Alfred Scott Warthin, who considered senile involution to be an entirely normal (and natural) process, commented in 1929 that "the so-called rejuvenation produced by the sex hormones of the transplanted testis or by ligation of the vas is no rejuvenation in any sense of the word, but is a re-erotization wholly. . . . It is another example of the dangers of putting new wine into old bottles as has been shown by the sudden deaths of a number of re-erotized old men."[230] Biologist Heinrich Poll speculated that the Steinach operation might actually *shorten* life: an elderly individual's already depleted nervous reserves might be squandered far more swiftly than expected because of the renewal of sexual activity, leading to death at an age earlier than would have been the case had he been left to decline naturally.[231] "While Voronoff's operation may perhaps be justifiable as an experimental procedure in young subjects in whom the testes have been damaged or destroyed by injury or disease, the treatment of senility by this method seems indefensible," expostulated physiologist Samson Wright in Britain. "Though the testicular grafts may stimulate physical activity and sexual desire, they cannot restore the worn heart, arteries and essential organs to their normal state. There is grave danger that excessive strain may be put on damaged structures, with disastrous results."[232]

Voronoff's supporters emphasized that mere arousal of sexual desire and restitution of potency were not the primary aims of the operation. The "restoration of mental and muscular vigour" was the chief goal. The restoration of potency, Leonard Williams claimed, was indeed common after the operation, but it was only a "by-effect which has not declared itself in all cases."[233] As for the dangers of putting new wine in old bottles, Williams countered that

> the human body is not a glass bottle, and Voronoff's results now suggest very strongly that this brand of new wine renovates the very bottles themselves. It is claimed that hard arteries become softened, high blood pressure is lowered, stiff joints become supple, and atrophied muscles grow plump. Slow cerebration is speeded up, long lost interests are revived, and the desire and capacity for work are restored.[234]

Rejuvenation was not a simple resexualization but a revitalization of the entire organism.

It was not entirely a question of saving frail older people from post-coital heart attacks that motivated the opposition of more traditional physicians. Morris Fishbein, the vociferous American opponent of all rejuvenative procedures, declaimed: "The young men, handicapped in competition because of lack of experience and of the world's goods, will be compelled to struggle against their rejuvenated fathers and grandfathers, while the latter, by repeated operations, maintain a permanent lead. It's a sad, sad outlook; but fortunately it isn't true."[235] Some critics were animated by their revulsion for the mating of the old with the young. "Who would conserve the proprieties if these gay blades of sixty and over were loose and eager with renewed appetites and the wisdom born of past experience?," asked one medical commentator.[236] In Germany, Hans Much proclaimed that an old man having intercourse with a young whore was so unnatural that it could only be counted as a perversion, and the Prague physician Alfred Kohn compared such a coupling with the "unnatural" marriage of a castrate masculinized by a testicular implant.[237]

Confining his attention to Steinach's rejuvenated laboratory rats, Kohn argued that the spectacular sexual revitalization of the rats—amounting to "sexual paroxysms"—did not indicate a return to the normal sexuality of youth but a morbid sexual excitement qualitatively different from youthful vigor. The Steinach operation may well have other rejuvenative results, but as long as the sexual consequences were so prominent and so very "unnatural" in intensity, the doctor of conscience must hesitate before recommending it to an elderly man. The question, after all, was not only of one elderly individual but of his family and the community they lived in. To turn a widower or grandfather, a teacher or a civil servant, into a pleasure-crazed womanizer (*weibersuchtiger Genußmensch*) was hardly a medically or socially praiseworthy achievement. Even the most amoral doctor must admit that a human being did not live in the freedom of Steinach's laboratory but in society. The unnatural lust aroused by the operation might, indeed, need to be repressed because of social reasons—and that repression itself might cause grave physical and mental problems.[238] Others remarked on the risk of encouraging venereal infections in old age and yet others emphasized the danger to the human species of permitting exhausted old men to procreate. This last charge kept recurring in spite of the Steinachites' eminently logical assertion that a bilateral "Steinach," by causing total sterility, would forestall any eugenic risk.[239]

The sexual rejuvenation supposedly brought about by the operation remained well known long after the operation itself had gone out of vogue. As late as in 1955, the behaviorist John B. Watson, then living in retirement on a Connecticut farm, wrote to his friend Karl Lashley, the renowned psychologist and director of Yerkes Laboratories of Primate Biology in Orange Park, Florida: "Several of my friends have had their vas tied back and claim increased sexual stirrings. Any truth in this? The removal of the fear of knocking a gal up might account for it."[240] Champions of rejuvenation were at least partly responsible for creating this impression. They defined senility as desexualization; although that desexualization did not merely refer to diminution in sexual *activity*, enhanced sexual activity and the alleviation of impotence were always presented as major benefits of the Steinach operation. No matter how hard they tried to surround the issue of sexual reawakening with other, ostensibly more important signs, their audience attended most closely to the sexual effects alone. The dilemma is evident from Steinach's own remarks about the effects of his operation on the libido and sexual potency. "This undeniable effect," he observed, "greatly hindered the dissemination and acceptance of vasoligature at the beginning. The marked generalized effects on the body and the mind were ignored, while critics and wits focused only on the stimulation of the sex gland and the psychosexual actions of the operation. Reactivation was besmirched by the odium of sexuality."[241]

For a few advanced thinkers, however, the connection of sex and new life was a convincing one. American psychologist G. Stanley Hall incorporated an extensive review of endocrine approaches to aging in his treatise *Senescence*, concluding that "the physiological dominance of sex glands and their products, and the immense role played by sex life, especially in man, suggest that it is in this field that the cure of his most grievous ills must be sought, just as the oldest and most persistent myths and legends have so long taught that it was in this field that the so-called fall of man took place."[242] Relatively few influential figures thought in those terms, however, and at the height of the controversy Steinach's friend and colleague, the socialist (and Lamarckian) biologist Paul Kammerer, put forward a more conciliatory argument. Even if the restoration of the libido was considered immoral, he explained, the other improvements that resulted from the Steinach operation "should be considered a sufficient compensation ... rejuvenation, by restoring mental faculties together with the love and ability to work, endows life anew with high, ethical riches."[243]

"Ability to work" is a key phrase in the history of rejuvenation research. In virtually all the case reports, "inability to work" was one of the major complaints of the patient—especially in males—and "enhanced energy and efficiency" were the earliest and most frequent improvements reported after the operation. The decrepitude of genuine old age, Steinach and the advocates of his operation emphasized, was unlikely to respond dramatically to the operation. The real target, instead, was *premature* senility, indicated by physical weakness, mental fatigue, and loss of libido affecting middle-aged people in the absence of physical disorders such as tuberculosis.[244] The fewer the symptoms and the sooner the operation was performed, the better were the results.[245] Rather than making old men young, the Steinach operation claimed to make aging people vigorous, energetic, and productive. It was not age, nor simply the libido, but energy that was the point at issue.

This, I believe, is worth pondering. As Anson Rabinbach has demonstrated, from the nineteenth century onward, religious conceptions of sinful idleness gave way to new preoccupations with the nature of labor and with ways to maximize efficiency. One way to enhance productivity was to reduce or eliminate fatigue, and in this task science, of course, was considered to be of paramount importance. In 1904, for instance, one German physiologist claimed to have discovered that fatigue was caused by the accumulation of toxins in the body and eventually produced a chemical that might counteract those toxins. In 1909, this physiologist and an assistant visited a Berlin secondary school and sprayed a classroom with the new chemical without informing the students about its actual purpose. That same afternoon, the students were tested with special exercises and their performance was extraordinary: the speed of calculation increased by 50 percent; the number of errors were reduced; and some students, usually tired and sleepy by that time of the day, were now fresher than in the morning.[246]

One of Steinach's most vociferous supporters, Peter Schmidt, wrote a popular medical book entitled *Don't Be Tired!*, in which he described fatigue as a "world-epidemic, beneath the burden of which the greater part of humanity is daily groaning more loudly." That book ended with a chapter on rejuvenation, in which Steinach's work was presented as the ultimate "medical method of combating old age," and which ended with the ringing cry, "Why do you want to be tired? Because you are old? Yes, but then you need not be old!"[247] Fatigue, as Rabinbach has shown, was not simply a malaise of individuals but a canker at the heart of contemporary civilization: only the creation of "a body without fatigue" would enable Western man to fulfil his cultural destiny. Glandular rejuvenation was far more than a purely medical issue: it was integral to a whole cultural project.

These implications were sharpened immeasurably in the aftermath of the Great War. The war had robbed Europe, particularly Steinach's Central Europe, of a large part of its young male population, destroyed the economy, and caused cataclysmic cultural effects. Paradoxically, although the labor market of the Weimar period was crowded with relatively young workers, the German birthrate was falling sharply from 1900, leading to pessimistic convictions, which came to a head in the 1920s, that the German nation was heading toward extinction (*Volkstod*).[248] Inflation, and then hyperinflation, destroyed the value of the mark, and there was little food available for purchase. Steinach himself observed with regard to his first "rejuvenated" patient that his "diet consisted mainly of inferior bread and vegetables, owing to the shocking conditions of living which prevailed generally in Vienna at that time as a consequence of the Great War." The operation, Steinach claimed, had enabled the patient to cope with such privations.[249]

Steinach's supporters presented the operation as a scientific response to the devastation of the war in far more direct terms. "When the last gun was silenced," remarked Peter Schmidt, "not only were the best men of every nation taking part in it decimated, but the survivors seemed to be absolutely at the end of their resources of endurance."[250] Elsewhere, he observed, "if we consider that in Germany the large *rentier* class was compelled, after the fall in the value of money, to begin earning their livelihood again, we can understand the real value of a measure which restores their working power."[251] A medical correspondent of the *New York Times* remarked that "in Continental centres, rejuvenation is of great interest because of the wastage of the war."[252] Young women, observed a Berlin gynecologist in 1924, had been so malnourished by the war that they frequently stopped menstruating and developed menopausal symptoms. The use of x-rays to stimulate their ovaries—in other words, the same treatment used for rejuvenating older women—would alleviate their symptoms.[253] In America, the German sympathizer George Viereck argued that the Steinach operation could help Germany recover speedily from its postwar economic destitution: since the operation added five to ten years to the life of an individual (a claim Steinach himself would have rejected), every German should use those extra years to further the reconstruction of Germany and the payment of reparations. In 1924, Gertrude Atherton was widely ridiculed in the German and American press for her impulsive suggestion that Germans should rejuvenate their aging elite and thereby gain a leading place in the world once again.[254] The rejuvenation of individuals, in short, would regenerate Germany and Central Europe.

Postwar Central Europeans regarded rejuvenation in general and the Steinach operation in particular as contributions to what Paul Weindling has termed "regenerationist biology." "Biology," according to Weindling, "inspired national values during a crisis when the state seemed to have lost its direction. . . . At a time when Spengler's categories of organic destiny gripped the imagination of the scientific community, biologists offered creeds of national regeneration, combining the scientific with resonating categories of nature, life and the race."[255] The regeneration of society could be brought about only by eugenics, whether the negative eugenics of eliminating the supposedly unfit or the more positive eugenics that encouraged the multiplication of the good and the desirable. Steinach's rejuvenation was undoubtedly consonant with the latter.[256] Paul Kammerer suggested that rejuvenation should be applied particularly to men who could guide humanity to a higher plane.[257] The journalist George Sylvester Viereck did not think rejuvenation should be confined to any particular group but observed, "if we save valuable human material by applying the process of rejuvenation impartially to men and women of ripe experience, our dreams of Utopia may come true at last. . . . Within ever expanding limits, biochemistry will hereafter direct the trend of eugenics and evolution."[258]

Simultaneously, the 1920s celebrated youth in all its forms. Youthful vitality, regardless of gender, was seen as a force of social and moral regeneration in Central Europe and as the fundamental expression of the American spirit.[259] Youth was no longer simply a phase of life but a state of being charged with unique aesthetic, moral, and even political meanings, and the 1920s, according to F. Scott Fitzgerald, represented the "peak of the younger generation."[260] The era was also marked by new concerns with the body and with lifestyle-reform in Central Europe. Enthusiasm for sport, nudism, sunbathing, hiking, vegetarianism, and quests for bodily efficiency, strength, and beauty were characteristic of Weimar culture. Relating these factors to rejuvenation research, historian Stefan Schmorrte has argued that the work of Steinach and others presented "youthful vitality" and a new, improved body as consumer articles. Aging and fatigue were caused by biological deficiencies: replenish those deficiencies with a fairly simple procedure and one would soon possess a rejuvenated body surging with the energy of youth.[261]

And in America, as Laura Davidow Hirshbein shows, "readers of popular literature were told that they should have a special interest in European rejuvenation procedures since American national identity revolved around youthful efficiency and energy."[262] Overwork, especially, was supposed to be the bane of American life. In 1926, Adolph Lorenz's interviewer asserted that everybody expected America to be

> the greatest field for this operation . . . because the United States has the greatest percentage of men and women who have lost vitality through overwork of any nation in the world. The largest group seeking restoration of the powers and "pep" of youth consists of the broken-down business men and women whose health has been undermined and who have aged prematurely through exhaustion of the glandular system.[263]

The Steinach champion George Viereck put it more lyrically: "America, the land where work is most intense, where the value of time is most keenly appreciated, America which consumes the vital energies of her children at a prodigious rate, America will receive the greatest benefit from Steinach's discovery. America is destined to advance the cause of rejuvenation and to champion its world-wide adoption."[264]

And in Central Europe at least, the energy of youth would not suffice: it was the energy of fully *virile* or fully *feminine* youth that was the desideratum.[265] This insistence on health being incompatible with incomplete sexual differentiation, of course, had long been a recurring motif. In the late nineteenth century, one of the cardinal features of degenerate bodies was their lack of adequate gender differentiation: degeneration, as Barbara Spackman has remarked, was "degenderation."[266] An effeminate man or a masculine woman was biologically degenerate. At the turn of the century, in Steinach's Vienna, the philosopher Otto Weininger (1880–1903) had published his massive work, *Geschlecht und Charakter* (Sex and Character) in which he condemned his epoch as "the most feminine of ages," while Weininger's admiring reader, the satirist Karl Kraus called it, simply, "the vaginal epoch."[267]

This makes the interpretation of Steinach's concept of sex-gland antagonism rather more complex than it might seem at first glance. Total sexual differentiation was a norm, an ideal that might, perhaps, have been a feature of the misty (evolutionary?) past, but it certainly was not a reality that turn-of-the-century thinkers and doctors could point to. The majority of doctors and cultural critics addressing issues of sexuality and gender worried over the sheer fluidity of gender that they perceived in bodies, minds, lives, and broader culture.[268] Whether they attributed it to degeneration, congenital developmental anomalies, or incomplete evolution of the human species, the phenomenon was virtually unquestioned. Equally unquestioned was the *desirability* of full differentiation between men and women: even theories of universal sexual intermediacy, as we have seen, considered incomplete differentiation to be pathological.

At one level, of course, Steinach's research further undermined traditional notions of virility and femininity by demonstrating that they were attributes generated by the sex glands, which could be altered by manipulating the glands or their secretions. At another level, however, his "cure" of homo-sexuality and rejuvenation promised that while gender might be a matter of secretions, those secretions could be easily regulated by medical science. To an age beleaguered by profound anxieties that men were becoming feminized and women virilized, Steinach's work offered reassurance that science could resolve the problem. Even if 100 percent men and 100 percent women did not exist, medicine was capable of creating them. Small wonder, then, that Karl Kraus, who hated suffragettes and journalists as archetypal symbols of the "vaginal era," imagined Steinach saving the world by changing feminist activists into maternal women and journalists into real men. Small wonder, too, that Steinach's wife claimed that her husband's research had validated the speculative theories of Otto Weininger.[269]

Thanks to biological expertise, gender would be clear and unambiguous in utopia: ever sharper gender differentiation, in fact, would *create* utopia. As Spengler and his epigones lamented the decline of the West, biologists dreamed of creating a new world, a new species, and a new, clearly sexed, unfatiguable body. The significance of the Steinach operation within that utopian project far transcended its actual clinical efficacy. Paul Kammerer's lyrical invocation of its future potential summed it all up:

> Man, who has subjected iron and stone to the power of his indomitable will, will at last become master of his own house. He will now be able to shape the course of his own life and career according to his wishes. And just as one can rule others only after mastering one's own will, so man, after he has acquired sovereignty over his own body, will achieve a heretofore unimagined new freedom to shape the future of mankind and attain true civilization.[270]

The Superhuman and the Bestial

Kammerer's vision of glandular rejuvenation leading to the regeneration of humanity was not universally shared. For some, glandular trickery could lead only to perdition. Arthur Conan Doyle's 1923 Sherlock Holmes story "The Adventure of the Creeping Man" provides a striking illustration of this motif.[271] This late episode in the career of the great detective revolves around the sixty-one-year-old Professor Presbury, a distinguished physiologist at the University of Camford, who had, in middle age, decided to marry the young daughter of a colleague. The young lady expressed some objections regarding their difference in age, and, after this, Professor Presbury suddenly disappeared on a mysterious trip, returning a changed man. His faithful wolfhound Roy now took to attacking him, he began to receive confidential letters marked with a cross under the stamp, and most unnervingly of all, took to crawling around his house at night "on his hands and feet, with his face sunk between his hands." Nevertheless, he now had "more energy and vitality" than ever before and his intellect seemed to be at its peak. "But it's not he—it's never the man whom we have known," cries his secretary.[272]

Holmes deduces that every nine days, the professor "takes some strong drug which has a passing but highly poisonous effect." On the next ninth day, Holmes returns to Camford and, as his deduction had predicted, he beholds the professor, clad in his dressing-gown, climbing up an ivy-covered wall. "From branch to branch he sprang, sure of foot and firm of grasp, climbing apparently in mere joy at his own powers, with no definite object in view. With his dressing-gown flapping on each side of him he looked like some huge bat glued against the side of his own house." The bizarre sequence ends with Presbury being attacked by Roy, who bites his master almost lethally on the throat.[273]

In Presbury's study, Holmes finds "an empty phial, another nearly full, a hypodermic syringe, several letters in a crabbed, foreign hand" and a more cultivated letter from one H. Lowenstein stating there were some risks associated with Presbury's "case." "It is possible," the letter explained, "that the Serum of Anthropoid would have been better. I have . . . used black-faced Langur because a specimen was accessible. Langur is, of course, a crawler and climber, while Anthropoid walks erect, and is in all ways nearer." The name Lowenstein reminds Holmes and Watson of "some snippet from a newspaper which spoke of an obscure scientist who was striving in some unknown way for the secret of rejuvenescence and the elixir of life. Lowenstein of Prague! Lowenstein with the wondrous strength-giving serum, tabooed by the profession because he refused to reveal its source." The mystery is solved: the professor was injecting himself with a monkey serum every nine days and turning himself, for all practical purposes, into a langur! "When one tries to rise above Nature," Holmes intones at the end, "one is liable to fall below it. The highest type of man may revert to the animal if he leaves the straight road of destiny. Consider, Watson, that the material, the sensual, the worldly would all prolong their worthless lives. . . . It would be the survival of the least fit. What sort of cesspool may not our poor world become?"[274]

In his 1939 novel *After Many A Summer*, Aldous Huxley showed us what that cesspool might be like. The aging millionaire Jo Stoyte—apparently modeled on tycoon William Randolph Hearst—finances a research team to find a means of rejuvenation. It is clear, however, that the age of Steinach and Voronoff is over. Dr. Obispo, the chief researcher, uses hormones on Stoyte, but the ultimate solution, he is sure, is not to be found in the endocrine system. "Brown-Séquard and Voronoff and all the rest of them—they'd been on the wrong track. They'd thought that the decay of sexual power was the cause of senility. Whereas it was only one of the symptoms."[275] After many twists and turns, Stoyte and Obispo, assisted by scholar Jeremy Pordage, find that the two-hundred-year-old Fifth Earl of Gonister and his mistress had become, for all practical purposes, immortal by eating the raw guts of a specific variety of carp. This immortality, however, was anything but blissful. Stephen Jay Gould has shown how Huxley borrowed from the hypothesis of Dutch anatomist Louis Bolk (1866–1930) that human beings were essentially apes whose full development had been retarded by an evolutionary alteration in the hormonal system. Humans grew old and died long before their simian features could be manifested.[276] As one of Huxley's characters put it: "You grow up; but you do it so slowly that you're dead before you've stopped being like your great-great-grandfather's foetus."[277] The Earl of Gonister, as Obispo, Stoyte and Pordage are horrified to discover, had lived long enough to turn into an ape. Inhuman and bestial, the rejuvenated earl lived on, a symbol of the futility and horror of eternal youth. Humans were human only because they died before growing up.

Gender, Biology, and Identity

What, if anything, united such apparently bizarre and, in presentist terms, wrong-headed clinical interventions as glandular rejuvenation and the glandular treatment of homosexuality? What could possibly be common to the biology of aging and the cause(s) of homosexual orientation? And if there are such commonalities, what insights might they offer into the history of medicine?

Glandular theories of sex, for all their uncertainties and for all the disagreements over mechanisms, transformed masculinity and femininity from immutable, inborn qualities into morphological and psychological attributes that were variable in nature and malleable in practice. Steinach's experiments, above all, showed that males could be feminized and females virilized by glandular manipulations, which, certainly, was a claim subversive of traditional social certainties. Subversion, however, was only part of the story. What Steinach took away with one hand he gave with the other. His research on rejuvenation as well as his "cure" of homosexuality had the same fundamental premise: the surgical or chemical restitution of masculinity. True males were not simply virile but also strong, energetic, active, courageous, creative, libidinous, and heterosexual. These, of course, were glandular phenomena rather than inborn, unchangeable qualities, but they were not trivial or so unstable as to be fictions. In the final analysis, hormonal theories of gender did not lead to any irretrievable breakdown of the categories of "male" and "female" or to any collapse of the normativeness of heterosexuality. Louis Berman encapsulated these themes with characteristic confidence: "In any attempt at measurement of men and women, the quality and quantity of the internal secretion of the interstitial cells must be respected as a fundamental consideration. The womanly woman and the manly man, those ideals of the Victorians, which crumbled before the attack of the Ibsenites, Strindbergians and Shavians in the nineties, but which must be recognized as quite valid biologically, are the masterpieces of these interstitial cells when in their perfection."[278]

What was novel, however, was that when *not* "in their perfection," those glandular cells could be assisted by Science. Gender, to be sure, was a matter of secretions, but since medicine could now control those secretions, an ideal world of virile, youthful, heterosexual men was well within reach. Magnus Hirschfeld, we should recall, used Steinach's findings to validate his own biomedical construction of a new homosexual identity that was neither wholly male nor wholly female, but was in no way a threat to conventional definitions of masculinity or femininity. And that was just one possible use of glandular theories of sex. As Steinach's own work on rejuvenation and Voronoff's monkey gland transplants showed, the chemical theory of gender had repercussions far beyond the domain of sexuality. Sex was not just a matter of certain acts or a set of physical attributes any more. In its fluid, malleable incarnation, sex was the very essence of life—an essence, moreover, that had yielded its secrets to science and could now be controlled medically.

"Every single feature of the individual," declared Knud Sand, "is in and of itself a sexual character," and those characters were now modifiable. Science was no longer confined to *studying* man as a sexual entity; science could now, almost literally, *make* men.[279]

The turn of the century was the best time to dream such dreams. It was an age of pervasive anxieties about gender, and Eugen Steinach's Vienna resounded with debates on the relevance of gender to civilization. Writers Arthur Schnitzler and August Strindberg, cultural critics Karl Kraus and Otto Weininger, feminist theorist Rosa Mayreder, and physician Sigmund Freud, in their own ways, constructed, deconstructed, hailed, and castigated the "new woman," the "new mother," and what James Joyce, in a somewhat different context, would immortalize as the "new womanly man."[280] The signs and meanings of masculinity and femininity were in flux: a heroic masculine age had been succeeded by Kraus's "vaginal" epoch.[281] This sense of apocalypse was greatly heightened by contemporary feminist activism. The successes of feminism may have been negligible as yet, but the very fact that women were beginning to demand such quintessentially male prerogatives as sexual freedom, political rights, higher education, and entry into the professions suggested to many thinkers of the era that women were becoming masculine and men effeminate and decadent. It was the end of civilization as the intellectuals knew it, and the specter of feminism encouraged much misogyny as well as new, anxious quests for the meaning of masculinity.

Not simply an age of antifeminism, the turn of the century was marked also by quasi-Nietzschean quests for a "pure, healthy, and authentic" masculinity, some of which would later, in Central Europe at least, flow quite naturally into nationalistic and fascist channels.[282] In the United States during the same period, nineteenth-century notions of manly self-control and continence were being transformed into a more open celebration of male sexual power and vigor.[283] Steinach himself was no believer in untrammeled male sexual adventurism, but his preoccupation with the replenishment of masculinity was entirely consonant with and shaped crucially by the social and existential concerns of his age.

As the old categories of gender came under threat, strategies evolved to construct new identities, while reinforcing or undermining old ones. Historians have not adequately appreciated the diversity of roles that biomedical discourse played in those redefinitions. While it has been shown repeatedly how biological arguments were used to negate demands for autonomy (most notably, demands for female autonomy), rather less attention has been bestowed on the historical fact that in other contexts, one could appeal to biology in order to affirm identity and demand autonomy without actually subverting the conceptual power of traditional ideas of male and female. *That,* I would suggest, is the fundamental significance of the outlandish experiments and treatments discussed in this chapter. They not only encouraged doctors and medical scientists to play flamboyant, invasive games with people's bodies, but they also allowed the sexual deviant or the aged to claim new, "normal" identities. In the days of Krafft-Ebing, too, a male homosexual had a clinical identity that, as Harry Oosterhuis has shown, was not entirely negative. But it was the identity of a degenerate, and it was fixed. With the dawn of the glandular age, effeminacy or decrepitude were no longer necessarily irreversible. The marvelous new techniques of glandular science promised to transform the effeminate, the homosexual, and the decrepit not into supermen or superwomen but into Everyman and Everywoman as represented in the cultural imagination of the era.

And that, ultimately, was the reason why the new chemical understanding of gender turned out to be less destabilizing than conservative. The theory was radical and the practices revolutionary, but they were based on norms of gender that were entirely traditional.

Chapter 4

Ending Gonadal Hegemony:
Sex and the Endocrine Orchestra

> The more you study the endocrine glands, the more deeply do you
> become impressed by this element of strife within the body. There
> is war between the male elements and the female in the product of
> conception; there is war between the thyroid and the pancreas; there is
> an armed neutrality between the thyroid and the suprarenal; it is pull
> devil, pull baker, between the pituitary and the gonads.
> —LEONARD WILLIAMS[1]

G landular research in the interwar years was not confined to
rejuvenation, although the mass media of the time did not
have much time for anything else. It was during those very years
that the simpler notions of the sexual body on which Steinach and
others had based their clinical interventions were reformulated
root and branch. By the end of the 1930s, endocrine scientists
and clinicians were working with ideas of the generation, maintenance, and modification of the sexual body that were immensely
more complicated than anybody could have imagined at the beginning of the 1920s. The drama that had begun with testicles
and the ovaries had now been invaded by an ever-expanding cast
of characters, whose interactions were subtle and unpredictable.
Apparently minor members of the cast seemed to be influencing the course of events far more profoundly than expected, new
subplots were mushrooming at a disconcerting pace, and anticipated denouements were dissolving into mere peripeteias.

For the generation of gland researchers active around the First
World War, male and female were mutually opposed forms, even

though it was a fact of life that many individuals were often mixtures of male and female. This sexual ambiguity was produced, it was claimed, by abnormally differentiated gonads secreting masculinizing as well as feminizing hormones. *Normal* individuals, however, were assumed to be exclusively, or at least overwhelmingly, male or female in their bodies and in their sexual orientation. A fully masculine male, needless to say, would be exclusively heterosexual. Deviations in masculinity or femininity, of course, were only too frequent, but no matter how common, they were always pathological, always to be identified as aberrant. It was the job of medicine to correct them. Steinach's glandular treatment of homosexuality, for instance, was founded on this conviction.

From the 1920s, however, this beautifully clear scheme was to be undermined by physiologists and biochemists. By the time of the Second World War, it had been established—at least among biochemists—that the testicles *normally* secreted hormones characteristic of the female as well as the male, and every ovary produced female and male hormones. And sex, by now, was more than a matter of the gonads alone. The latter notion, to be sure, was not exactly new. We have already seen how Blair-Bell sought to establish the polyglandular nature of femininity, and as early as 1916 physical anthropologist Sir Arthur Keith had declared that the sex glands were "not only seed treasuries, but also busy offices from which missives are being constantly dispatched to the adrenals, thyroid, pituitary, and other growth-controlling laboratories."[2] Such ideas were to become part of orthodox teaching in the interwar period. In 1922, zoologist Julian Huxley confidently asserted that "the gonads do not operate as independent organs, but in conjunction with the whole of the rest of the endocrine system—thyroid, pituitary, adrenal and the rest."[3]

Although the glandular constitution of the sexual body was getting complicated, there was as yet no implication that the categories of masculinity and femininity overlapped to any significant degree. Quite the contrary. "It thus comes about," Julian Huxley explained, quoting Blair-Bell, "that the relative proportion or relative activity of the parts of the whole ductless gland system is different in male and female."[4] In other words, male and female were still different categories and, perhaps, different states of being. That polarity, however, was now acknowledged to be generated by a combination of endocrine glands. "Masculinity in girls is not usually the result of disease of the ovaries" asserted Kenneth Walker in 1924, "but of the suprarenals; new growths of that gland . . . bring about precocity in boys and masculinity in girls. . . . It is not unlikely that other hormones besides those elaborated by the gonads play an important part in the development of the secondary attributes of sex."[5] Details remained obscure, but it was now certain that the sexual body was not produced or maintained solely by the gonadal hormones.

By the end of the 1930s, the hegemony and the sexual exclusivity of the testicle and the ovary had broken down irretrievably. They were known to secrete hormones "appropriate" to *both* sexes, and they were no longer considered self-sufficient and self-governing. Yesterday's virtuoso soloists were now demoted to being mere members of the endocrine orchestra (important members, to be sure, but no more so than their peers) and under the baton of the master gland, the pituitary. Sex was not a gonadal sonata any more; it was an endocrine symphony, or at least a concerto. This chapter, then, will tell two interwoven stories. The first is about the displacement of Steinachesque concepts of hormonal sex-polarity (what he himself called sexual antagonism) by more complex ideas of hormonal "bisexuality." The second is about the end of gonadocentrism and the construction of a polyglandular infrastructure for the hormonal sexual body. Neither event, as we shall see, amounted to more than partial revolutions, but together they represented the most profound reformulation of the fundamental principles of sexual biology since the days of Brown-Séquard.

The Rise and Fall of Sexual Antagonism

As we saw in the last chapter, Steinach conceived of the male and female sex glands as functionally opposed. The male puberty gland *stimulated* traits inhibited by the female puberty gland and *inhibited* those stimulated by the female. The male gland, for instance, induced a growth of coarse fur whereas the female gland inhibited that growth. The basic concepts of sex-gland antagonism did not change when Steinach began to speak of hormones instead of glands.[6] The idea of a functional antagonism between the male and female glands received powerful support from Chicago embryologist Frank Rattray Lillie's studies of the freemartin, a sterile, zygotically female but morphologically ambiguous twin of a male-female pair whose placentas were fused (leading thereby to the free mixing of their blood), occurring in cattle and sheep. The freemartin, although a biological rarity, was, as Anne Fausto-Sterling has emphasized, regarded as a key to the riddle of sex determination and development.[7] While acknowledging that the fundamental determination of sex was a genetic process, Lillie explained the occurrence of the freemartin as the "masculinization" of the female twin by male hormones from the other twin. "He thus demonstrated how the genetic view of sex worked in concert with the hormonal view. Genes started the sex determination ball rolling, but hormones did the follow-through work."[8]

The female twin of the freemartin was genetically determined to be female, but every embryo, regardless of its genetic sex, went through a sexually indeterminate phase in its development. The development of specific sexual characters began after the appearance of the sex glands. The female freemartin twin could become masculinized (in spite of its genetic femininity) for two possible reasons. First, the male hormone from its twin might suppress the action of its own female hormone because the male sex hormone was naturally more powerful than the female hormone. Lillie's studies suggested, however, that it was all a matter of timing rather than a straightforward struggle between two kinds of hormone. In the male twin of the freemartin, the gonads developed slightly earlier than in the female twin. Since the circulatory systems of the twins were in communication, the male hormones affected the still sexually indeterminate female, resulting in the masculinization of its reproductive organs.

The female embryo did not, however, become *totally* male, which would have been the case if hormones had been all powerful in sex determination and development. Although the external genitals of the abnormal twin were female, its internal reproductive organs were "intersexual" in type rather than unambiguously female or male. "The various cases," Lillie stated, "can be arranged in a series of increasing male-likeness, but the transformation of the female zygote does not . . . proceed all the way to the normal male condition."[9] This partial masculinization of the female freemartin twin, Lillie pointed out, was "analogous to Steinach's feminization of male rats and masculinization of females," although "more extensive in many respects on account of the incomparably earlier onset of the hormone action."[10]

Paradoxically, however, it was Lillie who was to be instrumental in the demolition of Steinach's concept of the antagonism of the sex glands, when he asked his junior colleague—and later his successor as departmental chair—Carl Moore to investigate whether the female freemartin twin was masculinized directly by male hormones or indirectly by the absence of the ovary.[11] Moore never in fact managed to crack that riddle. While trying to do so, however, he performed a series of gland-transplantation experiments attempting to replicate Steinach's findings and became more and more puzzled by the discrepancies that emerged.[12] Dorothy Price, officially Moore's research assistant at the time but actually the more imaginative researcher, recalled,

> Moore was not concerned in studying rejuvenation, but some of Steinach's experiments seemed to support Lillie's theory. However, these theories needed confirmation and Moore was just the man to do it. But, indeed, Moore did *not*!

> He repeated Steinach's experiments . . . and found no sex gland antagonism in rats and guinea pigs.[13]

Some of Moore's experiments "were carried out without previous removal of the normal gonad; ovaries were transplanted into male animals in which one testis had remained undisturbed, or testis material was placed in the body of a female one of whose ovaries were yet intact."[14] This, as Moore was well aware, was a procedure that, according to Steinach, could never lead to success.[15]

Steinach's assertion that gonadal grafts did not "take" if the recipient's own gonads were intact had followed from the so-called Law of Halsted that the transplant of an organ would "take" only if the same organ were not in situ in the recipient's body. Steinach, as we know, had conflated this supposed antagonism between the gonads themselves with the antagonism between the morphological effects of their secretions.[16] Moore sought to repudiate the entire concept of antagonism by disproving what was only one component of it, that gonads could not be grafted successfully in the presence of the animal's own gonad. Presenting fourteen successful instances of prolonged "persistence of sex-gland tissue grafted into an animal of the opposite sex with one of its normal gonads undisturbed," he swept aside Steinach's contentions with great confidence. There was no antagonism between the sex glands, and Steinach's results were simply not replicable.[17]

So how, asked Moore, could this finding explain the freemartin? In the freemartin, too, he asserted,

> there is no real indication of an antagonism between the sex glands. Hormone action is not characterized by an inhibition but by a stimulation, and the stimulus from the male hormone acting upon the indifferent gonad of the foetus exerts an influence quantitatively greater than the inherent influence toward femaleness, hence the resultant of the two forces is a type of development more nearly resembling the male than the female. . . . The degree of intermixture of the sexual apparatus would thus be a function of the quantitative difference between the two stimuli and not the result of the suppression of one gonad by the other.[18]

So, all that Moore could suggest was that the male hormone simply overpowered the "inherent" femininity of the female twin. This, of course, was similar to a hypothesis that Lillie himself had considered only to reject. Moore, in any case, did not offer any definitive proof for such a quantitative difference between the actions of the male and the female hormones.[19]

But weak as his explanation was of the freemartin, Moore's assault on Steinach's concept of antagonism was serious and did not go uncontested.[20] Alexander Lipschütz reported in 1925 on more than one hundred and fifty experiments with guinea pigs, which had shown that living ovarian grafts in the presence of intact testicles failed to feminize the animals. If the testicles were subsequently destroyed, then the ovarian graft came into its own, transforming many of the secondary sex characters of the hitherto masculine animals into feminine forms. Obviously, then, the testicular secretions had the power to inhibit the actions of the ovarian graft.[21] The next year brought a major riposte from Steinach himself. Acknowledging that he had been wrong about the direct antagonism between testicular and ovarian tissue, he emphasized that there could, however, be no doubt about the antagonistic actions of the male and female sex gland secretions. The real antagonism, in other words, was *hormonal* antagonism, expressed in the masculinizing or feminizing *effects* of the gonads.[22] Frank Lillie's studies of the freemartin had demonstrated that the hormonal antagonism between testicular and ovarian secretions was indisputable. The internal secretions of the sex glands, Steinach reiterated, were not only sex-specific in their effects but, contrary to what Moore claimed, they also actively suppressed the development of feminine sexual characters.[23]

Steinach was so unimpressed by Moore's arguments that he did not even bother to refute them point by point. He merely asserted that his earlier masculinization and feminization experiments, which were now beginning to be confirmed by others, demonstrated the reality of hormonal antagonism quite irrefutably. Steinach warned, however, that this antagonism should not be interpreted mechanically. In certain circumstances, it could be weakened or even eliminated. In his own experimental hermaphrodites, for example, each graft had seemed to bring about the development of the homologous sexual characters and inhibit the characters of the other sex only very modestly. That weakening of hormonal antagonism, in fact, was what enabled one to produce an experimental hermaphrodite: with absolutely equal and opposed antagonism, neither graft would be able to produce its sexualizing effects. The testicular graft would cancel all the actions of the ovarian graft and vice versa.[24]

Gonad transplantation, in any case, was no longer of much experimental importance, now that potent extracts were becoming available and the Allen-Doisy test had provided an accurate method to standardize the female hormones. In Steinach's own laboratory, a powerful female hormone had been elaborated from the placenta of sheep, which had proved to be identical in its effects with that obtained from the ovaries. Injections of hormones could now produce the same effects as glandular transplantations. In females, the injection of the placental extract regenerated the atrophied ovaries of senile animals, stimulating follicular activity and bringing about generalized revitalization of the organism. But it was equally effective in inhibiting masculine sexual characters. Unveiling a new series of experiments on infant male rats, Steinach claimed that injections of the placental extract inhibited the development of such male characters as the prostate and the seminal vesicles.[25]

Moore was not stumped by Steinach's new experiments, and the availability of hormone extracts expedited his demolition of Steinach's theoretical edifice.[26] In 1929, Dorothy Price recalled, Moore commenced another extensive series of experiments using testicular and ovarian extracts. The results were stunning. Ovarian extracts damaged the testicles and the accessory sexual organs such as prostate and the seminal vesicles "almost without exception"—exactly as Steinach would have predicted and as other researchers such as Fellner, Hermann, and Stein and the Laqueur group had demonstrated.[27] Was this proof of hormonal antagonism, however? When both hormones were injected into *castrated* males, the male accessory sexual organs—such as the seminal vesicles—developed normally. This showed "a lack of interference with the action of the testis hormone upon the male accessories by oestrin [ovarian hormone]. Oestrin is inactive upon male structures showing neither stimulation nor depression."[28] But when estrin was injected by itself, it did damage the male accessory organs. This, argued Moore, was not due to the antagonism of estrin and male hormone, but because of the absence or insufficiency of the male hormone consequent upon castration. When estrin and testis hormone were injected together into *intact* male animals, the testicles were damaged but the accessory organs remained unaffected.[29]

Estrin, then, could damage the testicles themselves but not inhibit the actions of the testicular hormone. The prostate and seminal vesicles of adult male rats shriveled after castration, but this castration-induced involution was in no way intensified by injections of estrin. So, it was likely that, contrary to Steinach's assertion, "oestrin is without effect upon the male accessories. It neither stimulates nor depresses them." Nor did spayed female rats show any response to injections of testis hormone, suggesting that "the testis hormone stimulates the homologous accessories, but is without effect upon heterologous ones."[30] Only the first half of Steinach's concept of sex hormone function—the stimulation of homologous sex characters—was correct.

The most innovative argument was still to come. This concerned the damage that estrin caused to the testicles. Such damage could, of course, be explained by some kind of direct antagonism, but Moore and Price rejected that possibility. Astoundingly, Moore had found that the testicles of young rats were damaged even by the administration of the *male* hormone. "No one," observed Moore and Price triumphantly, "has proposed that testicle hormone exerts a 'hormonal antagonism' effect in the male, and such a viewpoint would be incredible. And since similar results obtain from oestrin, we suggest that the same hormone antagonism conception is likewise untenable."[31] This was a devastating critique of Steinach's concept, but Moore and Price still had to explain how the sex hormone injections, whether male or female, damaged the gonads. That was far from easy. Moore was due to give a talk at the Second International Congress for Sex Research in London on sex hormone antagonism, and "there was great urgency for him to prepare a manuscript, but he could not interpret his findings." Pressed for time and ideas, he asked Price to reflect on the problem and, according to her, the solution to the puzzle occurred to her "quite suddenly."[32] Whether the solution did in fact appear so suddenly is not of great moment for our story; what is crucial is that Price's interpretation of Moore's perplexing results involved a radical remapping of the sexual body—or, as Moore put it pithily, it provided "a new explanation of so-called sex."[33] Before we can examine that explanation, however, we need to move away for a while from the gonads and peer into the hidden recesses of the brain.

The Riddle of the Pituitary

Diminutive in size (no bigger than a large pea) and tucked away at the base of the brain and protected by two layers of bone, the pituitary gland or hypophysis cerebri was, according to its great investigator Harvey Cushing, "like the nugget in the innermost of a series of Chinese boxes. No other single structure in the body is so doubly protected, so centrally placed, so well hidden."[34] Although the pituitary was known since ancient times, the sheer difficulty of exploring it ensured that its purpose and significance remained unclear until the early twentieth century.[35] Observers like Cushing, however, were convinced that Nature would not waste so much effort on protecting an insignificant organ and suspected the pituitary of being "a most important member of the endocrine series."[36] By the end of the 1920s, the pituitary gland (which has two parts: a posterior lobe derived from neural origins and a glandular anterior lobe) was known to be prolific in its secretions, many of which regulated the secretory activities of the other endocrine glands.[37] Physician Walter Langdon-Brown hailed it as the "leader in the endocrine orchestra," a status it retained until the question of its own regulation led to surprising answers.[38] Since the pituitary seemed to represent the meeting point of the neural and the chemical spheres of the organism, Kenneth Walker called it the "centre of our being."[39] Such descriptions did not seem too hyperbolic in the light of the fact that the anterior lobe of the gland seemed to be essential to life.[40]

Predictably, then, it was the anterior pituitary that fascinated the glandular speculators of the 1920s.[41] "Perhaps in no department of medicine has quackery flourished so much," observed Benjamin Harrow, "as in that dealing with the ductless glands, and more particularly with the pituitary. The over-enthusiastic scientist has joined hands with the pseudo-scientist, and both have provided ample material to the charlatan to advertise his goods and to ingratiate himself with an all too credulous public."[42] Harrow was not exaggerating. Louis Berman's readers would have learnt that milder varieties of pituitary insufficiency could turn human beings into "pathological liars with little or no initiative or conscience—amoral, not merely theoretically, but instinctively and unconsciously."[43] The posterior lobe governed the emotions while the anterior pituitary was the gland of courage and intellectuality. The administration of anterior pituitary extracts to the mentally retarded could lead to the enhancement of intellectual capacity, and the bravest men were usually pituitary-dominant.[44] Napoleon was the prime example of pituitary malfunction. Singularly bereft of feelings, the emperor had a highly developed intellect, suggesting posterior-pituitary deficiency and hyperactivity of the anterior pituitary.[45]

The popularizers, however, were not the only speculators. As notable an investigator of the pituitary as Harvey Cushing was not entirely averse to speculation, although his notions were diametrically opposed to Berman's. Cushing thought that the hyperactivity of the pituitary, as seen in gigantism, stunted the intellect. "Pathological giants, as a rule, have a low intellectuality," he remarked, adding that "it took Holofernes five years and three months to teach Gargantua his A, B, C's backwards."[46] Even more interestingly, neuropsychological symptoms were virtually invariable in pituitary conditions, "varying from nervousness to epilepsy and actual mental derangement." It was strongly likely, therefore, that deficiency of the pituitary hormones brought about mental instability. "There can be little doubt," Cushing concluded, "that many of the psychasthenias and neuroses of one sort or another will prove to be associated with ductless gland disturbances, more particularly with those of hypophyseal origin."[47] More than thirty years later, the London surgeon Lennox Ross Broster, a pioneer in the study of the adrenogenital syndrome, suggested without any sign of irony that the dinosaurs had suffered from pituitary gigantism and died out because of the low fertility that was known to accompany that condition.[48] The reality of pituitary research, however, was hardly less intriguing than these fantastical hypotheses.

"The Motor of Sexual Function"

As so often in our story, clinicians were preeminent in pituitary research. Harvey Cushing's classic monograph *The Pituitary Body and its Disorders* was based on surgical work conducted at the Johns Hopkins Hospital and the Johns Hopkins University School of Medicine. "There are few subjects in medicine," began Cushing, "which promise a wider overlap upon the fields of many special workers than this one of hypophyseal disease." Pituitary disorders could cause symptoms demanding the attention of ophthalmologists, neurologists, gynecologists, urologists, internists, general surgeons, otolaryngologists and pediatricians, not to mention radiologists, pathological anatomists, and physiological chemists.[49] But in spite of its obvious importance, the investigation of the pituitary was still "at the tail of the procession" of the ductless glands being studied by "experimentalists, pathologists and clinicians."[50]

Some things, however, were tolerably clear by Cushing's time.[51] It was accepted, for instance, that the extirpation of the pituitary led to inanition and death.[52] Partial removals of the gland were more interesting, because they seemed to cause symptoms of possible clinical relevance, including massive obesity, metabolic disorders, and "sexual inactivity or actual atrophy of the reproductive glands."[53] Infant puppies deprived of certain parts of the pituitary never attained sexual maturity—the females never went on heat and "the males were quite anaphrodisiac." The ovaries showed few follicles and the interstitial cells of the testicles degenerated. The other endocrine glands were also affected by partial extirpations of the pituitary, producing what Cushing called a "polyglandular syndrome." This experimentally produced complex of symptoms was very similar to a rare human condition called the adiposo-genital syndrome, characterized by stunting of skeletal growth, obesity with a feminine distribution of fat, and undeveloped genitalia. (Cushing's interest in the pituitary had, in fact, been aroused by a patient of this syndrome.)[54] Noticing that "fat amenorrhœics and impotent males" often suffered from pituitary disease, Cushing declared that "unquestionably the interrelation between hypophysis and testis or ovary appears, on clinical grounds, to be more intimate" than that between any other endocrine glands.[55] The influence of the pituitary on the gonads would indeed turn out to be one of its most fascinating aspects.

In the 1920s, Philip E. Smith (1884–1970) of the Department of Anatomy at Stanford University conducted a remarkable series of experiments on the pituitary's influence on sex that stamped Cushing's clinical deductions with the imprimatur of the laboratory, catapulting the pituitary to the forefront of sexual physiology. Smith began by showing that the removal of the pituitary in rats led to sexual atrophy, which could be reversed by pituitary implants. He then used pituitary implants to hasten the sexual development of immature female rats. The implants never failed in this task, but they could not reverse the genital atrophy brought about by oophorectomy. It was clear, therefore, that the pituitary exerted its sexualizing effects through the ovaries, and "daily pituitary transplants in the adult female," Smith reported, "usually leads to an increase in the size of some of the follicles of the ovaries."[56] The pituitary effect was not only anatomical but behavioral: pituitary transplants induced immature animals to mate. Only *anterior* pituitary transplants had any sexual effects, however, and there was some evidence to suggest that "the rapidity of maturing is correlated with the amount of pituitary tissue implanted."[57]

Hindsight shows us that a new model of pituitary interaction with the gonads was taking shape in the work of laboratory scientists like Smith and clinicians like Cushing. The novelty of the new concepts become clearer if we compare them with slightly earlier views of the link between the gonads and the pituitary. Eugen Steinach, for example, did consider it important to investigate how the gonad and the pituitary were linked, but left that task to one of his assistants, Joseph Schleidt. Schleidt found that in castrated animals, the histological appearance of the pituitary soon changed in ways acknowledged by scientists to be typical after castration, but castrates reimplanted with gonads possessed perfectly normal pituitaries. The masculinized or feminized animals, which had been implanted with gonads after castration, had histologically normal pituitaries. The conclusion did not really concern the pituitary but the gonads: since the implanted gonads always lost their germinal elements, it must be the interstitial cells—the "puberty gland"—that maintained the normality of the pituitary.[58] The integrity of the pituitary, then, depended on the gonads.[59] It was the gonad that was silently assumed to be the more important partner in the gonad-pituitary duo.[60]

Smith's extraordinary experiments apart, the nature of the pituitary influence on the gonads was clarified most strikingly by the dogged research of the German—later, Israeli—gynecologist Bernhard Zondek (1891–1966), based for many years at the Charité Hospital in Berlin. His work, some of it conducted with Selmar Aschheim, was central to the fall of the gonads from sexual omnipotence. Appropriately, it was also Zondek who was responsible for reporting one of the most dramatic pieces of evidence suggesting that the so-called male and female sex hormones were found normally in both sexes. On the occasion of his seventy-fifth birthday in 1966, the *Journal of Reproduction and Fertility* observed:

> Those of us who were at work in the late 1920s will not forget the string of discoveries coming from the hands of Zondek and Aschheim. . . . apart possibly from the brilliant work of the steroid chemists, no discoveries made at that time had a greater influence than these on the development of reproductive endocrinology.[61]

One of Zondek and Aschheim's most intriguing early findings was that although the genital tracts of infant mice could be artificially matured by injections of ovarian hormones or by implants of ovarian tissue, their ovaries remained immune to such stimulation and did not secrete any sex hormones.[62]

Artificial sexualization, therefore, did not amount to the induction of artificial puberty. What, then, induced the gonads to start secreting sex hormones at puberty? After failing to activate infantile gonads with numerous proteins such as organ extracts or fluid from ovarian cysts, Zondek and Aschheim concluded that only another endocrine gland could induce the ovary to secrete hormones.[63] They now tried to bring about premature puberty in infant mice with transplants or extracts of a range of endocrine glands—thymus, adrenal cortex and medulla, posterior pituitary and the thyroid. All failed except the anterior pituitary.

A pituitary implant, Zondek reported, gave rise to changes in the vaginal tissue of immature animals that were characteristic of ovarian follicular activity. The uterus grew in size and its mucosa proliferated.[64] This was the first phase of a complex reaction.[65] In the second phase, which commenced within a hundred hours of implantation, the ovaries were visibly enlarged and showed dramatic follicular activity, leading to hemorrhages within the follicle, visible as "blood points." This marked the end of the second phase. In the third phase, the follicles ruptured, the ovum was expelled into the uterine tube and a corpus luteum was formed.[66] Even in senile animals that had long ceased to go on heat, the anterior pituitary, as Zondek and Aschheim put it, "brought new life to the degenerating ovaries," an effect they were not embarrassed to describe as "rejuvenating."[67] Only spayed animals failed to respond to a pituitary implant.[68] Later, Zondek showed that with prolan (his name for the gonadotropic hormone of the anterior pituitary), one could even induce pregnancy in an infantile animal.[69]

After much more research, Zondek concluded that the anterior pituitary hormone, identical in men and women and also in humans and animals, was the supreme, general sex hormone, "the motor of sexual function."[70]

Supreme as it was, however, it was powerless without its underlings, the ovary or the testicle. The ovary could not secrete sex hormones without the pituitary stimulus but the pituitary could not work except through the ovary: it did not itself have the capacity to sexualize the body. Moreover, the gonads regulated the pituitary secretion of what we call gonadotropins today: in the absence of the ovaries, the pituitary poured out gonadotropins incessantly and pointlessly. "The motor of sexual function," Zondek asserted, "is regulated also by its own products. . . . Removal of the hypophysis ends all functional activity of the sex gland. Without a hypophysis, there is no follicle maturation, no impregnation. If, on the other hand, the sex gland is removed, the anterior hypophyseal lobe not only keeps on functioning but indeed, puts out more prolan."[71] Sex was no longer the business of any one gland: neither the pituitary nor the gonad was of any use without the other.[72]

Zondek and Aschheim were quick to realize that some of their findings had great clinical potential. Since the pituitary hormone was excreted in the urine, any elevation in its urinary level could be measured and this might provide important clinical indicators. If a patient's ovaries had been removed in part, for example, one could easily assess whether the residual portion was functioning adequately by measuring the prolan in the urine. If the ovary stopped functioning, then the level of the latter would rise.[73] Measuring urine hormones was of even greater importance in pregnancy.[74] The pregnant organism, in fact, was awash in the pituitary hormone: immediately after fertilization of the ovum, the anterior pituitary "exploded" into action and large amounts of its hormone were excreted in the urine.[75] Since anterior pituitary and ovarian hormones were excreted in large quantities in the urine of pregnant women, could they, then, be used for a pregnancy test?

Detection of the ovarian hormone alone by the Allen-Doisy test would not suffice.[76] That test would be positive in nonpregnant women and not very strongly positive in the first seven weeks of pregnancy, because the ovarian hormone was excreted in relatively high quantities only after the eighth week.[77] The anterior pituitary hormone, however, was a far more reliable indicator. Its level in the urine could be assayed by injecting a few cubic centimetres of urine directly into mice and then examining the reaction of their ovaries.[78] Of the seventy-eight pregnant women tested, seventy-six had a strong positive reaction. No biological test, Zondek and Aschheim claimed, could be more precise.[79] The timing, however, was crucial. In the first eight weeks of pregnancy, far higher quantities of the pituitary hormone were excreted than the ovarian hormone. Between the third and seventh months, the pituitary hormone level remained almost unchanged but the ovarian hormone levels rose steeply to overshadow it, a trend that continued until the birth of the child and for at least a month beyond that. The best time to perform the Aschheim-Zondek test, therefore, was early in pregnancy, which, of course, added to its value as a pregnancy test.[80]

There were many disagreements about the origin and properties of the hormone appraised by the Aschheim-Zondek reaction. Zondek himself argued that the hormone in the urine of pregnant women was the anterior pituitary gonadotropin, but the work of others suggested that the gonadotropic hormone in the urine of pregnant women was different from the pituitary gonadotropin, and there was no consensus, in any case, even about the pituitary gonadotropin itself. As late as in 1933, American experts emphasized that many researchers were still unsure whether the anterior pituitary secreted one or two gonadotropins. Considerable differences had also been noticed between species.[81] "The position with regard to gonadotropic hormones is decidedly involved," observed Walter Langdon-Brown in 1935. The true gonadotropin would have to bring about the maturation of the ovarian follicle as well as maintain the corpus luteum in experimental animals whose own pituitaries had been removed. The so-called prolan obtained from the urine of pregnant women "cannot do either under such conditions." It was also uncertain whether follicular maturation and the formation of the corpus luteum (luteinization) were governed by two different gonadotropins or just one "which produces these two effects according to the stage of the cycle," remarked Langdon-Brown.[82]

What these debates underlined, Zondek responded, was the centrality of the pituitary to the bodily economy and the multiplicity of its roles. Even the posterior lobe of the pituitary, he pointed out, had recently been shown to secrete two distinct substances, one that stimulated the uterus and another that raised the blood pressure and prevented urine production.[83] It was likely that the so-called prolan was not a single substance but a combination of two (or possibly even three) different hormones.[84] One (which he named Prolan A) induced ovulation and ovulation stimulated the anterior pituitary to secrete the other hormone (Prolan B), which brought about the formation of the corpus luteum. He had already isolated Prolan A, which, even on long-continued administration, did not stimulate luteinization.[85]

The postulation of two distinct prolans and the ability to isolate Prolan A in urine complicated an already complex picture. The Aschheim-Zondek pregnancy reaction, it was now evident, detected Prolan A as well as B, since the urinary levels of both were elevated in pregnancy.[86] But in nonpregnant women of reproductive age, the levels of Prolan A in urine were so low as to be undetectable at any stage of the menstrual cycle, even when the urine specimens were concentrated fivefold.[87] This changed during the menopause.

Zondek suggested that the menopause was not an event but a process with at least three successive stages. In the first stage, the level of the ovarian (follicular) hormone rose sharply. It then dropped rapidly during the second stage to levels so low as to be undetectable in urine. It was during this second stage that the well-known vasomotor disturbances associated with menopause made their appearance. The third and final phase was characterized by the definitive termination of ovarian function—the secretion of Prolan A was now elevated and the Aschheim-Zondek reaction would be positive at this stage of the menopause.[88]

Doubtless because of their gynecological focus, Zondek and Aschheim's research tended to skirt around male physiology. Nevertheless, they did elucidate some aspects of prolan action in males. Philip Smith had reported that repeated daily transplants of the pituitary stimulated the development of the immature male genitals in infant male rats.[89] Using their own aqueous pituitary extract, Aschheim and Zondek showed that multiple injections produced unambiguous stimulatory effects on the male genitals, evinced by the enlargement of the seminal vesicles and prostate. The pituitary extract did not, however, produce any effect in castrated male animals, thus suggesting that in males, too, the pituitary could act only through the gonads.[90] The separate functions of Prolan A and B in males, however, remained unknown.[91] Clinically, the level of prolan in urine was of use in diagnosing testicular carcinoma, but it was not reliable enough for a negative result to rule out testicular malignancy.[92]

While much about the gonad-pituitary link still remained to be deciphered, enough had been understood by the end of the 1920s for that leading gonado-centrist Eugen Steinach to reconsider his position on the pituitary. Stimulated by the reports of Zondek, Smith, and others, Steinach launched a new series of experiments investigating the role of the pituitary in sexual development. Steinach had long been content with identifying the gland that triggered puberty—now, at last, he was compelled to ask what triggered the puberty-gland itself.[93] First, Steinach replicated the findings of Zondek and Aschheim that the pituitary could induce precocious puberty in infant animals and then confirmed that sexually undeveloped rats became potent and sexually mature after injections of anterior pituitary extract.[94] Various grades of testicular insufficiency were seen often enough in human males: these, Steinach suggested, were likely to be caused by the inadequacy of pituitary stimuli and should be treated with pituitary extracts.

Expectedly enough, however, his chief interest lay in rejuvenation.[95] After about a dozen injections of anterior pituitary extract, senile rats, he reported, became more active, energetic, and sexually potent. Microscopic examination of their testicles revealed a proliferation of interstitial cells. Steinach still thought, however, that the gonads stimulated the pituitary. The reactivating influence of testicular transplantation or vasoligation, he argued, could well be due to the stimulation of a sluggish or inactive pituitary. The relation between the gonads and the pituitary, Steinach implied, was bilateral and harmonious.[96] This egalitarianism, of course, was quite different from Zondek's putatively hierarchical concept. The pituitary hormones, Zondek insisted, "represent the superimposed, nonspecific, general, sexual hormones. The hypophyseal hormones are primary, the specific sexual hormones are secondary factors."[97] To be sure, Zondek accepted that the gonadal hormones *regulated* the anterior pituitary, but he also emphasized that the pituitary was in no way dependent upon the gonads for its secretory functions. If the anterior pituitary was removed, the functional life of the gonad ended; if the gonads were removed, however, the anterior pituitary continued to produce its sexual hormones.[98] Even within a broadly polyglandular paradigm, then, there was still considerable room for different interpretations of the nature of the bond between the pituitary and the gonads.

The Pituitary and the Gonads: Feedback

It was this uncertainty that was resolved by Carl Moore and Dorothy Price. The original problem facing the Chicago researchers, of course, was to explain how injections of the sex hormones could damage the testicles and the ovaries.[99] When estrin damaged the testes, one might attribute that to sex hormone antagonism, but why did estrin damage the ovaries or the male hormone the testicles? The secretion of both male and female hormones, as Zondek and others had shown, was stimulated by the anterior pituitary. Instead of assuming, like Steinach, that the gonadal hormones, in their turn, stimulated the anterior pituitary, Price reasoned her way to a fundamentally different kind of link. "If, in the normal male, the anterior pituitary controlled the secretion of male hormone, and the male hormone, in turn, controlled the secretion of gonadotropin by the anterior pituitary, there would be a splendid scheme for a seesaw balance between production of gonadotropin and production of male hormone and thus, control of their levels in the blood."[100] Again, we do not need to decide whether Price's late-life reminiscences idealized the actual process of discovery. What we need to focus on is the crucial difference between her feedback model and the Steinachian network of glands running on a simple stimulation/inhibition mechanism. For Steinach, despite the expansion of his horizons beyond the gonads and simple sex gland antagonism, the sexual body remained a field of mutually opposed forces. In

Price's feedback model, however, the network ran on a different, quantitative basis: as long as the gonadal hormones remained above a certain level, they exerted no action on the pituitary; when they fell below that level, however, they still did not directly *stimulate* the pituitary. The pituitary detected that difference in the circulating blood and pumped out gonadotropins to restore gonadal secretion to their usual levels.

There was already some experimental evidence for such a feedback loop. It was known, for instance, that castration stimulated gonadotropic activity, which could then be reduced by the injection of sex hormones.[101] Sex-hormone-induced damage to gonads, then, might have nothing to do with Steinachian antagonism but be mediated through the pituitary and its gonadotropic hormones. "On such slender evidence I hung the Moore-Price theory that night," reminisced Price. When she explained her reasoning to Moore, "he thought for several minutes and then began to warm up; soon he began to consider it a brilliant idea."[102] A new series of experiments was launched and the findings were intriguing. When male rats were implanted with a pituitary graft and then injected with estrin, the testes as well as the accessory sexual organs remained normal. The female hormone failed to inhibit the development of masculine secondary sex characters, and the testicular tissue showed no damage. The same results were obtained when the sex hormones were administered with a gonadotropic extract obtained from pregnancy urine. When this latter extract was given alone to castrated animals, it had "no influence in warding off castration involution . . . hence any effect it may have had in other males must have been, of necessity, directed through the testicles."[103] This, of course, was what Zondek and Smith had demonstrated, and Moore and Price were well aware, too, of studies indicating that castration led to "a greater sex-stimulating capacity of the hypophysis" even though no stimulation could actually occur in the absence of the gonads.[104] What would happen to the pituitary and the gonads, they wondered, if the circulating levels of the gonadal hormones were raised artificially?

When they injected normal female rats with female hormone (estrin) for thirty days, they found that the pituitaries of those rats were "far inferior in their ability to stimulate gonads" than pituitary glands taken from untreated controls.[105] Since estrin damaged the testicles or the accessory sexual apparatus when administered by itself but not if accompanied by gonadotropic substances, it appeared logical to Moore and Price to conclude that in the former case, estrin suppressed the pituitary from "delivering into the organism a sufficient amount of the gonadal-stimulating substance [which] is required for the continuance of testis function." The testicles were damaged, then, by the *absence of gonadotropic stimulation*, and not by any direct antagonism of the female hormone on the male sexual system.[106] Excessive gonad hormones *of either sex* would depress the gonadotropic activity of the pituitary, which would eventually damage the gonads and inhibit their production of hormones. Here was the solution to the perplexing observation that even injections of male hormone could damage the testicles and female hormone the ovaries. It was not the antagonism between male and female but the consequence of a simple shutdown of the pituitary gonadotropin. Even if Steinach was correct in his claim that after his operation for rejuvenation, the testicle swarmed with hormone-producing cells, the additional hormones produced by those cells would merely inhibit the pituitary, thus *lowering* the production of gonadal hormones.[107] It was the quantitative feedback loop between the pituitary and the gonads that was the "real keynote to the solution of the hormone antagonism question": male and female hormones were not the simple, qualitative antagonists that Steinach had imagined them to be.[108]

"Sex hormone antagonism, in the Steinach sense, does not exist," pronounced Moore and Price.[109] Each sex hormone stimulated the development of what Steinach called the homologous sex characters but left the heterologous characters completely alone.[110] The higher the level of the gonadal hormone, the lower the secretion of the pituitary gonadotropin; the lower the former, the higher the latter.[111] Gonadotropic activity was *inversely proportional* to gonad hormone levels; the gonad and the pituitary were linked in the manner of "a seesaw, push-pull, or negative feedback."[112] The gonads, crucially, were not self-regulating hormone secretors: they remained "wholly non-functional" without adequate stimulation from the pituitary. But the pituitary was no absolutist monarch. When the levels of the gonad hormones rose, hypophysial activity was lowered. This ensured that "overloading of an organism with sex hormone cannot normally occur."[113]

It would be wrong, however, to assume that this was the end of the matter. Moore and Price's findings, although weighty, were open to contestation. Was the pituitary-gonad feedback *exclusively* negative? In 1931, Bernhard Zondek had emphasized that the follicular hormone of the ovary could reactivate the sexual organs of senile animals and also *stimulate* the anterior pituitary. Soon, there was some evidence to support the idea that the pituitary-gonad relationship might also involve a positive feedback. In the Schering laboratories in Berlin, Steinach's one-time assistant Walter J. M. Hohlweg (1902–92) noticed in 1934 that contrary to Moore and Price's observations, the administration of estrogenic substances in massive doses to sexually mature female rats did not cause ovarian atrophy.[114] Instead, the ovaries became full of numerous large corpora lutea. This was such an unexpected finding that Hohlweg initially assumed that it was due to an experimental error. With further experimentation, however, it seemed clear that large doses of estrogen stimulated the release of Luteinizing Hormone (Zondek's Prolan B) from the anterior pituitary. This phenomenon came to be known as the Hohlweg Effect.[115]

The most predictable ripostes to Moore and Price's work came, obviously, from Steinach and his supporters. Heinrich Kun, one of Steinach's prominent later associates, argued that all that Moore had done was to postulate a different, indirect mechanism whereby the sex hormones inhibited the heterologous sex characters. Whether injected ovarian hormone harmed the testicles directly or by inhibiting the pituitary gonadotropins, the fact remained that the effects of the former on the secondary sexual characters were antagonistic to the effects of male sex hormones on those traits. Steinach's original concept of hormonal antagonism, therefore, was intact: only its mechanism needed to be rethought.[116] There was a further, larger question to be answered. The inhibition of the male sex characters after injections of estrin would, if Steinach's hypothesis were correct, be a direct, active inhibition by the female hormone; not a consequence of testicular damage. Examining the testicular histology at the commencement of inhibition and comparing it with that when the inhibition had reached its full extent, Steinach found no abnormality whatsoever in the testicular histology in the first case—even though the male sexual characters of the rat were already stunted. In the latter, there was extensive damage to the germinal tissue but none to the interstitial cells. The female hormone, then, suppressed the male secondary sexual characters *before* affecting the testicles themselves—although it was not unlikely that this inhibition was later reinforced indirectly by inhibition of testicular secretion.[117] Experiments with pure hormones, in other words, supported every contention Steinach had made on basis of his earlier transplantation studies—which, he pointed out, had recently received further support from the report of Ernst Laqueur and his associates that their female hormone preparation, Menformon, exerted an "antimasculine" effect on male genitalia.[118] The nub of Moore and Price's critique of sex hormone antagonism—that testicular damage occurred even after the injection of *male* hormones—Steinach and his associates never addressed.

It would be fallacious, as Anne Fausto-Sterling has reminded us, to imagine that Moore's extensive series of experiments ended by refuting the concept of hormonal antagonism to everybody for all time. No doubt Steinach's experimental results were no longer taken to be as authoritative as they had been, for instance, by Lillie in his early papers on the freemartin. The cultural polarization of male and female ensured, however, that a rough-and-ready concept of hormonal antagonism seemed intuitively right to many (especially clinicians) and therefore, as we shall see in the next chapter, persisted in spite of much contrary argumentation in scientific papers. "Although defeated by hormone biologists, the idea of hormone antagonism," Fausto-Sterling emphasizes, "did not die."[119] Even the discovery of male hormones in "normal" females and of female hormones in males did not, in itself, spell an end to the concept of sex hormones as sexually antagonistic substances. As late as in 1945, Paul de Kruif—admittedly not a scientist but hardly an ill-informed figure—referred to the "curious antagonism, this chemical war between the male and female hormones," offering these examples:

> Take the male hormone, testosterone. It can repress the menstruation of girls entering puberty; it can inhibit menstruation in women and can bring on their change of life; it can stop milk secretion in the breasts of nursing mothers. Take one of the female hormones, the one called estradiol. Large doses of estradiol have wiped sexual desire clean out of a sexual criminal, have made him meek as a lambkin, have knocked out the ability of his testicles to form sperm, have brought about actual degeneration of the sperm-forming cells of his sex glands.[120]

As Fausto-Sterling puts it pithily, "a scientific fact, once established, may sometimes be disproved in one field, remain a 'fact' in others, and have a further life in the popular mind."[121] We shall encounter further evidence of this persistence of hormonal antagonism when we examine the clinical use of sex hormones.

At the same time, however, the persistence of sexual antagonism should not blind us to the larger, almost revolutionary implications of Moore and Price's work for the study of sexual physiology.[122] It had long been suspected that the sexual body was generated and governed by an oligarchy of endocrine glands rather than by omnipotent gonads. Moore and Price provided the first plausible, detailed, and sophisticated analysis of the structure and mechanics of that oligarchy. With their work, sexual physiology took the first real step toward the postgonadal age, even though it was hardly a firm step and the direction of movement remained far from definite.

The Stallion's Secret

The reconstitution of the sexual body did not end with the fall of gonado-centrism or the undermining of the idea of sex gland antagonism. Far more radical was the reconstruction of male and female bodies as hormonally androgynous. Again, there had been suspicions and hunches (at least from Steinach's day) that the gonads of both sexes secreted male as well as female hormones and the idea of universal sexual intermediacy was even older, but by the end of the 1930s it was no longer a matter of surmise. The categories of male and female were now undermined in that language of chemicals, numbers, and formulae that was sacred to the modern scientist and his audience.

"Those who have seen a fiery stallion," observed an expert on hormone chemistry in 1966,

> will hardly refute its masculinity. Nevertheless, its daily oestrogen output exceeds that of the non-pregnant female animal by a factor of 100 to 200. Ever since this discovery, the question of how a male may retain its masculinity in spite of such an enormous oestrogen secretion has remained one of the enigmas of biology. Recent research has added little or nothing to our understanding of the problem.[123]

The story, perhaps fortunately for the tranquillity of scientists, didn't begin with the sudden discovery of the stallion's secret. In 1921, the Viennese gynecologist Otfried Otto Fellner, after examining extracts of testicles, ovaries, and placenta, had found himself doubting that the ovaries and testicles produced totally different secretions. A careful study of his extracts revealed that they were identical in color, solubility, and odor. All three extracts, astonishingly, brought about pregnancy-like changes in the female reproductive tract. Male and female gonads, Fellner commented, were likely to be "hermaphroditic in a secretory sense." Indeed, it was probably the case that males and females did not exist as pure entities—all individuals were only predominantly male or predominantly female.[124] This, of course, would have been old news to readers of Otto Weininger or Magnus Hirschfeld but perhaps not to the majority of gland researchers of the time. Even Eugen Steinach, who would later change his views, was still a firm believer in the monosexual gonad, even though he acknowledged that many testicles did secrete feminizing hormone(s). That, however, was because they were incompletely differentiated. *Normal* male and female gonads were still polar opposites.

A more vigorous assault on the idea of the monosexual gonad would come from a slightly different quarter, and the source of assault indicated that the center of gravity of endocrine research was moving rapidly away from gynecology. "Although the presence of female sex hormones in male bodies was reported by gynaecologists and biologists, the biochemists," Nelly Oudshoorn has observed, "took a key position in the debate on the sex-specific origin of sex hormones."[125] First, the Dutch school of Ernst Laqueur reported in 1927 that they had found the female hormone in significant quantities in the urine of healthy men whose "'masculinity' was not to be doubted." The chemical identity of this substance to the hormones produced by the testes or the ovary still had to be proved, but it was nonetheless clear that normal adult men regularly produced a substance that possessed at least one characteristic attribute of the female sex hormone.[126] That characteristic attribute was the ability to cornify the vaginal epithelium, or, in other words, produce a positive outcome in the Allen-Doisy test.[127] Another laboratory team at the University of Dorpat found that the urine of all men contained female and male hormones.[128] The occasional clinician, too, contributed to the discussion. Gynecologist Hans Hirsch soon reported that male blood could bring about a positive Allen-Doisy reaction.[129]

Then came the news of the stallion. A letter from gynecologist Bernhard Zondek in *Nature* spelled out the sheer magnitude of the phenomenon. After demonstrating that the urine of pregnant mares was the richest known source of female hormones, Zondek had found "that in the urine of the stallion also, very large quantities of oestrogenic hormones are eliminated.... According to my analyses, the amount of hormone varies between 10,000 and 400,000 m.u. [mouse units] per litre of urine." A "sexually mature woman" excreted only 30–200 m.u. of the same substance in every liter of urine and even a pregnant woman reached only 10,000—the lower limit of the stallion's output. Even the mare did not excrete female hormones on such a large scale. The sex of the horse, then, could be detected from its urine but, as Zondek put it, "we find the paradox that the male sex is recognised by a high oestrogenic hormone content." Even more disconcertingly, the urine of the stallion was *not* rich in male hormones. Since there was no mass excretion of estrogens in the urine of geldings and immature colts, Zondek argued that "*the testis of the horse is the richest tissue known containing oestrogenic hormone.* According to my analyses, the hormonic content of both testes of the stallion is more than 500 times as great as that of both ovaries of a sexually mature mare."[130]

Even tiny quantities of stallion urine could stimulate uterine development and bring about "a retardation of testicular growth in the infantile male rat" and the growth of breasts in male rabbits, leading to lactation.[131] The big question now, of course, was not so much biochemical as one concerning the very biology of gender: "How can the occurrence of so great a quantity of oestrogenic hormone in a male organism be explained?"[132]

Zondek's answer can be seen as one culmination of a century of debate over sexual intermediacy:

> I believe that the female hormone which is regularly present in the male or-
> ganism represents a normal physiological product of the metabolism of the sex
> hormones . . . a conversion of the male hormone into the female one appears to be

quite possible. I am further of the following opinion: the metabolism of the sex hormones is, in the main, the same in both sexes. At first, the male sex hormone is synthesised from substances which are still unknown, and the male hormone is then converted into the female one.[133]

There was, in other words, no *absolute* chemical distinction to be drawn between male and female. At the hormonal level, the two sexes were effectively interchangeable.[134]

Another popular theory at the time claimed that the female hormones came from food—and, therefore, had nothing to do with the (male) body itself. The food option seemed less and less viable as it was shown, first, that in humans, urinary estrogen excretion was raised by the administration of pituitary gonadotropins and that the "oestrogenic activity of testicular vein blood from stallions was about twenty times higher than that of peripheral blood."[135] It was, however, suspected from early on that even if the testis did secrete female hormones, it may not be the *sole* source of such secretions: the adrenals, as we shall see later, would emerge as the "third gonad" by the end of the 1930s, further complicating the already byzantine government of the sexual body.[136]

Hormonal bisexuality became so incontrovertible in the 1930s that even Eugen Steinach virtually recanted his earlier conviction of the sex-specificity of gonadal secretions. He still considered the male and female sex hormones to be opposed in their actions but now accepted that sex glands of *all* normal individuals produced both.[137] In 1936, Steinach and Kun reported that the injection of the male hormone androsterone into male rats increased the excretion of estrogenic substances in the urine. This, they concluded, showed that the male organism converted male sex hormones into an estrogenic substance.[138] As always, however, Steinach refused to restrict himself to experimental animals. Could human males, too, convert male hormones into estrogenic substances? Injecting large quantities of male hormones—testosterone propionate or androsterone benzoate provided by the pharmaceutical firm of Schering-Kahlbaum, the corporate sponsors of Steinach's later work—into a group of six men, some senile and some young, Steinach reported that the levels of estrogenic substances in their urine climbed rapidly and substantially above the normal level. After the cessation of the injections, the levels returned to normal in a few weeks.[139]

The exact nature of the estrogenic substance remained unknown, but Steinach, characteristically, went off on a clinical tangent hardly related to the study. Chemically pure male sex hormones (such as androsterone or testosterone) did not produce hyperaemia in the brain, but the follicular hormone (oestrone), he claimed, did.[140] The Steinach operation, he re-minded his readers, also produced its effects through hyperaemia, especially of the brain, but it now seemed likely that this effect was not produced by the male hormone.[141] This, of course, carried the piquant implication that the Steinach operation, hailed as a surgical/endocrinological enhancement of masculinity, actually exerted its restitutive effects on thought and creativ-ity not through the male hormones but rather through female hormone-like substances secreted by the reinvigorated testicle.[142]

Such ironies were frequent at the time. Whilst scientists were still trying to explain the occurrence of female hormones in male urine, the urine of females was found to contain male sex hormones.[143] It was already well-known that certain ovarian tumors could produce defeminization—amenorrhea, loss of hair, and shrinkage of breasts, for instance—or masculinization, notably of the voice or of body hair distribution. Berlin gynecologist Robert Meyer speculated in 1930 that these effects were caused by the stimulation of latent masculine elements in the female organism by pathological secretions from the tumor cells. All cells in all individuals, he emphasized, were sexually bipotential, albeit to varying degrees.[144] In 1931, gynecologist Harald Siebke found substantial quantities of the male hormone—for which Siebke used the name androkinin—in women's urine, in menstrual blood, and in tissue extracts from female pelvic malignancies.[145]

In the 1930s, even Steinach argued that the stimulation of heterologous sex characters was possible without the transplantation of the heterologous gonad or the injection of the heterologous sex hormone. Reporting dramatic new experiments on female guinea pigs, Steinach and Kun asserted that the ovary could be directly manipulated to secrete male hormones, bringing about the development of male sex characters and the inhibition of female characters in the female animal.[146] Prolonged irradiation of the ovaries by x-rays could lead to the transformation of the guinea pig's clitoris into a large, penis-like organ. Behaviorally, the female guinea pigs now pursued normal females with all the enthusiasm of the rampant male.[147] Under the microscope, the ovaries of these animals were shown to be full of corpora lutea and lutein cells. The follicles were mostly atrophied.[148]

Why, however, should luteinization of the ovary lead to masculinization? The only possible answer was that the corpus luteum must secrete a hor-mone the actions of which were similar to those of the male hormone. The injection of luteal extracts into oophorectomized female guinea pigs did pro-duce sustained masculinization. Moreover, luteal extract proved capable of bringing about normal sexual development in infantile *male* castrates.[149] The extract, in other words, reproduced the results Steinach had obtained earlier by testicular transplantation. Other scientists had reported the presence of male hormones in the urine of females and in the placenta, but here, claimed Steinach's associate Kun, was the first experimental demonstration of the source—or, at least, one important source—of those male hormones. The corpus luteum, he explained, always secreted small quantities of a masculin-izing hormone in addition to the hormone that was now known to be active in maintaining pregnancy. Once the lutein cells had proliferated under the influence of radiation, the masculinizing secretion was produced in quantities large enough to bring about overt somatic and psychic virilization.[150] The normal state of the ovary, then, was one of "hormonal bisexuality."[151]

Other experiments, too, suggested that the ovaries themselves secreted masculinizing hormones. One important study was reported in 1937 by the Yale anatomist R. T. Hill.[152] Hill transplanted ovaries into castrated male mice. Expectedly enough, the mammary tissue of the grafted mice were stimulated by the ovarian transplant—but their seminal vesicles and prostate, too, were saved from castrate atrophy.[153] The injection of female hormones (progestin and estrin) into another group of castrated male mice resulted in mammary development, but the seminal vesicles and prostate underwent atrophy. It followed, therefore, that normal ovarian secretions did not simply contain female sex hormones but masculinizing agents too.[154] Alan Parkes hailed Hill's report for providing "conclusive proof of the potential andro-genic effect of the ovary." There was, however, another possible source for the male hormones in women: the adrenal gland.[155]

The "Third Gonad"

From the early twentieth century, there had been considerable interest in elucidating the connection of the adrenal glands with sex. The adrenals, evi-dently, were extraordinary organs. The adrenal medulla, of course, secreted adrenaline, and the adrenal cortex had long been shown to be essential to life.[156] Apart from those vital functions, however, they seemed to have con-siderable influence over the sexual domain.

Long before the secretions of the adrenal cortex had been chemically iden-tified, a classic 1905 report on eleven children—mostly girls—had linked vir-ilism with adrenal hyperplasia.[157] It was also appreciated early that the clinical picture varied considerably depending on the sex of the patient and the age of onset. A female baby with prenatal adrenal pathology was usually born with a large clitoris. "A little girl of 2, 3, or 4 years of age perhaps will come to ex-hibit the growth and appearance of a girl of 14," wrote Louis Berman. "She begins to menstruate, her breasts swell, she shoots up in height and weight, sprouts the hair distribution of the adult, and the mentality of the adolescent, restless, acquiring, doubting, emerges. A tot bewitched into puberty!"[158] Or, in the words of Sir Arthur Keith, "a child, scarce off its mother's lap, begins to assume, as a monstrous garb, the face, voice, and demeanour of sexual maturity."[159] In 1865, John Ogle had reported on a precociously pubescent girl of three who spoke "language worthy of a fishwife."[160] When the adrenal pathology commenced after puberty, the most usual signs were hirsutism, amenorrhoea, obesity, clitoral hypertrophy "and in some, but not in all instances, osteoporosis and a masculine mentality."[161]

At least from the time of Ernest Glynn's work in 1912, there was considerable speculation about the possible internal secretory dimensions of the defeminization and masculinization of patients suffering from adrenal tumors.[162] In 1933, S. Levy Simpson and others, stimulated by Womack and Koch's report of a masculinizing substance in women's urine, began to investigate whether the symptoms of virilism in women were due to excess male hormones.[163] After a series of experiments and biochemical tests on urine, they reported that most of their patients excreted substantially higher amounts of androgens in their urine than normal women. The source of the excess male hormones was harder to pinpoint, although, since it was known that "marked virilism could occur after ovariectomy," the ovary was unlikely to be the sole source of surplus androgens in virilized patients. The pituitary was another possibility, but they found nothing unusual when the urine of virilized women were tested for excess pituitary gonadotropic hormones. The adrenals, the authors concluded, "are suspect on clinical grounds but convincing biological evidence is still awaited."[164]

Curiously, adrenal hyperactivity was always masculinizing, regardless of the sex of the individual. "Adrenal hyperplasia or neoplasia," the pioneering investigator Ernest Glynn had observed in 1912, "is associated with a diminution of female and an increase of male primary and secondary sex characters. The converse does not occur; for in males, on the contrary, the 'maleness' usually increases."[165] In the early 1930s, the definitive American handbook, *Sex and Internal Secretions*, pronounced that the adrenal cortex was "primarily a male gland."[166] Lennox Broster, a London surgeon and authority on adrenal virilization, was not surprised by this gender disparity: "The masculinisation of the female is a far more common natural outcrop than the feminisation of the male. Upon this generality it may be assumed that the female make-up is more unstable than the male."[167]

Boys with adrenal tumors presented with the "infant Hercules" syndrome of rapid, disproportionate skeletal growth. This was accompanied by rapid (but transient) muscular development, luxuriant development of pubic and axillary hair, lengthening of the penis, and in some cases, "frequent erections and emission of semen." The mental state usually corresponded, however, to the actual age of the patient, although "manifestations of a psychoneurotic nature are prone to develop, apparently due to the conflict between the patient's sexual maturity and mental immaturity." Death came early, being preceded by wasting of the muscles and sometimes a sudden senilization of the appearance of the patient. One child, who died at the age of five, was thought to resemble "a venerable old man."[168] The treatment of choice was the removal of the affected adrenal in tumors or, in the hyperplasia of both adrenals, the removal of one.[169] Results were good, with the hairiness being affected first. Subsequently, frigidity improved, fertility was regained, and "cases of homosexuality have undergone a subtle change to normal heterosexuality."[170]

The adrenals, plainly, were male in their nature, "neutralising more or less the specifically feminine influences of the internal secretions of the ovary."[171] Early-twentieth-century writers on glandular matters often had a propensity to assign certain glands to either sex. Berman, for instance, had suggested that "ovary and adrenal medulla, and posterior pituitary and thyroid predominance constitute the feminine formula. Testis and adrenal cortex and anterior pituitary predominance comprise the masculine endocrine directorate."[172] Adrenal atrophy, on the other hand, was implicated in "retarded sexual development, or rapid loss of function" in women.[173] Julian Huxley speculated that "the cortex of the adrenal gland, if active beyond a certain measure, assists the development of male, prevents the development of female, characters. . . . One presumes that a slight preponderance of the adrenal cortex in the normal endocrine make-up will lead to a less feminine type of woman than normal. . . . We shall have to search for the finer shades of temperamental difference between man and woman not so much in differences in the quality of the secretion of testis or ovary as in differences of balance in what the Americans call the 'endocrine make-up.'"[174]

By the 1930s, however, convictions of the maleness of the adrenal were becoming obsolescent. As Humphry Davy Rolleston would point out in his widely read 1936 treatise on the endocrines, the situation was far more complex. In *postpubertal* males, for example, hyperplasia or tumors of the adrenal cortex did not lead to virilization but intriguingly, to bodily feminization, testicular atrophy, and gynecomastia.[175] "This obvious ambisexuality of the abnormally stimulated adrenal," exclaimed Alan Parkes at the end of the decade, "is undoubtedly remarkable."[176] Endocrinologist Julius Baur suggested that the adrenal sex hormone exerted "a protective action upon the normally suppressed characteristics of the opposite sex." All women, in other words, were partly male, and in women with "masculinizing" adrenal tumors, their innate partial masculinity was permitted to flower under adrenal hormonal protection.[177]

The adrenal secretions soon revealed other perplexing attributes. Crude extracts of horse adrenals were found to be rich in a substance that had effects similar to progesterone.[178] Estrogenic properties, too, were strongly marked, and so, of course, was androgenic activity. Embryologically, the adrenal cortex developed from the same rudiments as did the testes and the ovaries. In his 1934 Hunterian Lecture at the Royal College of Surgeons, Lennox Broster observed that the embryological study of normal fetuses had shown that between the eleventh and fourteenth weeks, the adrenal cortices of the female fetus showed a staining reaction identical to that of the cortices from adult patients of clinically significant adrenal virilization and to the staining pattern of the interstitial cells of normal adult testicles. "The female foetus may therefore be regarded as passing through a 'male phase.'"[179] Since the male phase seemed to end around the time when the acidophil cells of the pituitary developed, it was possible that the male potential in females was suppressed by the pituitary. At times of stress (such as puberty and pregnancy), that balance, however, might be upset and "it is reasonable to suppose that this latent adrenal influence will burst into activity and exert itself."[180]

Writing in the 1950s, a Spanish researcher emphasized that the adrenal cortices represented two glands in one: one the cortex proper, secreting cortisol and reacting to stimulation by pituitary adrenocorticotropic hormone (ACTH), and the other a sexual zone secreting male and female sex hormones, controlled by the gonadotropins of the pituitary, not by ACTH. This "sexual zone" of the adrenal he named the "third gonad," an accessory sex gland.[181]

The "third gonad," even before it received that name, was a culturally resonant organ, especially in those supposedly mild cases of hyperfunction where it merely made a woman more pushy, energetic, and ambitious. Louis Berman, expectedly enough, had much to say on the subject. Women with even mildly hyperfunctioning adrenals, he said, were "the ones who, in the present overturn of the traditional sex relationships, will become the professional politicians, bankers, captains of industry, and directors of affairs in general.... An adrenal type will probably be the first woman president of the United States." The less well-adjusted adrenal types would, however, cause havoc. Berman was confident that "the suffrage revolution" had been brought about by such women.[182]

In the London of the 1930s, Lennox Broster was even more profoundly concerned by the social consequences of adrenal virilization. Adrenal virilism did not make men out of women but produced something far more troubling—"the intersex type," a creature that was neither fully male nor fully female. In evolutionary terms, it opposed the progressive differentiation of the sexes and represented "a retrograde movement. . . . Is it stationary, receding, or increasing? If the latter, then we may be the innocent spectators of an evolutionary process drifting slowly and inevitably into the neuter state."[183] If the condition was on the rise, then the implications for humanity could be cataclysmic.

The adrenogenital syndrome could well be an indication of something rotten at the sexual core of mankind, a weakness that was manifesting itself in the shape of women demanding emancipation. The only solution was to be found in eugenics. "The conditions of life," said Broster,

> are rapidly changing. The emancipation of women has been a big change; there is over-population in many parts of the world; there are many factors in our social lives leading to race deterioration, and they must react on the individual. Too many minds, content with their lot, and unaware of these abnormalities in others, decry this so-called tampering with nature. Theirs is a selfish attitude; but in clinical science there is a spirit of altruism which aims at projecting our acquisitions to the betterment of the human race.[184]

Only altruistic eugenics could halt the further masculinization of women and thereby save the human species from perdition.

Whether Broster was acquainted with the work of Otto Weininger I do not know, but the fundamental features of his anxious reflections came straight out of the nightmares of the Viennese philosopher, modulated, no doubt, by the debates on race deterioration and eugenics in inter-war England.[185] The Woman Question, for both, united human biology with culture, revealing the future (or lack thereof) of the species. The sexual biology Weininger knew had now changed beyond recognition, of course. For Broster's generation, sex was constructed, shaped, and maintained not by some inborn essence, nor by the secretions of any one gland, but by the concerted, tensely balanced and potentially fragile interaction of several. This new sexual body, for all its intellectual novelty, was still, however, as much an object of anxious cultural meditation as the old one had ever been.

"The Glands of Childhood": Thymus, Pineal and Sexual Development

Although we have so far been concerned only with mature sexual characters or deviations therefrom, sex, needless to say, is also a developmental phenomenon. The sexual body, in an obvious, physical sense, begins to take shape at puberty, and gland scientists devoted considerable thought to identifying the biological basis of that transition. Eugen Steinach, as we know, was convinced that puberty began when the "puberty gland" began to pour its secretions into the blood. Somewhat later, the great awakening of the gonads was attributed to the activating influence of the pituitary. Such hypotheses explained the onset of puberty but not its absence during childhood. What, in other words, held the interstitial cells back from secreting their sexualizing hormones before puberty?

Debates on this question, most of them long forgotten, show just how complicated endocrine theories of the sexual body were becoming in the early twentieth century. Maturity and senility had already been argued to be glandular phenomena; the research we are now going to explore drew infancy and childhood into the endocrine realm. Just as sexual maturity and involution were brought about by specific hormonal stimuli, the *absence* of sexual maturity, too, came to be linked with the glandular system. The rare phenomenon of sexual precocity suggested that puberty, rather than being a static, passive condition, might need to be maintained by a dynamic process. The usual sexual immaturity of childhood was a consequence of the *absence* of gonadal hormones, to be sure, but this gonadal quiescence might actually be the result of active restraint by forces that broke down in the rare cases of precocious puberty. What, then, were those forces? Having learnt much about the puberty gland, investigators now looked for what one could call the glands of childhood.

One major clue was provided by a brief report of a case of pineal tumor in a five-and-half-year-old boy published in 1909 by Lothar von Frankl-Hochwart, Professor of Medicine at the University of Vienna. At the age of three, this boy had suddenly begun to grow unusually fast. When five, he was as big as a child of seven, and the precocity was not simply physical. Mentally, too, he was older than his years, showing great interest without any prompting from his family in questions of immortality and life after death, interests that, Frankl-Hochwart remarked, were common enough during puberty but hardly ever in childhood. What truly amazed his doctors, however, was the boy's sexual development. A month before death, his voice had deepened, pubic hair had grown to a luxuriance rarely seen before the age of fifteen, the penis had attained adult dimensions, and he had regular erections. He had also developed various symptoms suggestive of a brain tumor, and eventually the postmortem revealed a teratoma of the pineal gland. A survey of the literature located a few earlier cases where sexual precocity and the more typical compression symptoms of a brain tumor had been ascribed, on postmortem examination, to a tumor of the pineal gland. Only one of the recorded cases was of a female and that report had not mentioned any genital symptoms or menstrual irregularities. Nevertheless, for boys at least, it was clear that when the typical symptoms of brain tumors occurred with accelerated physical, intellectual, and genital development, the first diagnosis to consider was of a pineal tumor.[186] So characteristic were these features that a distinct "pineal syndrome" was postulated, with such jaw-breaking synonyms as "macrogenitosomia praecox" or "praecocitas psychosomogenitalis."[187]

The pineal gland is found in all vertebrates and had been known since classical times. Phylogenetically, the pineal was supposed to represent a "third eye," but perhaps the most famous hypothesis about pineal function was that of philosopher René Descartes, who believed that man's rational soul resided in the pineal. Later, theosophists taught that in the distant future of humanity, the pineal would acquire "the power of divine insight."[188] By the 1920s, some scientists regarded the gland as important in determining the body's response to light. Dismissing all such theories as speculative, von Frankl-Hochwart asserted that the pineal was an endocrine gland, but since it underwent involution by puberty in humans, it was probably of importance only in early childhood. Since the destruction of the pineal by tumors caused genital hypertrophy, while the destruction of the pituitary by tumors was known to inhibit sexual development, it was, at least, plausible that the pineal functioned as a physiological antagonist of the pituitary. The pituitary stimulated sexual development, while the pineal retarded it.[189]

Harvey Cushing, citing von Frankl-Hochwart's paper with approval, remarked in 1912 that "the first information concerning function" of the pineal, and indeed of most endocrine glands, was not to be expected from laboratories but from "clinical observations of . . . patients suffering from proven hypertrophies or tumor of the gland."[190] Since pituitary deficiency in childhood, as Cushing knew well from his clinical studies, did retard sexual development, he agreed provisionally with von Frankl-Hochwart that there was "some antagonistic action between pineal and hypophysis in relation to the sexual glands."[191] In Central Europe, Artur Biedl asserted that until the age of seven, the pineal "exercises a definite and apparently inhibitory influence upon the development of the sexual glands, and it is probable that it has a secondary effect upon mental development. . . . That there is an antagonism between the activity of the pineal gland and that of the hypophysis is certain, for we know that pituitary insufficiency produces hypogenitalism."[192] Another medical scientist proclaimed that if the interstitial cells of the testicle constituted the "puberty gland," then the pineal was the "anti-puberty gland."[193]

Clinically, it all made sense, but experimental investigations produced confusing results. The excision of the pineal or the administration of pineal substance did not produce any straightforward consequences in animals.[194] Perhaps the most striking results were those of L. G. Rowntree and his team, who reported the acceleration of sexual development in young rats given a special pineal extract.[195] These results, of course, opposed the original theory that the pineal *retarded* the development of sexual maturity. Between 1934 and 1941, at least three different experimental teams conducted pinealectomy in rats for as many as six successive generations without any impact "upon weight, rate of development or age of sexual maturity."[196]

In spite of this confusion on the experimental front or perhaps because of it, many competing theories came to be offered to explain the sexual precocity in children with pineal tumors. Surveying them in the mid-1930s, Humphry Davy Rolleston found no reason to accept von Frankl-Hochwart's simple schema of an antagonism between the pineal and the pituitary. Since precocious puberty could occur in tumors of a whole range of glands and organs—adrenal cortex, hypothalamus and midbrain, gonads and, of course, the pineal—it was likely that sexual precocity was brought about by "a stimulating influence exerted on the anterior pituitary through the hypothalamus which is influenced by chemical or other stimuli originating in the adrenal cortex, gonads, pineal, or locally."[197] Research on the pineal continued, but its stellar role in sexual physiology as "the anti-puberty gland" proved transient.[198]

The career of the other supposed "gland of childhood," the thymus, followed a rather different path. Located in front of the neck and named thymus by Galen because it resembled thyme flowers, this gland was long thought to play some role in the lymphatic system.[199] It had also long been believed that the thymus did not persist into adult life, but there was no agreement on the tempo and mode of this involution. As recently as in 1954, the celebrated British surgeon Geoffrey Keynes exclaimed, "It seems to be extraordinarily difficult to get at the truth concerning even the size of this enigmatic organ."[200] Clinicians generally believed that the thymus began to shrink around the time of puberty and that this involution continued gradually over life.[201] This issue was considered vitally important not because of endocrinological reasons but because an enlarged thymus was supposed to be the cause of sudden unexplained death in infants and small children. Some children, it was suggested, were born with a constitutional hyperplasia of all lymphoid tissue, including the thymus gland. This condition was named status thymico-lymphaticus.[202] "The idea developed that children with status lymphaticus were well fed, pale, pasty, flabby and rather inert and effeminate, with large thymus and tonsils." In the early twentieth century, the removal of an enlarged thymus in infancy was supposed to be the only preventive treatment but soon had to be abandoned because of high mortality figures. Quickly, the grand new technique of irradiation occupied the place vacated by surgery, and the irradiation of an enlarged thymus was recommended, Ann Dally writes, by "virtually every textbook of the period. . . . There were even suggestions that all newborn babies should be irradiated as a prophylactic measure."[203]

Meanwhile, the endocrine importance of the thymus remained uncertain. In 1936, Humphry Davy Rolleston could refer only to "divergent answers" to the fundamental question whether the thymus secreted any hormones and if so, what those hormones actually did. Experimental removal of the thymus was reported to produce disturbances in calcium metabolism, while other studies found the removal to produce no consequences. More intriguingly, castrated animals were reported to show thymic enlargement—a fact, according to Rolleston, "long familiar to butchers."[204] It was claimed, therefore, that the thymus inhibited gonadal development in immature animals.[205] This hypothesis, however, was not universally accepted. It was criticized on statistical grounds, and others argued that thymectomies caused no changes in the gonads.[206] In 1913, William Blair-Bell summed up the rather equivocal situation. It was well known, he remarked, that "the gonads and uterus remain under-developed until the thymus atrophies at the time of puberty. At this period of life, owing to the withdrawal of the thymus secretion, as some authorities think, the genital organs begin to develop. Other investigators, however, believe that it is the development of the gonads which causes retrogression in the thymus. The probability is that both views are correct."[207]

Even this kind of semi-certainty was sufficient to spur the likes of Louis Berman on to speculative heights. For healthy sexual development, Berman insisted, "there must be a certain atrophy and retrogression of the thymus gland, and there must likewise be a similar atrophy and retirement of the pineal gland. Both of these involutions of the glands of childhood must occur before the normal hypertrophy and development of the sex glands and their secretions can start."[208] He went on to paint a striking portrait of what he called the thymo-centric personality. Children whose thymus glands remained active throughout life instead of disappearing after puberty, he claimed, were very beautiful to look at "but somehow unfit for the coarse conflicts of life." They were far more susceptible to tuberculosis, meningitis, and childhood diseases than other children.[209] Sexual abnormalities were common—the persistence of the thymus was nothing short of "partial castration" and the latent traits of the other sex manifested themselves in the individual. "Thymocentrics," warned Berman, "are often homosexual." Prone to every variety of moral irresponsibility, they took naturally to alcohol, narcotics, and crime, ending as "the hopeless misfits of society."[210]

The functions of the thymus were so enigmatic that even the serious scientists could not entirely resist speculation. On the thymus's relation to the gonads, that great critic of endocrine hype Swale Vincent commented that

> the experiments . . . tempt one to the hypothesis that the thymus furnishes an internal secretion of some kind which ministers to the needs of the economy before the reproductive organs are fully developed . . . if castration is performed, the thymus maintains its original structure and functions. The internal secretion must, of course, be of a different nature from that which determines the development of the secondary sexual characters, as these do not become manifest in castrated animals.[211]

And in Vienna, gynecologist Richard Hofstätter, following a persistent model of the thymus as the "anti-puberty gland," recommended in 1934 that "hypererotic" pubescent girls be treated with pineal and thymus extracts.[212]

By then, experimental studies had begun to challenge this simplistic theory of thymic function. Some studies claimed that the experimental administration of thymus extracts actually *accelerated* sexual development. Even castration, it was now claimed, did not prevent the involution of the thymus.[213] The beautiful symmetry of the old scheme was destroyed not only by such studies but more importantly by the gradual waning of the concept of status lymphaticus. As Ann Dally has shown, the concept came under increasingly negative scrutiny from the 1930s. Along with its occlusion from the realms of internal medicine and pediatrics, the thymus also retreated from the domain of glandular research, assuming a prominent position in immunological discourse and in explanations of conditions such as myasthenia gravis. It remained a fascinating organ but no longer to the investigators of sex, death, and moral degeneracy.

What Fell? What Rose?

It is easy enough to outline what was eliminated from sexual biology as a result of glandular research during the late 1920s and 1930s. First, the gonads lost their primacy. They were now seen as team-players: crucial to the production and maintenance of the sexual body, but not of much use without the other endocrine glands and especially the pituitary. The powers of the pituitary were chronicled with awe, but it was also recognized that those powers were less than absolute. The pituitary, although the controller of the endocrine system, could not work without feedback from the gonads. The adrenals, moreover, appeared to make or, at least, possessed the potential of making more than marginal contributions to the chemistry of sex. Most importantly, neither sex had a monopoly over the male or female sex hormones: each produced both. These reformulations led, as Nelly Oudshoorn has demonstrated, to a weakening of the concepts of "male" and "female." "By the end of the 1930s," writes Oudshoorn,

> scientists supported the idea that male bodies could possess female sex hormones and vice versa, thus for the first time combining the categories of male and female into one sex.... They defined female and male sex hormones as closely related chemical compounds, differing in just one hydroxyl group, which could be detected by chemical methods in both sexes. In this manner, they broke with the dualistic concept of male and female as mutually exclusive categories.[214]

It is important to note, however, that this demolition of the male/female boundary remained largely confined to the laboratory. Although it had some impact on clinical practice, endocrine therapeutics, as we shall see in the next chapter, was only partly revolutionized by the concept of universal hormonal bisexuality. In this as on many other issues, the laboratory and the clinic responded to one another's findings, but the lines of communication were often incomplete and rarely straightforward and wide open.

Chapter 5

The New Hormones in the Clinic

The Lewis Carroll of today would have Alice nibble from a pituitary mushroom in her left hand and a lutein one in her right and presto! She is any height desired.
—HARVEY CUSHING[1]

R esearch on the sex glands was driven by many forces but none, arguably, was more powerful than the hope of curing ancient ills and enhancing human life. Those aspirations, as we have seen, were at their most flamboyant in the 1920s. The standard historical view is that by the end of that decade a remarkable change had begun to sweep through the world of hormonal research. The hegemony of clinicians and "whole-gland" physiologists—the Fraenkels or the Steinachs—waned and a new, biochemical phase commenced, which soon flooded the medical marketplace with potent, standardized hormonal preparations. The clinician (especially the gynecologist), hitherto a high-flyer in laboratory research on gonadal functions, now became largely a consumer and tester of hormonal preparations.[2] The involvement of clinicians in endocrine research, while by no means over, would never again be as intimate as it had been in the days of Halban and Fraenkel.

This conventional narrative is perfectly true within its limits and will figure frequently in my argument in this chapter. What it encourages us to overlook, however, are the continuities between the old and the new. The clinician-researcher may have been been less prominent in experimental endocrine research than formerly,

153

but the clinical contribution to endocrine discourse actually multiplied in extent as well as importance during the inter-war decades. As the biochemists produced more and more therapeutic products, the clinicians filed report after report on their efficacy—or their lack of efficacy. Unlike the laboratory scientist Fraenkel producing an extract, which was then tested on his patients by gynecologist Fraenkel and the results reported as addenda to a largely physiological paper, glandular therapeutics of the late twenties and thirties evolved a clearer division of labor. The biochemist ruled *one* roost and depended to a very great extent, in any case, on the support of the pharmaceutical industry. The clinician retreated from frontline research but gained a different kind of power—that of authenticator. The most precisely standardized and rigorously purified extracts would not bring any glory to their producers if practitioners found them to be of little use in therapy. This happened repeatedly; in spite of the availability of dozens of hormonal products by the time of the Second World War, clinicians, as we shall see, were rarely hyperbolic in their praise and sometimes almost dismissive. That does not, of course, mean that the products in question were not prescribed by many but merely that endorsements from the higher echelons of clinicians were far from easy to obtain. The practitioner, in short, was more than a rubber stamp even during the heroic years of biochemical endocrinology.

"The Very Essence of Eve"

It was not just the biochemists of the 1930s who were involved in the elaboration of potent hormonal preparations. The exciting, complex story of the development of standardized ovarian extracts that began in the late 1920s involved clinicians, physiologists, and also eventually biochemists. Even before the 1920s, researchers had sought to obtain potent ovarian extracts, and some of those extracts may even have contained hormones in effective quantities. George Corner identified three gynecologists as having obtained such potent extracts: Henri Iscovesco in Paris, Otfried Otto Fellner in Vienna, and Edmund Herrmann, also in Vienna.[3] What differentiates these early attempts from the biochemical enterprise of the thirties is the former's lack of interest in isolating a particular set of chemicals with specific formulae and specific physical properties. The first potent extracts were judged by their biological actions, rather than by their chemical purity.

As so often in the history of sex gland research, the career of Eugen Steinach provides us with an excellent illustration. One of Steinach's last major projects was the development of a potent ovarian extract, of which we heard briefly with regard to rejuvenation. Having failed to induce significant sexual effects on experimental animals with commercially available extracts, Steinach and his associates produced an extract from ovaries and placentas of sheep. They began with a weak oil-based extract but then developed an aqueous, highly concentrated version, which, they claimed, was completely nontoxic, even to humans.[4] No details of its manufacture were published, presumably because of commercial secrecy. The research was sponsored by Schering-Kahlbaum, who soon marketed the extract under the name "Progynon."[5] Steinach made florid claims for the efficacy of his extract. It was powerful enough to induce normal sexual development in animals oophorectomized in infancy and to restore the estrus cycle in oophorectomized adults.[6] In the latter case, the induced estrus was so much like the real estrus that these oophorectomized females were pursued by males without any hesitation.[7] Moreover, the new extract could restore normal adult functions in senile animals whose reproductive organs had atrophied and who had long ceased to go into estrus.[8] Often, the senile ovary itself was reactivated and the cycle continued for months without injections.[9] Microscopically, the hitherto atrophic ovaries now showed active follicles and corpora lutea.[10] The reactivation of the senile ovary led, predictably enough, to a reactivation of the entire senile organism, indicated by rise in appetite, weight gain, improvement of the fur, general liveliness, and in a few cases, normal coitus. After many injections, the ovaries of the reactivated animals returned virtually to normalcy, with ovulation and fertilization occurring as in youth.[11]

Steinach himself never published any account of his experience with Progynon in humans, but from a report by Viennese gynecologist Erwin Last it is clear that its actions in humans, although noteworthy, were not even remotely as dramatic.[12] Reporting on four years' experience with Progynon in all available forms (oil-based injection, aqueous injection, suppositories, and tablets in a range of doses), Last observed that the aqueous solution gave the best results and treatment was most useful in women who had stopped menstruating for unknown reasons but who had no uterine atrophy. In many of these cases, Progynon reestablished the normal cycle, which persisted indefinitely in many without further treatment. It was also successful in reestablishing menstruation in menopausal women, especially when supplemented with irradiation or diathermy of the pituitary.[13] Supplementation of Progynon with pituitary extracts gave good results, although Last provided no details.[14] The bad news was that Progynon had failed to establish the menstrual cycle in women who had never menstruated or to reestablish it in women who had uterine atrophy. Steinach had found Progynon to work on rats in comparable conditions, but humans did not seem to respond similarly. Apart from the obvious fact that human physiology was infinitely more complex than that of castrated animals, Last emphasized that the human menstrual cycle had recently been shown to be governed by two hormones, the follicular hormone contained in Progynon and the hormone of the corpus luteum. Supplementation of Progynon with the latter, he was confident, would improve results significantly, and he was eager to do so as soon as a suitable preparation was available.[15]

Although physiologists and gynecologists were not displaced at one stroke by biochemists, a crucial shift did occur in the mid-1920s. One important landmark in this context was the work of American scientists Edgar Allen and Edward Doisy. George Corner, who lived through this period and participated in some of the pathbreaking research, recalled in old age that "at the beginning of 1923, before Allen and Doisy's first paper appeared, the literature of ovarian endocrinology was in a very confused state."[16] The American duo succeeded because they "did not," as Corner drily remarked, "know enough about the earlier history of the subject to be confused by it."[17] Edgar Allen was an anatomist and Edward Doisy a biochemist (both then at the Washington University in St. Louis, Missouri), and Doisy was far more interested in 1922 in insulin than in ovarian secretions. Nevertheless, when Allen drew him into helping him with some ovarian extracts, their collaboration led to the identification and subsequently the crystallization of an ovarian hormone from ovarian follicular fluid. This was named Theelin.[18] That Allen and Doisy's work led to a glorious new epoch in sex gland research is universally acknowledged, and their first joint paper of 1923 is justly celebrated as a classic contribution to the subject.[19] Allen and Doisy did not simply produce a reliable ovarian extract—they changed the very nature of ovarian research.

Take, for instance, their indicator of hormonal potency. They relied on Stockard and Papanicolaou's new test for detecting estrus in an animal from smears of its vaginal cells.[20] If and when the administration of the extract changed the appearance of the vaginal cells in a certain way, then the extract was taken to be a potent one. Soon, this change in the character of the cells would come to be known as the Allen-Doisy Reaction. The new smear test was quick, reliable, and did not require the animal to be killed and dissected. If, however, the two Americans had chosen to assess the potency of their extract from traditional indicators like its effects on the growth of the uterus, the outcome of their study may have been very different. At the very least, it would have required greater time and biological expertise to perform and, therefore, would have been difficult to standardize and popularize. Soon, the smear test was so precise that Allen and Doisy could define one rat-unit of their hormone as the smallest amount of the extract that could induce cytological changes characteristic of estrus in the vagina of a spayed female rat.[21] George Corner pointed out that the British physiologist Francis Marshall had also used estrus induction as a yardstick for extract potency, but since Marshall was using bitches and did not have the Stockard-Papanicolaou test at his disposal, he had been compelled to wait and "watch for the slow onset of ill-defined signs of oestrus in the bitch."[22] Allen, on the other hand, "could read the results of his tests in a day or two" from vaginal smears.

Harald Siebke, one of the last gynecologists to engage seriously in experimental research, declared that it was only the availability of the Allen-Doisy test that had ensured the extraordinarily rapid progress of research on the follicular hormone.[23] In 1929, gynecologist Robert T. Frank, whose research on matters glandular had commenced in 1904, recalled that

> methods of recognizing the active principle involved in the production of "feminineness" were more difficult, time-consuming and expensive than they are today since the Allen and Doisy reaction permits at least an easy qualitative recognition of the female sex hormone without sacrificing the test animal. Before this method was introduced, this field of investigation was but sparsely peopled. Since 1923 the subject has attracted innumerable workers who are elbowing and jostling each other and jockeying for position in the neck and neck race to isolate and synthesize the much desired and long sought for hormone, which is bound to relieve many of the ills from which women now suffer.[24]

Indeed, if one wanted to be perverse, one could argue that Allen and Doisy's work was no more than an imaginative and topical extension of Stockard and Papanicolaou's cytological research.

Animals, then, remained indispensable, but their glands were no longer the invariable objects of hormonal research. While animal glands had once been used to *produce* knowledge, they would now increasingly be regarded as the sources of extracts. Preparing the extract and testing its potency were the new challenges. As Harry Benjamin wrote sadly to his aging mentor (who, being half-Jewish, was in exile in Switzerland after the Nazi takeover of Austria) after failing repeatedly to find him a post in an American research institution, "hormone research in America is no longer in the hands of physiologists and biologists but almost exclusively in those of chemists . . . no Steinach is needed for whatever animal experimentation is required in that research."[25] There was, then, a major shift from the physiological to the biochemical, but it was tied to the emergence of a technique that, although itself essentially physiological, made physiological expertise with whole glands less crucial than it had been before 1923.

It is important to appreciate, however, that the physiologist did not thereby become subservient to the biochemist. Physiologists welcomed the new extracts and hormones as reliable research tools. Even a dyed-in-the-wool physiologist like Steinach gave up experimenting with new methods to stimulate the ovary once he had developed a satisfactory extract, which a lay admirer of his hailed as "the very essence of Eve."[26] Steinach's great adversary, Carl Moore, who had begun his investigations of sex gland function in the whole-gland phase of the science but moved with great enthusiasm to the more biochemical phase, observed with reference to the new testicular extracts developed by his Chicago colleagues McGee and Koch: "The success of the biochemists in obtaining effective preparations of the male hormone, and the development of effective and quickly applied procedures for determining its presence or absence in the organism . . . unite in providing instruments with which real advances may be made in the study of the production and the function of the internal secretion of the testis."[27] As Moore's (and Dorothy Price's) research on the pituitary eloquently demonstrated, however, biochemically produced extracts facilitated the kind of physiological research that no biochemist could conduct. Their work relied on biochemically obtained extracts, but their conceptualization of the findings was a milestone in the history of physiology, not biochemistry.

The real change was in the location and role of the clinician. The age of the clinician-physiologist was now definitely on the wane. The new physiological research, no matter how innovative, was conducted in laboratories far more distant from the clinic and larger society than were the laboratories of, say, Knauer or Fraenkel.[28] Physiologists would now leave their own shores very rarely, and the borders of their specialized domain would progressively be closed to clinical adventurers and research-minded gynecologists. The division of endocrinological labor, in short, became clearly established, but the biochemist, the physiologist, and the clinician all retained considerable authority over domains that, although separate, were far from insubstantial. It was the old-style generalists who disappeared.

Gynecology in the Age of Hormones

The basic narrative of the standardization, nomenclature, and manufacture of ever purer hormones and subsequently of their synthesis was recounted with panache by physiologist Alan Parkes. He looked back on the years between 1929 and 1935 as a period of "spectacular chemical and physiological advances" leading to the identification and characterization of "the chief naturally occurring oestrogens and androgens, as well as progesterone," together with the discovery of the "hypophysial, placental and endometrial gonadotrophins." It was, simply, "the heroic age of reproductive endocrinology."[29] What I would like to examine is what exactly happened in the clinic once the sex hormones were available as potent, standardized and reliable therapeutic preparations. Did they bring about the therapeutic revolution prophesied by so many in the 1920s?

It was the gynecologists who had expected most of hormonal treatment, but those hopes proved hard to fulfil. In 1935, gynecologist Emil Novak summed up the hope and the disappointment concisely: "Every one feels that gynecologic organotherapy is, on the whole, extremely disappointing, and yet the feeling is universal that it is in this direction that one must look for improvement in methods of treating the functional gynecologic disorders."[30] The German gynecologist Franz Siegert observed in 1934 that although the importance of the sex hormones in female life was beyond dispute, the human efficacy of the new hormonal preparations was still very uncertain. Clinicians continued to use hormones but in an intuitive manner. This was evident in the treatment of the "functional" gynecological disorders. These conditions were unaccompanied by any structural lesions and difficult to differentiate from one another with absolute clarity; they were, however, very common and hard to treat.[31] Hormones were often used in such disorders on the assumption that since there was no gross organic abnormality to be found, the symptoms must be caused by the dysfunction of ovarian or pituitary hormones. That, Siegert emphasized, was not only a most unsatisfactory rationale for treatment but overlooked the possible role of the other endocrine glands such as the thyroid or the adrenals.[32]

Novak accepted that in the current state of knowledge it was impossible for clinicians to be sure about what the ovarian hormones could do. The major indication for treatment with estrogenic substances was amenorrhea, even though the results were "notoriously unsatisfactory."[33] And yet, amenorrhea was one of the commonest complaints for which a woman saw her gynecologist or indeed, any doctor.[34] "This," remarked Novak, "is particularly true because of the prevalent belief of the public that menstruation is essential to health and well being, and that absence of the function, with its supposed retention of harmful substances in the system, may bring about serious results."[35] Although estrogens, by definition, could induce estrus in animals, that, Novak sadly commented, was "very different from menstruation in women." The physiology of the human menstrual cycle had been clarified sufficiently for gynecologists to realize that estrogenic substances alone, no matter in how high a dose, could never bring about menstrual bleeding. Menstruation occurred when the uterine endometrium had been built up by the sequential action of estrogen *and* progestin. It was irrational to hope that even if the luteal hormone was lacking, large doses of estrogen might build up the endometrium sufficiently. Secondly, ovarian hormones, Novak stressed, could not stimulate inactive ovaries. As Price and Moore had established, the administration of sex hormones depressed the secretion of pituitary gonadotropins, thereby further depressing an already inactive gonad.[36] Therapy with estrogenic substances, therefore, could only be replacement therapy; it could compensate only for one part of the menstrual cycle and should never be expected to reestablish "the regular menstrual rhythm."[37] Although it might be ideal to treat amenorrhea by "a really gonadotropic preparation," the currently available pituitary preparations "do not fill the bill."[38]

Moreover, amenorrhea could be due to a number of reasons having nothing at all to do with the endocrine glands. Sometimes, the cause might be endocrine but not ovarian: administration of the thyroid hormone restored the cycle in some cases.[39] The Viennese gynecologist Richard Hofstätter agreed. Many cases of amenorrhea, frigidity, dysmenorrhea and sterility, treated with indifferent success by estrogenic hormones, often responded brilliantly to treatment with thyroid hormone.[40] Novak did not distinguish between primary amenorrhea (where the woman had never menstruated and might suffer from general ovarian insufficiency expressed through other anomalies such as undeveloped breasts) and the secondary variety, where menstruation had ceased during the reproductive years but not because of pregnancy. Most gynecologists in Central Europe emphasized the importance of this distinction: primary amenorrhea, it was widely agreed, was almost impossible to relieve while the prognosis for the secondary variety was often better.[41] But even in the latter, the prospects of success with hormonal therapy were far from brilliant, especially if menstruation had not occurred for more than a year. By then, the atrophy of the reproductive tract had advanced too far for hormones to have any effect.[42] Novak summed up the endocrine treatment of amenorrhea

> very unenthusiastically, because it will usually be unsuccessful. . . . I know of no gynecologist of standing who is not unenthusiastic about the treatment of amenorrhea by estrogenic substances. On the other hand, I know of few who do not often resort to organotherapy for amenorrhea because they know of no treatment that offers any more prospect of success.[43]

Things obviously hadn't changed much since 1929, when New York gynecologist Robert Frank, having failed to alleviate amenorrhea by ovarian preparations, had refused to place much faith in "the gratifying results reported by a number of clinicians who gave female sex hormone preparations."[44] He had hoped, however, that "in the near future . . . by means of potent preparations we may be able to load the endometrium to such a degree that sudden withdrawal will be followed by menstruation."[45] Six years later, at the height of Parkes's heroic age of reproductive endocrinology, that dream was still as distant as ever.

Gynecologists quickly appreciated that the findings of physiologists—erected on the foundation of animal experiments—were seldom applicable in the clinic, and the judgments of individual clinicians on dosage and treatment regimes varied widely.[46] As Friedrich von Mikulicz-Radecki, the director of the Gynecological Clinic in Königsberg, explained, the experimental physiologist investigated the function of single glands under carefully controlled laboratory conditions, whereas the clinician had to deal with the complex, interlinked, and still largely mysterious endocrine system of individual patients. A woman with a menstrual dysfunction may well benefit from hormone treatment, but she was unlikely to be suffering from a disorder that was sharply localized to one endocrine gland. The condition was far more likely to implicate both the ovary and the pituitary, not to mention the thyroid and other glands, and to variable degrees. The involvement of the nervous system could not be ruled out either. This extraordinary complexity of human sexual physiology explained why clinicians had not yet had any very remarkable successes to report on the hormonal treatment of gynecological disorders. The human being, he exclaimed, was not a retort![47]

When one compared the results of ovarian transplantation with those of hormonal treatment, the former, admitted Franz Siegert, seemed far more impressive.[48] The secretions of an actual gland were still superior to the pure hormones. The achievements of hormone biochemists, Siegert acknowledged, were enormous, but those achievements did not necessarily count for much at the bedside.[49] But there was no simple way around the issue. Ovarian transplantation, obviously, was far too serious an operation for the routine treatment of functional gynecological disorders.[50] At the very least, hormone treatment might eventually reduce the necessity for serious pelvic operations.[51] All therapeutic innovations faced doubt and uncertainty in their early days, remarked Carl Clauberg, particularly because the "striking discoveries . . . made in animal experiments" were invariably "extrapolated to humans with exaggerated expectations." It was especially important, Clauberg warned, not to expect anything that the animal experiments had not suggested to begin with; others added that it was equally vital not to promise one's patients more than what hormones had ever claimed to deliver. Harboring hopes for any "lasting replacement of missing ovaries in oophorectomized women by hormone therapy" was to invite disappointment, observed another gynecologist. Yet another counseled his colleagues against promising such grand and unattainable results as "rejuvenation" to their patients.[52] The optimists, of whom there were a few, continued to claim that hormones would yield excellent results, if only gynecologists would lose their timidity and prescribe them in high enough doses.[53]

Patchy therapeutic success was not the only problem facing hormone treatment in gynecology. Practical problems abounded. Dosage remained totally uncertain and was largely empirical.[54] The units in which hormones were measured varied from nation to nation and from pharmaceutical company to pharmaceutical company. Although the League of Nations Conference on the Standardisation of Sex Hormones had determined an "international unit" in 1935, this did not immediately come into regular *clinical* use. Emil Novak explained how in America the unit of dosage was still the Allen-Doisy rat unit (the amount of hormone needed to induce estrus within three days in oophorectomized mature rats of standard weight), which was equivalent to about three times the new international unit. There were other "rat units" too—not to mention "mouse units"—and many manufacturers were slow to standardize their products in the new "international units." For Schering products, for example, a rat unit was almost five international units; for Parke-Davis, a rat unit was 3.3 international units.[55]

Considering the speed at which hormonal science and therapeutics moved in the years before the Second World War, one might expect these early concerns to have been resolved by the end of the thirties. That, however, was far from the case. In 1939, gynecologist Hans-Ulrich Hirsch-Hoffmann acknowledged that despite some undoubted contributions of hormone therapy, the early enthusiasm with which practicing physicians had greeted the sex hormones had already evaporated.[56] Also, the gynecological profession, it seemed from Hirsch-Hoffmann's paper, still needed to be reminded that the endocrine glands comprised an interactive system: amenorrhea did not necessarily indicate an intrinsic ovarian dysfunction or even an endocrine problem at all.[57] The prices of hormonal products were still so prohibitive that neither Hirsch-Hoffmann's patients nor their insurers could pay for their long-term use in any case. Appalled by the expense involved in long-term treatment with sex hormones, another gynecologist remarked that while that kind of cost might be acceptable in the treatment of life-threatening disorders such as diabetes mellitus, it was less immediately justifiable in ovarian dysfunctions, which rarely endangered the patient's life.[58]

Hirsch-Hoffmann was concerned, moreover, that so much attention was being paid to the dosage of the hormone that the importance of its source was being neglected. It seemed quite implausible that a hormone obtained from urine could be as potent as one obtained from the gland itself.[59] It was quite probable, moreover, that the female sex hormones were not meant to act in isolation but only in synergy with other specific substances secreted by the ovaries.[60] Often, in the vasomotor menopausal symptoms, low-dose whole-gland preparations yielded far better results than megadoses of pure, standardized follicular hormones extracted from urine.[61] This could only be because the whole-gland extracts provided the unknown synergistic agents that were found in the ovary and which were essential for hormone action.[62] The very purity of the new hormone products might be impeding their efficacy. Single symptoms that were clearly due to single hormonal deficiencies— Hirsch-Hoffmann mentioned dysmenorrhea, which Novak had considered to be a troublesome condition to treat with follicular hormones—should certainly be treated with the newer hormone preparations. When the ovarian insufficiency was expressed in more generalized ways—as in climacteric symptoms—then organ extracts were much more appropriate.[63]

Fertility and the Sex Hormones

In an age of relatively simple assisted reproduction and reliable contraception, it is easy to forget that it took a while to learn how to use hormones to enhance or diminish human fertility.[64] From the early days of organotherapy, there was widespread interest in treating infertility with glandular extracts. As Naomi Pfeffer has emphasized, an enormous range of "pluriglandular preparations for the treatment of functional sterility could be bought in chemists' shops without a doctor's prescription" in the early twentieth century.[65] The results were poor, and the situation did not improve with the introduction of a pure follicular hormone. Novak found it so unsatisfactory in the treatment of sterility that it seemed "scarcely worth while to theorize" on it. While deficiency of follicular secretions might be responsible for occasional cases of infertility, treating every case with the estrogenic hormone, he asserted, was no more than firing a "shot in the dark."[66] In Central Europe, Friedrich Geller observed pithily that the hormonal treatment of sterility was a "thankless" task.[67]

There were, as usual, some positive reports too. In 1937, Hans Otto Neumann of the University Gynaecological Clinic of Marburg reported the successful treatment of infertility in six women who had been sterile for three to eight years.[68] One twenty-one-year-old woman who had scanty periods for three years and an undeveloped uterus became pregnant after two courses of treatment with high-dose Progynon and gave birth to a healthy girl eight months later.[69] Another patient, thirty-four years old and in a childless marriage for eight years, conceived after four courses of Progynon, but aborted the fetus after four months. However, her periods, which had been scanty but regular before treatment, returned to normal after the miscarriage.[70] In each of these cases, the husband's semen had also been examined and pronounced normal.[71]

Were these successes to be attributed solely (or at all) to hormone treatment? Neumann was well aware that the miracles may have been brought about not by the biological actions of the hormone but by associated "psychosexual moments." All gynecologists knew that women could conceive suddenly and unaccountably after being infertile for many years.[72] But even if it had been the hormone treatment that had helped in his own cases, the mechanism of action remained quite unclear.[73] Many practitioners continued to emphasize the polyglandular basis of infertility. In Britain, for instance, gynecologist Eardley Holland found infertility to respond exceedingly well to treatment with thyroid and anterior pituitary preparations, as did Cedric Lane-Roberts, who also considered insulin injections before breakfast and dinner to be useful in ovarian failure.[74] Even in the 1950s, things had changed only slightly as far as the utility of estrogen was concerned. In cases of infertility, a British gynecologist advised the use of estrogen only in some cases of secondary amenorrhea and occasionally in patients with excessively viscous cervical mucus. In the latter patients, the estrogen would merely thin the mucus, facilitating the sperm's entry into the uterus.[75]

As for the control of fertility, the history of the contraceptive pill and its foreshadowing in the research of the Innsbruck physiologist Ludwig Haberlandt (1885–1932) has been dealt with so exhaustively by Hans Simmer, Elizabeth Watkins, and Lara Marks that we do not even need to dip our toes in that complex ocean of facts, interpretations, and sociocultural debates.[76] In order to bring the story of the corpus luteum to some kind of culmination, however, I would like to look very briefly at Haberlandt's work. For reasons which remain unclear but may have been eugenic, he sought a way to sterilize women temporarily. "He reasoned," explains Simmer, "that hormonal excess should cause anovulation."[77] In this hormonal scheme, it was the corpus luteum that was the key.[78] As we have seen, the corpus luteum had long been suspected of preventing ovulation. Veterinarians knew, furthermore, that cows with persistent corpora lutea were infertile—although they did not know *why*—and they had evolved a technique to render these infertile cows fertile by squeezing out the persistent corpora lutea through the rectum.[79] The removal of luteal cysts in humans, Josef Halban would demonstrate in the 1920s, could sometimes cure amenorrhea.[80] Interestingly, however, all the research on luteal functions around 1900 had tended to focus on the corpus luteum's role in sustaining pregnancy rather than in suppressing ovulation.[81]

Also of importance to the development of hormonal contraception was the concept of the "interstitital cells" of the ovary and the report of their hyperplasia during pregnancy. It was speculated that the hyperplastic interstitial tissue acted in the same way as the corpus luteum when the latter regressed. It followed, then, that the secretions of this hyperplastic interstitial "gland" might act in the same way as the corpus luteum and prevent ovulation and conception. Haberlandt, moreover, was familiar with Steinach's claim that in transplanted ovaries, the follicles perished, whilst the interstitial cells proliferated.[82] Combining these isolated insights and hypotheses, Haberlandt tried to induce sterility in rabbits by transplanting the ovaries of pregnant rabbits into sexually active, nonpregnant rabbits. The first recipient became infertile and remained so "for about two and a half months after the transplantation despite frequent coitus."[83] By 1921, Haberlandt could report on several more experiments on rabbits and guinea pigs. There were many failures, but eight animals were indeed rendered temporarily sterile by the transplants.[84] If ovarian grafts from pregnant animals prevented conception, then *extracts* of such ovaries should also be effective.[85] Haberlandt also used placental extracts in his sterilization experiments since placental extracts were supposed to potentiate corpus luteum activity. Results were variable but often quite good, and placental extract, of course, was far easier to obtain than extracts of the corpus luteum.[86] Finally, Haberlandt succeeded in achieving the same end in mice through oral administration of ovarian and placental extracts.[87]

Initially, Haberlandt was inclined to restrict any future human use of temporary sterilization to patients with medical problems that might make pregnancy unsafe, but by the time he spoke to the World League for Sexual Reform in 1930, he referred quite explicitly to human birth control (*Geburtenregelung*).[88] In collaboration with the pharmaceutical firm of Richter of Budapest (the German company IG Farben was not interested), he evolved an oral preparation for sterilization called Infecudin, and it was reportedly sold on a restricted basis for experimental use.[89] Even before it came on the open market, however, a theologian of Graz lamented in 1928 that it was the height of scientific irresponsibility to introduce a birth-control pill when the populations of Austria and Germany were in precipitous decline. Not hormone tablets, but more children—*that* was the need of the hour.[90] The death of Haberlandt in 1932 resolved the issue for the time being, and Infecudin was never widely marketed.[91]

Simmer has dealt in meticulous detail with the confirmations and extensions of Haberlandt's work, and there is no need to pursue them here. What is noteworthy is that it was subsequently found that the administration of pure *estrogenic* extracts also interfered with the implantation of the fertilized ovum in the uterus by preventing progesterone from preparing the mucosa for the implantation.[92] The contest between estrogen and progesterone and the modulation of those processes to influence fertility provide us with further illustrations of the ways in which the sexual body came to be conceptualized as a product of hormonal *interactions*. And the use of hormones in contraception as well as in the treatment of sterility reunited the sexual and reproductive spheres that had been torn asunder by earlier gland researchers. That reunion of the sexual and the reproductive, of course, remained incomplete until the introduction of hormonally assisted reproduction and the contraceptive pill, events falling well outside the temporal boundaries of this book as well as of what Alan Parkes called the golden age of reproductive endocrinology. The availability of potent, standardized hormones was, no doubt, worthy of celebration, but as far as the clinical arena was concerned, those hormones did not immediately bring about anything like a revolution in the gynecological clinic. It would be mistaken, however, to assume from this that hormonal treatment was of no worth at all. We can appreciate the usefulness, even if limited, of the new hormonal therapeutics by examining the treatment of menopausal symptoms.

The Hormones and Menopause

The many concepts of the menopause, their unities and divergences, and, above all, the contexts influencing their construction and their use are topics that historians have not yet explored exhaustively.[93] For us, menopause is a crucial subject because its symptoms constituted the one category where hormonal treatment was found useful even by its most stringent critics.[94] The ever-skeptical Emil Novak acknowledged that menopausal symptoms could, "in a small number of cases," be so severe as to justify the use of hormones.[95] Apparently, some gynecologists still found irradiation of the pituitary to be useful in treating distressing menopausal symptoms, but most, according to Novak, preferred therapy with follicular hormone, especially since Bernhard Zondek had identified the specific phases of menopause when there was a genuine estrogen deficiency.[96] Although it could be difficult to rule out "the ever-possible psychic factor," it was clear that estrogenic therapy was of very real, if sometimes unpredictable, utility in relieving many of these symptoms, especially the vasomotor ones.[97] Since the treatment could be stopped as soon as the symptoms disappeared, the expense was not as severe a problem as with long-term hormonal treatment.[98] The follicular hormone was, so far, the sheet anchor of treatment, but although luteal preparations still remained far too expensive to be of practical use, they, too, had their champions. Ludwig Fraenkel continued to preach the virtues of his Lutein tablets into the 1930s and recommended them as enthusiastically for the treatment of menopausal disturbances as in other gynecological conditions.[99]

So far, I have referred to menopause or its treatment as if there was a clear consensus that menopause was akin to a disease and needed to be treated. This was not entirely the case. For some clinicians, especially in Britain, menopause was a natural process—hormonal treatment was permissible but only to help women over the climacteric. Long-term use was to be avoided. In 1937, Peter M. F. Bishop, endocrinologist at Guy's Hospital and the Chelsea Hospital for Women, asserted that although the vast majority of menopausal women did suffer from some climacteric symptoms, these were not always due to hormonal factors. The examination of vaginal smears could establish the patient's hormonal status reasonably reliably: if the smears were of "the oestrous type," then no estrogenic treatment was required, no matter how florid the symptoms.[100] "It is fortunate," Bishop observed, "that the human vagina is a delicate indicator of oestrogenic influence, for one is enabled to estimate the correct dose of ovarian extract for each individual patient by determining the amount necessary to induce an oestrous type of smear." Treatment must be stopped as soon as the smear became estrous.[101]

Bishop agreed that estrogen therapy was useful in the so-called vasomotor symptoms and in some other symptoms associated with menopause, such as atrophic senile vaginitis and skin lesions (patchy keratosis) of the palms and soles.[102] However, he advised against removing every symptom. The clinician's aim was not to restore the hormonal level of the premenopausal years but simply to palliate the patient's passage into the post-reproductive phase of life. A complete elimination of the hot flushes, for instance, was not the goal: one must aim, instead, to reduce them to "one or two every day or every other day."[103] Another British gynecologist would assert in the 1950s that the ideal dose was the one that was small enough to "just fail to relieve the symptoms."[104] These practitioners had no interest in keeping their menopausal patients youthful or feminine with hormones. As the British gynecologist Kenneth Bowes commented, "It has perhaps been a pity that they [hormones] have been made so easily available, and particularly that many women regard them as almost a necessity at the climacteric."[105]

But the seemingly miraculous effects of hormone therapy were not to be denied, and from the 1940s the availability of the synthetic estrogen, diethylstilboestrol (DES)—later withdrawn because of its carcinogenicity—made hormone replacement easier.[106] The psychological effects of estrogenic treatment of aging women led, it was reported in 1950, to "a definite improvement in memory and motivation measured by psychometric tests."[107] "The dramatic response of the senile female sex organs (except the ovaries) to estrogen administration," wrote one gerontologist in the 1950s, "has in many respects the character of a rejuvenation." Nor were the changes confined to the reproductive tract. The skin, too, showed considerable regeneration, the cosmetological importance of which was not overlooked.[108]

Soon, it was appreciated that the sex hormones might also be able to prevent the bone loss of old age. Treatment with estrogens *or* androgens could prevent the loss of bone by increasing the formation of organic bone matrix, suggested Fuller Albright in the late 1940s. At the time, it was a speculative hypothesis, but a review in the *Journal of Gerontology* welcomed it as "most interesting, since senile osteoporosis is one of the most important gerontologic diseases."[109] Research on osteoporosis would, in fact, be so fruitful that eventually, *routine* hormone replacement therapy in menopause would be justified by its prevention of senile bone loss. In its journey into medical and cultural prominence, however, senile osteoporosis was to become associated almost exclusively with menopausal women: relatively little research was done on bone loss in aging men, and in spite of suggestions that testosterone supplementation might be of use in the condition, it was never to become standard clinical practice.[110]

Androgens, however, were tried in menopausal women, partly because estrogenic substances never proved to be of much use in enhancing the libido.[111] Some clinicians found testicular extracts to enhance sexual desire in women.[112] It was initially presented as a paradoxical clinical finding. In less than a couple of decades, however, small doses of testosterone were being prescribed to older women for preventing senile vaginitis and also for better control of the hot flushes and depression.[113] The hormones were not given in sequence as in hormone replacement therapy today, but in limited courses. One British practitioner, for instance, recommended daily doses of ethinyl estradiol and methyl testosterone for about ten days and half those doses for another ten days. A second course was prescribed if necessary, but "after an interval of not less than three weeks, preferably with the dosage again reduced." After this, only "infrequent short maintenance courses with the smallest possible doses of these hormones may be given."[114]

Hormone replacement therapy became routine for older women only during the later decades of the twentieth century and only in a few national contexts.[115] Although it would take us far beyond the scope of this book to explore that topic in detail, it is nonetheless important to emphasize that it was not merely the progress of hormonal science that was responsible for this shift in clinical practice. In 1963, American gynecologist Robert Wilson and his wife Thelma Wilson described menopausal women as "castrates" who "exist rather than live"; their low oestrogen levels not only caused physical problems (osteoporosis, for instance) but also had grave psychological consequences, ranging from depression to "a vapid cow-like feeling."[116] In his 1966 bestseller *Feminine Forever*, Wilson called for the routine replacement of estrogens in *all* menopausal women. Aided by funds from the pharmaceutical industry and publicity from the press, Wilson became the prophet of a hormone replacement therapy that, as Elizabeth Siegel Watkins has emphasized, was miles removed from the older (especially the older British) approach, which merely prescribed estrogen *temporarily* to *some* women who could not cope with the transitional symptoms of the menopausal period.[117] Wilson was not solely concerned with the health of the menopausal woman: he also saw her as a social threat. "Estrogen-starved women," he proclaimed, were vulnerable to alcoholism, drug addiction, and divorce.[118] Rather than a mere palliative for a transition that could be difficult for some, routine hormone replacement was a tool to ensure social peace and harmony.

We should not, however, assume from Wilson's forceful words that HRT was imposed by a patriarchal medical profession (aided and abetted by the pharmaceutical industry) on a compliant and ignorant female population.[119] As Elizabeth Watkins has shown, the HRT phenomenon was powered by consumer demand to a very large extent. It was seen as "one more weapon in an arsenal that included vitamins, hair dyes, and cosmetics" and, one might add in memory of Gertrude Atherton and her generation, ovarian irradiation and pituitary diathermy. From 1966 to 1975, the annual number of estrogen prescriptions almost doubled, and Premarin—a commercial estrogenic preparation obtained from the urine of pregnant mares—became "one of the top five prescription medications in the United States."[120]

There were ups and severe downs after that, too, and debates over the safety of routine hormone replacement therapy continue to rage to this day. Crucially, we must keep in mind that even today, HRT remains a fairly empirical therapy and there are few hard and fast rules. As Germaine Greer emphasizes, what we refer to as HRT is actually "a multiplicity of regimes using a multiplicity of products in various combinations and strengths. No single individual can find his way around the whole gamut. . . . Selection of patients suitable for treatment is governed by the subjective impressions of the practitioner, and selection of the treatment regimen is a matter of serendipity."[121] The role of different pharmaceutical companies in marketing hormones is, needless to say, a vital component.[122] Moreover, it is not unlikely that the "scientific" understanding of menopause is in fact shaped significantly by culture. Is menopause a deficiency syndrome or a natural decline of life? As Saffron Whitehead remarks, it is primarily in modern Western society that life after menopause is seen as an "old, sexless, and useless" state, from which a woman can be "saved" only by HRT.[123] Cultures exist in which women acquire new power and status after the end of their reproductive years, and the Western construct of menopause may well be less of a purely biological matter than the proponents of HRT allow. Such statements are necessarily generalizations, but they do serve to emphasize the need for further research on the historical and sociocultural contexts of menopausal hormone replacement therapy.

The Anatomy of Menopausal Melancholy

Doctors over the years have identified middle-aged women as being particularly vulnerable to severe, suicidal depression. "Insanity occurring in women at the climacteric period, connected with, and dependent upon the physiological changes then taking place, has been noticed by most writers on female diseases as one of the gravest and most important of the morbid conditions which are incident to that time of life," wrote the Edinburgh asylum physician Francis Skae in an 1865 report on two hundred cases obtained from the case books of the Royal Edinburgh Asylum.[124] "There is scarcely an arrow in the armory of pain," observed American gynecologist William Conklin in 1889, "that is not unsheathed at this period" of a woman's existence. "Sunshine and shadow rapidly chase each other over her daily life."[125] The predominant symptoms were profound depression, convictions of guilt, suspiciousness bordering on what we would call paranoia, "fear of undefined and impending evil," paroxysms of excitement and violent agitation and "the most dangerous, and at the same time one of the most prevalent of the symptoms of climacteric insanity, namely, suicidal tendency."[126] Fortunately, this "involutional melancholia" was also "one of the most curable forms of insanity," and if suicide was guarded against, then patients usually recovered within six months.[127]

For most early-twentieth-century psychiatrists, the likeliest cause of involutional melancholia was the physiological change occurring during menopause. Even before fully standardized hormonal preparations were available, there was considerable interest in treating menopausal psychiatric symptoms with glandular extracts. In 1927, British psychiatrist Charles Molony pointed out that the frequency of psychotic episodes at puberty and menopause suggested that the inception and termination of significant sex hormone secretion destabilized mental functions, albeit in opposed ways. At puberty, "with the mind at the threshold of its development, states of excitement will be the rule," whilst around the menopause, "with the whole system, and more particularly the mental powers, on the down grade, depression will predominate." Glandular extracts were almost always useful in such cases—he reported recovery rates of 80 percent.[128] One had to remember, however, that endocrine disturbances causing psychiatric symptoms were almost always polyglandular and required polyglandular organotherapy. For climacteric insanities, for instance, a "combined thyropituitary-ovarian extract" was far more efficacious than ovarian extract on its own.[129]

When estrogenic hormones became available in chemically reliable preparations, there was considerable enthusiasm for their use in treating involutional melancholia. Convinced that menopausal symptoms, including melancholia, resulted from the "instability of the autonomic nervous system secondary to endocrine imbalance" (the imbalance being due to ovarian insufficiency), St. Louis internist August Werner treated twenty-one women presenting with what he considered to be involutional melancholia with Theelin, an estrogenic hormone supplied by its manufacturers Parke-Davis.[130] All of them improved, while seventeen controls treated with saline injections did not. Theelin was "curative" in uncomplicated cases but brought about some improvement even in virtually every case. Moreover, estrogenic treatment served as a reliable diagnostic test: if the symptoms were not relieved by estrogen, then the patient was unlikely to be suffering from involutional melancholia.[131] Within a few years, American investigators were reporting 90 percent recovery rates in involutional melancholia treated with Theelin; that hormone, "for all practical purposes," was fast becoming "a specific in involution melancholia."[132] "The melancholic outlook, which sometimes appears at the menopause and descends like a black cloud blotting out the joy of life," popularizer Ivo Geikie Cobb assured his readers in the 1940s, "may recede when oestrone therapy is employed."[133]

Others were less enthusiastic. Without denying the importance of the endocrine aspects of involutional melancholia, a British psychiatrist emphasized the importance of heredity and the personality of the patient. The "premorbid" personality of those who developed involutional melancholia seemed to present certain characteristic features, such as "strong obsessional character," anxiety, shyness, and stubborn, jealous traits.[134] Indeed, it appeared from a study of twenty hospital patients treated with Theelin that the more stable and "normal" the premorbid personality and the happier the patient's family environment, the better the results of endocrine therapy. The more obsessional types with unhappy lives were often significantly less responsive to hormonal treatment, and the least responsive patients were often suffering from pluriglandular deficiencies that were *exacerbated* by exclusively estrogenic treatment. The endocrine factor in involutional melancholia, in short, was "of considerable importance" but for many investigators of the condition, it was only one in "a constellation of factors" that also included heredity, personality traits, and other physical conditions.[135]

Yet others had even less patience for endocrine explanations. In 1937, the very year of the Werner group's enthusiastic report of successful hormonal treatment, another team of researchers from Boston reported that of ten patients of involutional melancholia—four men and six women, as opposed to the all-female groups we have examined so far—there was "no evidence of any type of improvement in the mental condition."[136] Indeed, 30 percent of cases showed an exacerbation of symptoms.[137] "There may," they conceded, "be cases of involutional melancholia in which theelin is of therapeutic value.... The type which we treated, as judged by the severity of symptoms, was not, we believed, of this caliber."[138]

Still other endosceptics (if one may coin a term) argued that while hormone treatment might improve physical symptoms associated with the menopause and even produce a general sense of well-being, it had no specific action on depression.[139] A 1941 review of the evidence even denied that menopause was *the* cause of involutional depression in an endocrine sense. The problem, instead, lay in the inability of rigid personalities to cope with the physical, emotional, and social demands of the involutional period.[140] "The new emphasis on ECT which was developing at this time," observed Saul Rosenthal in his 1968 historical review, "cast estrogen treatment [of involutional melancholia] further into the background."[141] In 1944, an American research team reported that in seventy five patients diagnosed with involutional melancholia, estrogenic treatment had failed to produce any improvement but those same patients had responded quite quickly to electroconvulsive treatment.[142] A handbook of endocrine treatment addressed to the general practitioner observed in 1953 that full-blown involutional melancholia "invariably fails to respond" to hormonal treatment but responds very promptly to electroconvulsion.[143] The very existence of involutional melancholia as a distinct psychiatric entity "with its own etiology, natural history, and clinical picture" came to be doubted, and the depression of menopausal women came to be seen simply as psychotic depression in middle age.[144] Yet another supposed triumph of sex hormone therapy had proven unfounded.

Sex Hormones Beyond Sex and Reproduction

Estrogenic substances were used not solely to influence what I have called the sexual body but also in conditions that, for us at least, bear no relationship to hormones. Hemophilia is a good example. Since it occurs only in men but is transmitted through women, the Chicago physician Carroll Birch speculated that "there must be something in the female organism which holds the disease in abeyance." The normal male, moreover, had been "proved" to be "part female," and it had been "definitely established that the female sex hormone can be isolated from the urine of normal males."[145] Birch claimed that hemophiliac males, unlike normal males, did not excrete any estrogenic hormones in their urine.[146] Hence, it was plausible that the hemophiliac tendency was expressed only in the absence of female hormones. Birch had found that treating hemophilia with extracts of whole ovary was beneficial: coagulation times were reduced, hemorrhages became less severe and the patient's general condition improved greatly.[147] These claims were soon dismissed by another study, which failed to find any reduction in coagulation time or indeed, any hemostatic effect at all from the use of any estrogenic agent.[148] I am not arguing that hemophilia was ever a major indication for the use of estrogens in men, but this debate, in itself quite minor, does show how physicians sought to explain a whole range of sexual differences with endocrine concepts and how such explanations could lead quite easily to attempts at hormonal therapy. Moreover, Birch's theory of hemophilia offers another illustration—albeit a smaller one than Steinach's cure for homosexuality—of the sexually intermediate body of the nineteenth century becoming a therapeutically accessible entity under the auspices of the new hormonal science.

More generally, it was appreciated quite early that the sex hormones were really growth hormones.[149] "Concentration upon the growth factor of the anterior pituitary, which is at present rarely of clinical value, has deflected attention from the fact that testosterone is a potent stimulus to growth of many tissues of the body, especially of the muscles, of the hair and, until the union of the epiphyses which it hastens, of the bones," observed physician Raymond Greene in England.[150] Earlier, Ludwig Seitz, director of the University Gynecological Clinic at Frankfurt, had argued that since the follicular hormone of the ovary induced the hypertrophy, hyperplasia, and differentiation of the cells of the uterine mucosa during the regular menstrual cycle and brought about the development of the uterus during pregnancy, it should be regarded as a sex-specific growth hormone.[151] American gynecologist and pioneer hormone researcher Robert Frank had also regarded the female sex hormone as "a growth substance ... that has gradually become more and more specialized, until it has developed into the main factor causing 'feminineness' in all dimorphous species."[152] It is unclear, however, whether such statements reflected the realities of clinical practice. It is true, of course, that new clinical uses were still being proposed for the estrogens in the 1940s—to prevent osteoporosis for instance—but such proposals do not seem to have been too numerous.[153] The primary and virtually exclusive indications for estrogenic treatment were sexual.[154]

"The Most Secret Quintessence of Life"

A fervent lay admirer of Steinach's work had once hailed Steinach's Progynon as "the very essence of Eve."[155] Not to be outdone, journalist Paul de Kruif proclaimed that the elaboration of demonstrably effective testicular extracts and hormones in the 1930s had constituted the discovery of "the most secret quintessence of life."[156] The testicular essence had proven rather hard to find. Writing in 1943, the American physician Gerald Newerla observed that although the testis was "the first gland subjected to animal experimentation,"

> approximately a century—from 1848 until 1935—was needed before the androgenic hormone was isolated and synthesized in pure crystalline form. This fact is still more surprising when comparison is made with the internal secretions of other glands, such as the thyroid, the parathyroid, the pancreas and the adrenals, whose endocrine activity was not suspected until much later, yet whose active principles were isolated chemically, and even synthesized, much earlier than were those of the testis.[157]

Ever since Brown-Séquard, commercial preparations of testicular extracts had, of course, been widely available, but their physiological potency was always open to doubt.[158] By the 1930s, however, there was a more precise criterion. In order to be considered potent, a testicular extract now had to pass the capon comb test, i.e. it had to stimulate the growth of the shrunken comb and wattles of a castrated cock. The stimulation of growth was not enough: the comb and wattles had to shrink again after the injections were stopped.[159] The first testicular extract to pass this double test had been obtained in 1929 by biochemists Lemuel McGee and Fred Koch from bull testicles at the University of Chicago. The extract appeared to be nontoxic, but its chemical nature remained impenetrable—Koch and Gallagher were not even sure whether their extract had one active principle or more and declined to suggest a name for the extract. "Too often," they declared righteously, "a name gives a false sense of security as regards the purity of the product, a fact we wish to emphasize, for it is our firm opinion that the extract is as yet grossly impure." An even greater obstacle to clinical use was the low yield. In order to obtain the "minimal daily dose," one needed forty-three grams of frozen bull testicle![160] As Paul de Kruif put it with his customary brio, the hormone was "not stored up at all. . . . It is as if God wished to hide testosterone from the curiosity of questing men who, if they found it, might be bold to use it to make mankind happier than God intended."[161]

But God hadn't been clever enough after all. By 1930, it had been found that the so-called male hormone was also present in the blood of cocks.[162] It, therefore, "seemed logical to conclude that since the male hormone is in the blood, it would probably also be found in urine."[163] Male urine, therefore, was well worth exploring as a source of the male hormone. This was done by the vitamin pioneer Casimir Funk and his associate Benjamin Harrow. The extract—or sometimes, concentrated urine itself—was tested by the capon comb test: "in every instance injections caused the masculinising effect."[164] In 1931, Adolf Butenandt in Germany obtained a crystalline substance from 25,000 liters of male urine, which he called androsterone "in the belief that it was the essential male hormone."[165] By 1934, Leopold Ruzicka in Zurich had synthesized androsterone from cholesterol, and the pharmaceutical industry's interest in sex hormones reached fever-pitch. The next year, Ernst Laqueur and his team in the Netherlands, funded by Organon, obtained a crystalline form of active male hormone from the testicle, to which they gave, according to Alan Parkes, "the dreadful name 'testosterone.'"[166] Butenandt and a colleague, funded by Schering, synthesized testosterone in 1935, narrowly beating Ruzicka, who was sponsored by Ciba.[167] The synthetic hormone was made available in the form of injections (testosterone propionate was the most favored injectable), pellets for subcutaneous implantation, and subsequently as tablets of methyl testosterone for oral use.[168] The androgens immediately entered clinical practice and not always in the most expected ways either.[169] They were tried with some enthusiasm in the late 1930s, for instance, in the treatment of angina pectoris.[170] But the greatest clinical interest, undoubtedly, focused on using androgens to influence male sexual function.

Impotence, predictably, was a popular indication, but sadly for many, testosterone, in spite of the high hopes associated with it, proved to be of very little use in treating it in otherwise normal men.[171] Testing the hormone (supplied by Schering and Ciba) on twelve impotent patients aged between twenty-seven and sixty with no evidence of pituitary deficiency, two American surgeons found in 1940 that "while most of the patients felt less depressed mentally and thought their spirits were better, there was no improvement in the impotence." Even the psychic benefits, unfortunately, were replicated when sesame oil was injected instead of the hormone, and the authors inferred that "the benefits of androgenic therapy in the impotent male without hypogonadism are chiefly psychic."[172] Across the Atlantic, A. W. Spence at St Bartholomew's Hospital published a negative report on the use of testosterone propionate in six impotent patients without "any obvious endocrine disorder."[173] Spence noted that the absence of testicular hormone did not inevitably cause impotence, "for it is well known that occasionally some eunuchs are potent," a phenomenon explainable, he suggested, by the production of androgens by the adrenal glands. While the male hormone was essential to sexual function, it was not sufficient in itself; if the "psychogenic stimulus be absent or disturbed, excess of the hormone does not compensate for this deficiency."[174] With the coming of effective hormones, doctors and patients discovered (or perhaps rediscovered) that the testicles and the pituitary, for all their powers, constituted only one element of the sexual sphere. The psyche was equally if not more important, and purely psychogenic impotence was the commonest category of male sexual dysfunction.[175] No endocrine treatment was necessary or useful in such cases.[176]

The hormone, however, was quite dramatically useful in arousing sexual desire and potency in sexually undeveloped (eunuchoid) men and in cases of pituitary deficiency. If the endocrine function of the testicles failed before puberty, the secondary male characters did not develop, the genitalia remained infantile—and impotent—and the body developed into a neuter form: "long spindle limbs with a small trunk" and "a wide pelvic to shoulder girdle ratio."[177] If the testicular hormones failed in adult life, then, too, the effects were noteworthy. Obesity, coarsening of features, and impotence were invariable, and the mental and behavioral changes could be remarkable. As the British surgeon Lennox Broster summarized: "Usually these people tend to become inert, helpless recluses and avoid society, especially female. They tend to be unstable emotionally, and either through bad home influence or evil associations they are apt to develop vicious habits. Like women with virilism they suffer from a sense of inferiority and social ostracism."[178] Administration of testosterone produced marked benefits in such cases, although not the reversal of infertility.[179] In the younger eunuchoid, puberty could be induced by hormone treatment. So, testosterone did establish itself as a reliable therapeutic agent in some relatively rare conditions, but that, surely, was poor consolation for those who had shared de Kruif's vision of it being the "quintessence of life."

The Male Climacteric

The efficacy of the androgens in increasing muscle mass, promoting a general sense of well-being, and relieving a host of age-related frailties—almost exactly the same list Brown-Séquard had proffered in the late nineteenth century—led to their use in a condition labeled male climacteric. The concept has an interesting history, which deserves deeper exploration than I can attempt here.[180] That aging produced a diminution in sexual desire was, of course, an old idea.[181] The concept of a male climacteric, however, did not always emphasize the sexual sphere. In 1831, the President of the Royal College of Physicians, Sir Henry Halford, had published a discourse on "the climacteric disease."[182] He had argued that the "falling away of the flesh in the decline of life, without any obvious source of exhaustion" was not simply a manifestation of senescence but a distinct disease.[183] Tiredness, loss of weight, insomnia, a bloated countenance, and whitened tongue all featured on its list of symptoms, but interestingly none of them was sexual and the condition, Halford emphasized, had no clear analogue in women. "Though this climacteric disease is sometimes equally remarkable in women as in men, yet most certainly I have not noticed it so frequently nor so well characterized in females," he declared. "Perhaps the severe afflictions of their system, which often attend the bearing of children, or, what is more likely, the change which the female constitution undergoes at the cessation of the catamenia, may render subsequent alterations less perceptible."[184]

Menopause and the "climacteric disease," then, were entirely separate entities, and if there was a sexual component to the latter, then Halford did not consider it worth mentioning. Although his description of the climacteric would be quoted frequently by later medical writers as a pioneering statement on a controversial condition, clinicians a century later would focus on the sexual dimension and on the analogy to the female menopause, both of which Halford himself had emphatically excluded.[185] Obviously, Halford wrote long before the dawn of the hormonal age, but it is tolerably clear that he was conceptualizing the disease in a desexualized way that could, perhaps, be forced into a whiggish history of the concept of neurasthenia but not, without actual disregard of facts, into the prehistory of any sexual and endocrine concept of the male climacteric.[186] Despite this, his name and concept would often be cited by the many studies of the male climacteric that appeared in the early twentieth century.

In 1910, Archibald Church, Professor of Nervous and Mental Diseases at Northwestern Medical School in Chicago, reminded American doctors that it was "a well-established anatomic fact" that the early human embryo was sexually undifferentiated and potentially hermaphroditic, and it was a truism that boys and girls were "physically much alike" before puberty. It was too often forgotten, however, that "old men and old women again become alike both mentally and physically through the loss of sexual differences."[187] If the two sexes were so similar at the beginning and end of life, then was it not likely that even in their years of maturity, they were more similar than one imagined? One such unsuspected similarity might relate to their physiological periodicity. Men, to be sure, did not have a monthly flux, but many scientists agreed that they, too, operated on a biological cycle. Some of the abnormalities to be found across the sexes—migraine, periodic drinking, epilepsy—seemed to follow regular monthly rhythms. Speaking broadly, then, periodicity was not confined to females, and "there also exists in men a cyclic wave which is really a manifestation of a general law of vital energy."[188]

If there was a monthly cycle in males, was there also something like menopause? Women as well as men of the involutional age, Church emphasized, showed similar kinds of mental turbulence, which could range from a full-blown anxiety neurosis to vague, neurasthenic symptoms.[189] In the physical sphere, headaches, weight loss, and gastrointestinal torpidity (expressed through indigestion and diverse intestinal complaints) were common. The patient was almost always easily fatigued. Fortunately, however, most patients recovered after "a variable course of from eighteen months to three years."[190] Management of the male climacteric, therefore, required tact and patience rather than any energetic treatment. The patient's workload should be reduced, and he might be advised to travel, but by and large, nothing much needed to be done because the symptoms were not by themselves "of serious import" and the condition was essentially temporary.[191]

Shortly afterwards, German neurologist and psychiatrist Kurt Mendel published a paper on what he called the *climacterium virile*. It was, Mendel asserted, a very real affliction of aging men but was rarely taken seriously by doctors or, indeed, by many sufferers.[192] Since puberty was hardly less turbulent for males than it was for females, there was no reason to presume that the involutional years were easier for men than for women. True, the climacteric developed more gradually in men than in women, but that was no reason to doubt the existence of a natural male climacteric or that its likeliest cause lay in the sex glands. The woman going through a natural menopause exhibited symptoms similar to, but less intense than, an oophorectomized woman. Having studied men undergoing the change of life for more than a decade, Mendel was confident that their mental and behavioral peculiarities mirrored those of castrated men.[193]

In every case of the male climacteric, the essential, recurrent feature, Mendel stressed, was an unprecedented emotionality, a tendency to weep at the slightest provocation, or at no provocation at all.[194] Well-built, masculine, and hitherto healthy individuals would turn tearful and effeminate in middle age and in addition report hot flushes and sweats, anxiety attacks, palpitations, insomnia, and other symptoms characteristic of the menopausal woman.[195] Men in the climacteric became as capricious in temperament and as apathetic toward their previous interests as menopausal women, losing all their previous zest for life. Withdrawing from society, they developed a childlike dependency on their doctors. Although all of Mendel's patients were happily married and had never suffered from any sexual dysfunction, they had begun to lose their libido at the onset of the climacteric.[196] The condition, however, seemed to be as self-limiting as in Church's cases. Although the symptoms could linger for as long as four years, the usual range was between one and three years, and the patient recovered fully, albeit often with a permanently diminished libido. But that, Mendel remarked, was hardly unexpected by the patient at his age.[197] The prognosis for a climacteric male was generally excellent, no matter how severe the symptoms.[198]

Since the female and the male climacteric occurred at ages that were equivalent and had similar symptoms, course, and prognosis, it was only reasonable to assume that the male climacteric was brought about by the cessation or at least by the sharp diminution of testicular secretions.[199] There was, Mendel acknowledged, a problem with this hypothesis. Organotherapy with testicular extracts had not given particularly impressive results in the male climacteric.[200] Although hydrotherapy and electrotherapy were both of some use, psychotherapy was all most patients needed—and psychotherapy, for Mendel, did not seem to mean much more than reassuring the patient about his good prognosis.[201] Here, then, was a curious entity: a putative state of glandular deficiency that responded only to psychotherapy, but Mendel does not seem to have found it contradictory in any way.

It was also possible to accept the concept of a male climacteric without fully sharing Mendel's gonadal perspective. Following Mendel's report, London physician Bernard Hollander reported that he, too, had long noted identical symptoms in men of a certain age and offered some of his own findings to supplement Mendel's.[202] In both sexes during the climacteric, the physical machinery was re-tuned for a new (presumably lower) level of functioning.[203] Agreeing with all of Mendel's observations, Hollander declared that the climacteric male was, in a word, "effeminate."[204] The loss of sex gland function, he acknowledged, could entail the diminution of sexual as well as broader mental powers. Did mental energy depend, then, on the gonads and their secretions? Hollander—who was an enthusiastic phrenologist—demurred. The gonads probably did stimulate creativity by means of their *nervous* connections with the brain, but it was all a neural matter having nothing to do with the internal secretions. To treat a man suffering from severe climacteric symptoms, one must treat not his gonads but his nerves.[205] This relative lack of interest in gonads as endocrine organs was not an uncommon feature of contemporary concepts of the male climacteric, especially in Britain. As late as in 1919, Guthrie Rankin, a consulting physician of London, wrote in the *British Medical Journal* that while the female climacteric was undoubtedly due to "the functional death of the ovaries," the male climacteric probably had nothing to do with the testicles.[206] Rankin had little interest in matters endocrine and presented a virtually Halfordian picture, although far more negative than Halford or the Germans as far as the prognosis was concerned.[207]

In 1922, with the medical world abuzz with glands, Kurt Mendel returned to the male climacteric. His colleagues, he was convinced, were exaggerating the importance of the endocrine glands, but he still thought it wise to reevaluate his original concept in the light of new findings in glandular physiology, particularly those reported by Steinach.[208] As far as the symptomatology was concerned, Mendel had little to add to his 1910 discussion. Even then, he had suggested that the syndrome of the climacterium virile followed from the decline in sex gland function.[209] In the light of recent glandular research, however, it seemed likely that the thyroid, the pituitary, and the adrenals were also involved. Although the precise mechanics of their involvement remained to be elucidated, it was possible that the male climacteric resulted from the involution of particular endocrine glands and the compensatory hyperfunction of others. The details were complex, but the male climacteric was ultimately an *endocrine* disorder. Mendel had little to say on what caused the glandular dysfunction. His sole suggestion was poignantly unique to the era: the impairment of endocrine balance by war.[210]

As for treatment, there was no resolution of the old paradox of an endocrine condition that did not respond to endocrine treatment. Reassurance remained the best treatment of all. Organotherapy, Mendel reaffirmed, was not of much use, at least with the commercially available extracts.[211]

Interestingly, he refused to climb on to the Steinach bandwagon of the 1920s. In theoretical terms, the operation seemed to be rational enough, but he was not an admirer of its clinical effects. (In a separate paper, Mendel had recently implicated the Steinach operation in a case of irreversible psychosis.)[212] The only endocrine therapy that Mendel did not reject was testicular transplantation, although even that, he pointed out, was unlikely to produce lasting benefit.[213] Although seemingly paradoxical, however, Mendel's views could be shared even by champions of the Steinach operation. In Britain, for example, Kenneth Walker was a rare champion of monkey glands as well as vasoligature. In 1931, at the height of the endocrine era, he told the Sunderland Division of the British Medical Association that the male climacteric was a genuine clinical entity but not clearly analogous to menopause in women. The female menopause was "primarily a genital episode," while in the male the climacteric represented "his descent from maturity to old age."[214] "Whether it be pleasing or not to her political aspirations, it is a fact," stressed Walker, "that everything in woman is sacrificed to the reproductive function."[215] In man, however, procreation was not of primary importance, and although testicular insufficiency did blight the life of the average fifty-plus man, the male climacteric, unlike menopause, was not *purely* a manifestation of gonadal decline.[216] Neither the Steinach operation nor monkey gland implants would be of much use in treating it.[217]

Although the conceptual relationship between the male climacteric and the "premature senility" treated by the Steinach operation remains unclear, it is a fact that as surgical rejuvenation went out of vogue in the late 1930s, the male climacteric reemerged.[218] A. P. Cawadias claimed in 1947 that "the climacteric . . . is the opposite of puberty and indicates the involution of the gonads. . . . This occurs in men as well as in women, and to deny the climacteric of men is a negation of biological principles."[219] Kurt Mendel's endocrine hypothesis had now become a firm principle, although nobody seemed to remember Mendel's name or his skepticism about endocrine therapy. In America, the St. Louis physician August Werner proved to be a great champion of the male climacteric concept in all its endocrine dimensions.[220] A volume published in his honor in 1952 recorded:

> How Dr. Werner became convinced that [the male climacteric] is a definite clinical entity, necessitates a confession, which is always good for the soul. After having obtained many thousands of histories from women of all ages during the menacme who had ovarian afunction or hypofunction . . . he had many male patients who complained of decreased libido and potency and the characteristic symptoms as

described for women. . . . About 1937 Dr. Werner, almost suddenly, developed all of these symptoms in a severe form, which left no doubt in his mind that the male of the species, is also subject to this condition.[221]

The syndrome in men, Werner emphasized, was not enormously different from women, although the hot flushes that affected virtually all menopausal women were less common in men. The nervous and mental symptoms were the most frequent features of the male climacteric, including crying for no reason, severe depression (by now being referred to as involutional melancholia), and even suicide.[222] All this, of course, was fairly Mendelian, but Werner, unlike Mendel, was an enthusiast for endocrine treatment.[223]

By Werner's time, there was, of course, ample reason for such faith. Gone were the days of unreliable extracts or the Steinach operation: testosterone propionate had now been shown to be almost miraculously effective in enhancing the potency and elevating the mood of castrate patients who were "disturbed, anxious, and broken in spirit."[224] As well as restoring erectile functions, the hormone injections restored the "capacity to respond with the proper emotions not only to intercourse but also to other acts such as kissing or embracing."[225] Although Werner did not cite it, a German report published a few months before his own had waxed eloquent over the action of testosterone propionate on depression, delusional ideas, suicidal thoughts. and other mental symptoms in twelve middle-aged men.[226] If the climacteric male was analogous to the castrate, then there was no longer any reason to hold back from hormonal treatment, and, upon failing to find any organic problem in a fifty-year-old man with the symptoms of the male climacteric, Werner confidently prescribed testosterone proprionate injections for three months. This led to the patient's complete recovery, and there was no looking back for Werner after that.[227] The testosterone injections would now be his first choice in the treatment of uncomplicated cases of the male climacteric.

Although they did not occur in every case of the male climacteric, mental and nervous dysfunctions of various degrees were, of course, integral to it. The climacteric male, observed Werner, was "easily aggravated or excited to anger by word or deed. . . . They are hard to please, and frequently their family or associates can hardly get along with them." Disturbed sleep, headaches, loss of memory, delayed cerebration, and depression were virtually universal. Vasomotor symptoms like hot flushes, palpitations, and cold extremities were common, as were the classic symptoms of fatigue and decline of potency. Testosterone propionate (injections of 25 mg every other day "omitting Sunday," then 25 mg twice a week for a month and finally,

25 mg once a week for variable periods depending on the patient's condition) relieved the symptoms in virtually all cases.[228] The clinical and biological conclusion was clear and irrefutable: "That man is subject to varying degrees of sexual function and does have a climacteric is now an established fact."[229]

The old psychiatric dimension was not overlooked either. Although they did not mention (and may not even have known of) Kurt Mendel, mid-twentieth-century investigators of the male climacteric were retracing his path. Whatever the depth of their convictions about the endocrine pathogenesis and treatment of the male climacteric, most clinicians paid close attention to the psychic symptoms, especially depression. Indeed, for some experts, the male climacteric almost became a synonym for involutional melancholia in men. Two American investigators confessed in 1940: "The term 'male climacteric' is being widely used at present, both in the clinic and the experimental laboratory. Because of such usage we here subscribe to the term, but in a limited way. The cases described were diagnosed as involutional melancholia, thought to be caused by a lack of sufficient male hormone elaboration in the individual for mental stability. In that sense we use the term 'male climacteric.'" The report on two patients treated with moderate amounts of testosterone propionate claimed remarkable improvement lasting for almost a year after the last injection.[230] Mendel's male climacteric had been an endocrine disorder unresponsive to endocrine treatment; involutional melancholia was turning almost into its mirror image—a psychiatric condition that was best treated by hormones.[231]

Some established clinicians disagreed with Werner regarding the effectiveness of testosterone in treating the impotence of senility. In the discussion after one of Werner's papers, Victor Lespinasse of Chicago—a pioneer in testicular transplantation—observed that the male climacteric might well be due to deficiencies in blood circulation through the gonads. He also emphasized the importance of potency: "The men that I see are mostly interested in sex. They come for impotence. They don't lose their libido. . . . Testosterone works occasionally with a fair result."[232] Werner denied that testosterone restored potency even to the degree Lespinasse claimed, but their exchange is actually more important for what it reveals about the ways in which glandular physiology had changed since the 1920s. The inevitable decline of sexual function with age, Werner asserted in response to Lespinasse, was due in man as well as woman to the "failure of gonadal response to the gonadal hormone of the anterior pituitary, and this inherent failure of response is the condition that inititates the interglandular disturbance, with a consequent autonomic nervous system imbalance and the production of the symptoms described."[233]

While Lespinasse still thought in terms of self-contained glands acting in isolation, the endocrine body had changed into a less compartmentalized entity for Werner. The sex glands, if anything, were now considered *more* important than they had been earlier—even the vasomotor symptoms of the climacteric were now endocrine symptoms—but they were no longer considered to be as autonomous as they had been a couple of decades earlier. Testicular deficiency was no longer autochthonous: it was the testicle's lack of response to *pituitary* stimulation that was primary, not its lack of vascularization. The *symptoms* of the male climacteric, of course, were attributable to the deficiency of androgens and were essentially identical to those of "eunuchism, [bilateral] cryptorchism, hypogonadism, castration" and the immediate cause of those symptoms did lie in the testicles.[234] The deeper, biological foundations of the syndrome, however, were likely to involve dysfunctions affecting the entire glandular economy rather than the testicles alone. While Lespinasse was still focused on the testicle and its nutrition, Werner, like others of his generation, had begun to think of the endocrine glands as constituting a tense and occasionally disharmonious system, imbalances within which could be transmitted to—and through—the nervous system.[235]

Much of this, clearly, was only of theoretical import. No matter what the cause of the androgen deficiency, the clinical symptoms were due to that alone, and the only causal therapy advocated by Werner was androgen replacement. In strictly clinical terms, Werner's reasoning could even be described as circular. Whatever responded to androgen replacement was the male climacteric and, therefore, the male climacteric must be caused by androgen deficiency. Some of his contemporaries rejected this pharmacotherapeutic formulation. Calling for the development of a "specific sensitive index of testicular androgen secretion," American internist Richard Landau pointed out in 1951 that "while androgen deficiency in the castrate cannot be disputed, that of middle-aged men complaining of nervousness, fatigue, flushes and diminished libido requires substantiation by some more objective indicator than the beneficial results of the administration of testosterone."[236] Landau was not even very impressed with a 1944 study that had shown that patients suffering from the male climacteric excreted higher amounts of pituitary gonadotropins in their urine than "normal" men, suggesting that their testicles were secreting less androgen than the pituitary regarded as sufficient.[237] Describing these results as "almost unbelievably clear-cut," Landau emphasized the difficulty of "evaluating the efficacy of testosterone therapy" and pointed out that "an elevated excretion of urinary gonadotrophins may not inevitably reflect a deficient secretion of testicular androgen."[238]

The symptoms of the male climacteric, Landau emphasized, had a "potential emotional overlay," and it was crucial to eliminate any "purely psychologic effect" of the treatment. The "shift from maturity to old age" went smoothly for most, but some inevitably developed "neurotic reactions," manifested most frequently by problems of sexual potency. Most of these would respond to suggestion, especially the suggestion surrounding an expensive, potent, radically new form of treatment such as hormone injections. Until the concept of the male climacteric was supported by definite quantitative evidence of androgen deficiency, it was best, he asserted, to assume that the majority of cases of male climacteric were psychologically rather than hormonally induced. The diagnostic label of male climacteric itself, he urged, "should be abandoned."[239] Other critics argued that some of the symptoms of the so-called male climacteric could well be due to changes in individual organs and systems, rather than to diminished levels of androgen secretion.[240] In Britain, endocrinologist Raymond Greene dismissed the symptoms as "more easily explicable by arterial degeneration or by plain boredom and dissatisfaction." But sceptic as he was, even he had to acknowledge that "the administration of androgens in such cases to be occasionally justifiable.... There is no doubt that in the man whose testicular function is failing, albeit physiologically, the male hormone may have a noticeably tonic effect, increasing the patient's energy and sense of well-being."[241]

By the end of the 1940s, one can sense the emergence of a professional consensus that "the increased muscular strength, alleviation of mental depression, and improvement of sexual function occasionally seen after testosterone treatment of old men is usually of considerable value, as are also the effects of estrogen administration on the many postmenopausal complaints in the women."[242] Tiberius Reiter, a German-trained British physician and an upholder of the androgen-deficiency concept of the climacteric, added a new twist to the debate in 1951 by pointing out that since men and women produced estrogens as well as androgens, a proper replacement regime for the climacteric male had to combine both.[243] (Compare this to Steinach's suggestion that the rejuvenative effects of his operation might be replicated more reliably with hormone injections if a small quantity of the female hormone were added to the larger doses of the male hormone.)[244] That same year, clinical endocrinologists Max and Joseph Goldzieher, reiterating the androgen-deficiency explanation of the male climacteric, advised the American general practitioner to combine testosterone with estrogen to maximize therapeutic results.[245]

Details apart, much about this sunny optimism was quite reminiscent of the 1920s, as were some of the concerns. It was dangerous, physicians were reminded once again, to encourage older men with weak hearts to plunge into sexual extravagance.[246] Overall, however, the news was excellent: gerontologists could do quite a bit to restore the functions lost in old age.[247] This enthusiasm was not, of course, to last, but since this book ends around 1950, we can leave the aging male in a world that the rejuvenators of the 1920s would have welcomed.[248] Few midcentury gerontologists uttered the name of Voronoff or Steinach, but it was only in the more technical aspects of their principles and practice that they differed from those now-forgotten pioneers. In their larger vision, they were virtually indistinguishable. For the rejuvenators of the 1920s, sex was not just a matter of copulation or of anatomical characteristics. Sex was life itself, and sex was controlled by the sex glands: the decline of sex gland function lowered the vitality of the entire organism. Aging, sex, and the endocrine secretions were linked in one causal chain. To be sexless was to be old, to have healthy sex glands was to be youthful, or at least, vigorous. Steinach's generation was convinced that one could, within limits, revitalize the old if one could resexualize them. Much had changed over the next two decades in the *why* and *how* of gonadal secretion, but even in the 1940s, sex and sex hormones remained in conceptual terms exactly where Steinach himself had placed them: at the very heart of life itself.

Hormones and Homosexuality: *Plus ça change...*

Nor did endocrine theories of homosexuality disappear with Steinach. With the coming of the hormonal age, all that really disappeared from discourse was the gland: the focus shifted to the hormones, their levels in the body, and the ratios between different hormones. The possibility of "curing" homosexuality by hormonal treatment continued to be debated energetically. In the late 1940s, Ivo Geikie Cobb informed the lay world that while attempts to diagnose or treat homosexuality by endocrine means had yet to be unsuccessful, it was "distressing to realise that a few milligrammes of the wrong sex-hormone may be responsible for a maladjustment which affects an individual's entire being."[249]

One early attempt to determine the characteristic hormonal profiles of homosexuals was undertaken in Magnus Hirschfeld's Berlin: the urine of nine "genuine" homosexuals was assayed by the Allen-Doisy test for female hormones, but none showed any elevation of the latter.[250] The story did not end, however, with that negative study. In twentieth-century America, there was much medical and psychological interest in explaining the nature of homosexual desire and reforming the homosexual predisposition.[251] Endocrine theories and hormonal therapeutics attained considerable prominence in that project. In 1929, Robert Frank had suggested that the "anomalous discovery" of the female sex hormone in male urine might reflect "a quantitative and fluid transition from male to female, and that such individuals possess hidden female biological qualities, thus again offering, with due reserve, a possible explanation of homosexuality."[252] Frank did not pursue the connection, and it was Clifford A. Wright (1882–1961), co-founder of and senior attending physician at the Psychoendocrine Clinic of the Los Angeles County General Hospital, who was to become the leading American proponent of a hormonal explanation of homosexuality.

Wright revitalized the endocrine approach to homosexuality by moving away from older style of reasoning, which sought evidence of feminization in cells or in the levels of female hormone in male urine. He focused instead on the alteration of the normal *balance* between male and female hormones. All individuals, he accepted, excreted both kinds of hormone in their urine. In "normal" males, however, the ratio between the male and female hormones was markedly different from the ratio in "normal" females. In conditions of sexual inversion, it was this ratio that changed toward the norm characteristic of the other sex. A homosexual male would excrete male and female hormones in a ratio that approached or matched that in normal women.[253] The older model of "one sex, one kind of sex hormone" was now history, but even within the new and supposedly more "bisexual" paradigm, the male homosexual could still be presented as a hormonally feminized individual.[254] The new model, in fact, made the identification of feminization even easier than in the days of Hirschfeld: it was now a simple matter of biochemical measurement. The body itself no longer had any diagnostic utility, and there was no need to catalogue the feminine physical characteristics of male homosexuals.[255] All that was required to establish the feminization of an individual male was to show that "the usual balance or dominance [of the male sex hormone in urine] is *definitely* altered."[256] Negative diagnoses were equally important: "Hormone assays of the urine are important in helping to disprove homosexuality in a normal individual where arrest has been made because of an alleged overt act." Wright recalled one of his cases: "The patient was a girl of seventeen, whose teacher, a homosexual woman, made love to her. The girl denied homosexual inclinations and her hormone assay was normal.... Subsequent investigations confirmed these negative findings."[257] Sexual orientation, even in the old days, was not determined from sexual *acts* but from a whole range of psychological and morphological characteristics. Now, however, there was a simple chemical test for separating the "homosexual" from the "normal."

In some cases, of course, the ratios did not match up to Wright's expecta-tions. When "normal" people showed homosexual-type ratios, he categorized them as latent homosexuals. Avowed homosexuals who had normal-type ratios were obviously more problematic. Wright doubted whether such peo-ple were "of the true constitutional type," although he did not classify them, as Magnus Hirschfeld may have done in his place, as pseudohomosexuals.[258] Perhaps the greatest challenge, Wright and his collaborators acknowledged, was the finding that in the same homosexual person the urinary hormone ratios could vary widely across time. Consequently, one had to test the urine more than once and refrain from "attributing too wide significance to the result of individual determinations."[259] Despite these problems, however, Wright was sanguine that a firm diagnosis of homosexuality was possible on the basis of urinary hormone estimation.

"Homosexuality of itself is a vice, not a crime," Wright asserted, "though it may lead to criminal acts; it is a medical and social problem."[260] Punishment was useless: "It is no more logical to expect to cure a homosexual's abnormal sex instinct by eighteen months' stay in jail, than it would be to change the nor-mal sex instinct of a normal man or woman."[261] Treatment had to address the hormonal biology of the individual. Wright's attempt to cure homosexuality began with a twenty-four-year-old male, who "had in several instances almost uncontrollable desire toward his own sex." His urine showed an elevation of female hormones and a lowering of the level of male hormone. The patient was treated with pituitary gonadotropin injections. His mother soon wrote gratefully, reporting that her son was now "interested more or less in three girls, is better tempered and less nervous."[262] The use of gonadotropins was the logical treatment because they had "a tendency to stimulate the existing sex glands." Even if there was no deficiency of gonadotropins, administra-tion of them "would tend to readjust the hormone balance." Additionally, the use of androgens or the adrenocortical hormones in males might be ex-cellent adjuvants. Initially, Wright avoided using a "potent testicular extract" in homosexuals—following Carl Moore's principle, he thought the use of an androgenic substance would further depress the patient's own testicular function by lowering the secretion of pituitary gonadotropins.[263]

At this stage, there was little serious criticism of Wright's work and the approach received considerable support from other researchers.[264] In 1939, the magisterial American compendium *Sex and Internal Secretions*, in spite of claiming that homosexuality was primarily a psychological and behavioral condition with some possible morphological and physiological foundation, stated that "assays of urine for the male and female sex hormones have shown,

in some instances, a disturbance of the normal balance . . . substitution therapy for the deficient hormone has been reported with varying degrees of success. The procedure is worthy of future investigation."[265] In 1940, Abraham
Myerson and Rudolf Neustadt reported that independently of Wright and
his associates, they had studied "26 true male homosexuals" over two years,
to find that "23 show a strikingly characteristic disproportion between the
male and female hormones."[266] The urine of heterosexuals who had been
compelled into homosexual acts—there was the inevitable allusion to prisons and boarding schools—did not show any such disproportion between
the male and female hormones. Consequently, it seemed evident to Myerson
that urinary hormone assays offered a "direct, measurable, and scientifically
worthwhile approach to the sexual constitution of man."[267] Echoing older
notions of universal sexual intermediacy, Myerson and Neustadt observed
that "an original hermaphroditic nature in man has become lost in the shuffle
of evolution, but there still remains a bisexuality which is the most important
of all the sexual phenomena that can be separated from essential reproduction
itself."[268]

Such faith would soon be undermined. In 1941, a paper in the journal
Psychosomatic Medicine reported the case of a forty-six-year-old African-
American institutionalized for two decades in a mental hospital with "constitutional psychopathic personality without psychosis."[269] Upon admission to
the Worcester State Hospital in Massachusetts, he had been described in the
diagnostic summary as "a negro of passive homosexual type with feminoid
make-up, without evidence of psychosis." He was put on sex hormones provided by pharmaceutical companies such as Schering and Squibb. Treatment
commenced with the synthetic estrogen, stilboestrol. Then, he was implanted
with "a 150-milligram tablet of Testosterone (Schering)." This was followed
by injections of gonadotropin obtained from pregnant mare serum, which
was then substituted by pituitary gonadotropin and accompanied by testosterone propionate injections. Thyroid hormone was given simultaneously
to "enhance the responsivity to sex hormones." This astounding course of
treatment culminated with two kinds of estrogen. Considering the number of hormone preparations used, the physicians must have experienced
considerable chagrin to find that "none of the drugs of the entire series gave
rise to any detectable change of behavior or attitude." "Had the personal
psychodynamics of the patient been primarily dependent upon hormonal
factors it seems highly probable that at least minor shifts of preoccupation
and attitude would have been detectable with changes of medication," they
reflected.[270]

Two years later came another study, this time of eleven "overt homosexuals" between sixteen and thirty-four.[271] Four of the patients had "accepted organotherapy by compulsion"—court order in one case, parental insistence for three—and the remaining seven because "of a desire to live heterosexually." Urinary analysis revealed the pattern "usual in male homosexuality" and treatment was pursued with testosterone propionate, chorionic gonadotropin, and, in some cases, methyl testosterone tablets, for periods varying from three to twenty months.[272] Eight cases did not respond but three reported positive results—amounting largely to the disappearance of homosexual fantasies.[273] Of the eight who remained firmly homosexual, five "complained of an actual intensification of the homosexual drive so that further treatment was withheld or abandoned by them."[274] Since androgens led to "a generalized augmentation of susceptibility to sexual arousal," the authors concluded that it was logical that the administration of androgens to homosexuals resulted in an intensification of homosexual urge rather than an intensification of "normal" masculinity. (They did not, however, show any awareness of Moore and Price's theory that the administration of androgens would ultimately depress testicular functions.) As for treatment with gonadotropins, it was likely that they stimulated the secretion of all sex hormones—the levels of female hormone, therefore, were unlikely to be lowered by gonadotropins. "The organotherapy of homosexuality," the authors, therefore, conceded, "has not reached a stage where one may simply administer androgens and gonadotropins to all cases with confidence of success. The fact is that sometimes instead of helping one gets a worsening of the condition."[275]

This skepticism was soon to be reinforced by others. In his trenchant critique of hormonal theories of homosexuality, zoologist and sex expert Alfred Kinsey (1894–1956) rejected the assumption that homosexuality and heterosexuality were totally separate entities.[276] His own extensive collection of interviews suggested that although the supposedly characteristic physical stigmata *were* present in some homosexuals, innumerable others were "physically as robust and as athletically active as the most 'masculine' of men. There are, in short, intergradations between all of these types, whatever the items by which they are classified."[277] If the category of the homosexual was so uncertain and so fluid, then did it make any sense to delineate the supposed hormonal peculiarities of *the* homosexual?

Some of the upholders of the hormonal explanations ridiculed Kinsey for his nit-picking, but the nay-sayers continued to multiply.[278] Thomas Moore reminded his readers in 1945 that one Dr. Fischer of Germany had transplanted a slice of testicular tissue from "a normal, strongly heterosexual man" onto a eunuchoid masseur in order to enhance the latter's masculine development.[279] The graft had worked, but not quite as expected. The formerly hypogonadal masseur had become sexually active—but actively *homosexual*, and had returned to hospital to have his graft removed.[280] In a major review of "the clinical uses of testosterone in the male," published in 1947, the authors followed up a detailed critical analysis of the work of Wright with a summary of Kinsey's reservations, concluding that while "the *power* of the human sex drive is largely dependent on . . . the amount of circulating androgen (or estrogen in the female)," "the *direction* of the human sex drive seems to be largely dependent upon *psychological* factors, which are conditioned by the early environment and sexual experiences of the individual."[281] The idea of assaying sex hormones in the urine of homosexuals gradually faded away.[282]

It was the new awareness of the distinction between sexual drive and sexual orientation that led to a different approach to the hormonal "treatment" of homosexuality. If raising the androgen levels led to the intensification of desire rather than its transfiguration into masculinity, then, it seemed, the only way to extirpate homosexuality was to kill desire itself by *reducing* the androgen level. Although this could be done by surgical castration (and sometimes was), chemical castration with estrogen was far more acceptable.[283] This treatment was based firmly on the principle established by Moore and Price: the administration of estrogens inhibited the production of gonadotropins by the pituitary, which in turn led to the inhibition of androgen secretion by the testicles.[284] Despite the apparent resort to the old Steinachian concept of "sexual antagonism"—male hormones can be countered only by female hormones—the treatment of male homosexuality with estrogen was actually incompatible with Hirschfeld and Steinach's "female gland in a male body" hypothesis of homosexuality. From *their* perspective, treating a male homosexual with estrogens would have enhanced homosexual desire by interfering with the male hormones and accentuating femininity. Steinach had aimed to change the direction of desire itself by glandular and hormonal treatment; less than fifty years later, medical scientists had conceded defeat on that front, claiming merely the ability to *suppress* desire with hormones.[285]

What, however, of the biology of female homosexuality? Early-twentieth-century researchers had never explored the issue, and even clinical sexologists like Hirschfeld had not delved into the question of lesbianism. It was male homosexuality that was the riddle; female inversion was merely a phenomenon to be noted. Even studies of hormonal levels in lesbians were not undertaken until the 1970s, when some lesbians were reported to have showed "decreased estrone excretion and elevated testosterone excretion."[286] Other studies followed, with results almost as contradictory and confusing as for those of male homosexuals during the same period.[287] Hormonal masculinization seemed an unlikely explanation for female homosexuality since women virilized due to adrenal problems tended to be heterosexual.[288]

Hormonal theories of homosexuality did not merely encourage doctors to diagnose and treat homosexuals with sex hormones. There is some evidence to suggest that hormonal approaches to sexuality may have provided some foundation for a new kind of subjectivity that we might call transsexual, and a justification for modifying the body in the light of that subjectivity. In 1946, Michael Dillon had remarked in his strangely titled book *Self: A Study in Ethics and Endocrinology*, "Surely, where the mind cannot be made to fit the body, the body should be made to fit, approximately, at any rate to the mind."[289] Born a woman, Laura Maud Dillon was convinced that he was a man and in order to bring his body in line with his outlook, he had persuaded a doctor to prescribe testosterone to him and a plastic surgeon to remove Laura Maud's breasts.[290] Dillon then re-registered his sex at birth as male and enrolled as a male medical student at Trinity College, Dublin.[291] The surgeon Sir Harold Gillies eventually constructed a penis and scrotum for him, although Dillon's book was published before that operation. Perhaps the most interesting point about Dillon's case is not his claim that the body could be born with an incorrect sex but rather his conviction that the mind, although supposedly influenced by glandular secretions, necessarily spoke the truth about one's sex. It was not Laura Maud's mind that was incorrect in believing that she was actually male. What was wrong was her female body, which, therefore, had to be changed. As Bernice Hausman has emphasized, this belief animated later transsexual discourses on the individual's right to seek a sex-change to make the body "fit" the "true sex."[292] Endocrine science and plastic surgery helped transsexuals to "correct" their bodies, while their minds were not considered to merit any correction or even serious investigation.

Unisex Hormones?

By 1940, it was universally recognized that each sex produced male as well as female sex hormones.[293] Some authorities continued to harbor doubts about the origins and biological significance of crossed-sex hormones (to use a convenient, if uneuphonious shorthand expression), but that did not prevent clinicians from experimenting with their therapeutic use.[294] The use of "female" hormones in men was not embarked upon very swiftly, but uses were found quite quickly for androgens in women. An authoritative 1947 survey put it:

> For approximately a decade androgens have been utilized as therapeutic agents in women. The too rigid early concept of androgens as male hormones seemed at first to lend merely an academic interest to their application to women; but with the growing awareness that they were capable of profoundly altering the sexual physiology of women their potential value as therapeutic agents began to be appreciated.[295]

In Central Europe, Viennese gynecologist Hans Husslein recalled in 1954 that it had first appeared senseless or even potentially dangerous to treat women with a substance that was "foreign" to them (*artfremd*). Initially, the use of androgens in women was often termed "paradoxical hormone treatment."[296] By the time he was writing, however, it was clear that one had to accept that androgens played distinctive, if still incompletely known, roles in female physiology.[297] It was likely that in human beings, what was sex-specific was not the sex hormone, remarked Husslein in the spirit of Clifford Wright, but the *balance* of "male" and "female" hormones.[298]

Androgens were first found useful in combating the vasomotor and psychological symptoms associated with menopause. Although the estrogens were the preferred treatment for such complaints, they stimulated endometrial proliferation, which could result in severe bleeding.[299] In those cases, androgens were of great value, whether as a supplement to the estrogens or even as the exclusive treatment. Perfectly capable of combating sweats and faints on their own, androgens were also particularly valuable in enhancing the menopausal woman's physical strength and sense of well-being. Postmenopausal osteoporosis was also prevented. In this serious problem, the actions of estrogens and androgens were similar and the two could be used in combination to potentiate one another at doses which would be inadequate for one hormone.[300] There was little real analysis of possible mechanisms of action, but there was virtually universal agreement by the mid-1950s that the "male" hormones were indeed effective in revitalizing the menopausal female, especially when combined with estrogens.[301]

But menopause was not all that androgen treatment was useful in. Just as the removal of the ovaries had been found to stop intractable uterine bleeding (whether purely "functional" or associated with fibromyomas) in the late nineteenth century, androgen treatment was found useful for the same conditions in the mid-twentieth.[302] The male hormones, it was suggested, acted like "a 'medical curettage' achieved by the inhibition of ovarian hormonal activity via the pituitary."[303] Since androgens suppressed menstruation, they were also found useful in endometriosis and in pelvic inflammatory diseases associated with dysmenorrhea and even premenstrual tension.[304] With the increasing use of male hormones in women, it was discovered that "they appreciably enhanced libido."[305] The mechanism, once again, was unknown, but it was suspected that the male hormones increased the sensitivity and vascularization of the external genitals. Their enlarging effect on the clitoris was especially striking, and they were soon being recommended for the treatment (preferably "by the gentle inunction of an androgenic ointment" on the clitoris) of frigidity due to clitoral insensitivity.[306] "Testosterone," wrote an American specialist, "appears to improve libido partly by increasing the vascularity and sensitivity of the clitoris, and partly by its general 'tonic' effects." Estrogens, however, were "of little or no value."[307]

Central European gynecologists agreed heartily with the recommendation to use androgens in frigidity but attributed their aphrodisiac action more to the stimulation of cerebral sexual centers rather than, as in the Anglo-American literature, to stimulation of clitoral growth. Androgens, argued Hans Husslein, enhanced libido in doses that were too small to affect clitoral development.[308] Husslein recommended the general use of androgen supplementation in menopause to ensure that the libido remained healthy. Although androgens would later be used in post-menopausal hormone replacement regimens for this very reason, there was little inclination among mid-twentieth-century Anglophone clinicians to do so.[309] For them, in fact, the rise in libido was a major problem. This particular effect of androgen treatment was every bit as undesirable as some of the other side-effects, such as hairiness, deepening of the voice, or acne.[310] "It should be borne in mind," warned the 1947 review cited previously, "that the effect on libido may not be desirable in all instances, since increased libido without normal channels for satisfaction may become a serious individual problem and lead to psychic trauma."[311] In 1953, endocrine specialists Max Goldzieher and Joseph Goldzieher called it an "undesirable side-effect of testosterone in the female" that could be so troublesome as to compel termination of the treatment.[312]

There was, in short, no real consensus on the subject of endocrine treatment—just plenty of theories, studies, and opinions. At least up to mid-century, the use of sex hormones in gynecological practice remained more or less ad hoc. It involved much more than guesswork, of course, but the precision and level of consensus that had been reached by this time in the biochemical study of sex hormones were not to be matched in the clinic. At the beginning of the twentieth century, the gynecologist was showing the way while the chemists were not even on the scene, but by 1950, gynecologists were no longer extending the frontiers of hormone research. To be sure, they still had unquestioned authority in pronouncing on the clinical utility of hormonal products evolved by biochemists and the pharamaceutical industry, but unlike Halban, Fraenkel, or Blair-Bell, they could no longer claim to be at the leading edge of endocrine science.

Cancer and Sex Hormones

We saw earlier that some early practitioners strongly suspected a link between the sex glands and certain kinds of cancer. Those pioneering investigations did not, however, lead to any sustained research, and it was only in the 1940s that physicians came to accept that "organs such as the breast, uterus, and the prostate gland, which are common sites of cancer, are under endocrine control."[313] It was at least plausible, therefore, that "a causal relationship might exist between the hormones and the genesis of tumors of these organs."[314] Plausible, perhaps, but still not proven to the satisfaction of physiologists. Hormonal treatment in the palliation of various kinds of inoperable cancer—breast, prostate, uterus—was to become almost routine before there was "an exact, irrefutable explanation" of the "action of hormonal administration or deprivation on cancer."[315] The place of hormones in the history of cancer research and therapy is a crucial one, but its history is too vast and complicated to be covered here even in outline. Let us, therefore, focus only on the development of endocrine theories and treatments of breast cancer. That will not only round out the earlier discussion of the work of George Beatson but also help illustrate how the elaboration of new theories of hormonal action and the availability of pure hormones transformed concepts of the sexual body, its pathologies, and their chemical correction.

There was much consternation when it was reported in the 1930s that high doses of estrone over a prolonged period could produce malignant breast tumors in hitherto healthy mice. The new biochemical perspective on hormones offered the "tempting argument" that the sex hormones and the carcinogens had similar or even identical chemical structures—that sex hormones were, in fact, carcinogens.[316] Many investigators insisted, however, that similarities in chemical structure did not entail any physiological similarity. As the well-known biochemist E. C. Dodds pointed out, there was "much closer similarity between male and female sex hormones chemically than there is between estrone and, let us say, dibenzanthracene [a carcinogen]. Yet the clinician who is worried about the possibility of a carcinogenic action of estrone is not at all worried about the possibilities of producing masculinization by giving a woman estrone."[317] But whether the female sex hormones were directly carcinogenic or not, it was quite clear that they did have a stimulatory effect on breast malignancies.[318] In 1947, the American physician Ira Nathanson observed that it was possible as well as probable "that they are indirectly responsible, inasmuch as they may produce precancerous changes or may provide a suitable substrate so that another agent may act."[319] Although this, of course, was before the age of routine hormone replacement therapy, many women were already "receiving estrogens over long periods for various endocrine disturbances," and Nathanson urged that estrogenic hormones should be avoided in patients with "chronic mastitis, cancer, or any form of neoplasm either before or after surgical or radiation treatment."[320]

Hormones might cause breast cancer—but could they also cure it? "The problem," Dodds complained, "bristles with experimental difficulties, since there are so many factors influencing the rate of growth of tumors."[321] There were also many different rationales for treatment. Some physicians argued that because many of the carcinogens produced some cancers and halted the growth of others, the estrogens might also have palliative as well as stimulatory effects in cancer.[322] Other, completely different justifications for the use of the estrogens in breast cancer held that large doses of these hormones would inhibit the secretion of pituitary gonadotropins, and once the gonadotropins were suppressed, the secretion of ovarian hormones would taper off.[323] Dodds remained unconvinced about the worth of such an approach, but by 1950 Edith Paterson, radiobiologist at the Christie Hospital in Manchester, could report that the estrogenic hormones—especially the synthetic products like stilboestrol—had been used with impressive success in the treatment of breast cancer: about 25 percent of patients had showed improvements, usually within the first month of therapy.[324] Three years earlier, the American Medical Association's Council on Pharmacy and Chemistry commented that estrogen treatment worked best in the "soft tissue manifestations of the disease in older women," which agreed with Paterson's own findings.[325]

The center of gravity of hormonal treatment of breast cancer, however, lay elsewhere during this period. The consensus was for androgenic or anti-estrogenic approaches. Beatson and Boyd's use of oophorectomy in advanced, inoperable cases of breast cancer was remembered with respect by some, and in men suffering from breast cancer, castration was supposed to lead to similar improvements.[326] The therapeutic use of androgens in breast cancer, therefore, may seem to have been virtually inevitable. Actually, however, it happened rather less predictably. In 1939, Alfred Loeser of University College London prescribed androgenic treatment for two women with recurrent breast cancer after mastectomy. He had not aimed to treat the cancer at all but merely to stop the excessive menstrual bleeding in his patients. He noticed purely by chance that the tumors had regressed after androgen treatment.[327] After some focused research, Loeser argued that certain strains of mice and women from certain families were unusually vulnerable to breast cancer. "Apart from inherited susceptibility, one of the carcinogenic factors which may transform a normal cell into a cancerous one is said to be a surplus of oestrogenic substances in the body," he observed. "It appeared, therefore, worth while to try to counteract the female hormone by male hormone." Using implanted pellets of testosterone ("in order to maintain a permanent source of counteracting hormone in the tissues") in a strain of mice that was particularly predisposed to breast cancer, Loeser found that the implants protected a significant percentage from cancer. The mice which had received implants early in their lives did best. Once the cancer had developed, however, the hormone did not seem to exert any protective function.[328]

Loeser then tried the pellets on six women sufferers, all of whom came from families with histories of breast cancer and had already been treated with surgery and radiotherapy. "Male hormone was given to these patients to see whether when given over a long period it would inhibit severe metastasis and influence recurrences." The three patients who did not already show metastatic deposits remained free of them for three years after the hormone treatment and were still under observation at the time of publication. The general condition of the patients who had already had metastatic deposits when the treatment began also improved after the implantation "but the cancer itself remained unaltered and their decline could not be averted."[329] Even though his study was a small one, Loeser advocated the implantation of testosterone "in the operation site when the breast is removed for carcinoma, and implantation should be repeated when signs of masculinisation disappear." He also tried using progesterone pellets because that hormone could be "regarded as antagonistic to the follicular hormone," but it seemed to be of no use against cancer. All the women treated with testosterone developed overt signs of masculinization—"enlargement of clitoris, hoarse voice, increased sexual urge, increased growth of hair, gain in weight"—but no other side effect was reported.[330]

Rather than being a mere side effect, the masculinization, Loeser suspected, was fundamental to therapeutic success. Although the testosterone, he was sure, exerted a direct antagonistic action on the estrogenic hormones (the latter being the hypothetical cause of the cancer), "it is my impression that it was not the antagonistic action alone but the alteration in the patient's constitution—the masculinisation—which had an effect. The crescendo of

masculinisation coincided with improvement in their condition; the dimin-
uendo with progress of the disease."[331] Although there was no immediate,
widespread acceptance of androgen treatment in cancer, androgens had been
used as "*prophylactic*" agents to "prevent recurrence in patients who have had
a radical mastectomy," and one investigator from Brazil claimed a massive
rise in survival rates—close to 100 percent—of women treated with prophylactic testosterone propionate after radical mastectomy for breast cancer.[332]
Not all reports were very enthusiastic, however, and one study of seven cases
reported glumly that "six of the seven patients treated have died of the disease
within 16 months of the institution of treatment."[333]

In 1952, one of the most experienced American investigators of the
endocrine dimensions of cancer wrote: "Ovariectomy...is preferable if a
rapid clinical effect is desirable....When this was not feasible, we have used
testosterone in conjunction with radiation to hasten suppression of ovarian
function....Testosterone therapy may be used solely to suppress ovarian
function, but there is no proof that it is as efficacious as castration by surgery
or radiation."[334] In Britain, too, endocrine treatment of breast cancer was not
recommended for patients who could still benefit from surgery or radiotherapy. "This statement," asserted Edith Paterson,

> is based on the evidence that no patient has been "cured" by oestrogens or andro-
> gens, even in the limited sense that five years after treatment, disease cannot be
> detected....In cases which are beyond surgical treatment or radiotherapy, even
> in a palliative sense, the sex hormones play a role as palliative agents in about
> 25 per cent. of late cases, and are also of considerable theoretical interest in the
> study of malignant disease of the breast.[335]

Within this restricted field of palliative therapy, however, the androgens had
the pride of place. Younger, premenopausal patients responded better to androgens, but even those whose tumors were unresponsive gained something
from the therapy, especially a heightened sense of well-being.[336] New York
surgeon Frank Adair exclaimed that "in the occasional case with painful bone
metastasis this agent was nothing less than miraculous for the relief of pain
and for giving the patient needed rest without narcosis."[337]

Although no consensus was to emerge quickly, it is instructive to look at
some of the possible mechanisms of sex-hormone action in breast cancer
that were discussed in the literature of the 1940s and 1950s. The American
Medical Association's influential Council on Pharmacy and Chemistry
declared in 1947 that it was "not now prepared to discuss the mode of action" of hormones in inhibiting the growth of malignant tumors.[338] For some

researchers such as Loeser, however, the palliative action of androgens on breast cancer was due to generalized masculinization. The so-called male hormone was not only of use in boosting masculinity but in countering femininity. "The best responses of breast cancer," observed Ira Nathanson, "are intimately associated with the androgenic activity of a compound, irrespective of purely metabolic capacities."[339] It was not simply a matter of countering femininity. In the infrequent cases of male breast cancer, surgical castration was supposed to give excellent results.[340] At this point, then, one important approach to the endocrine treatment of breast cancer sought merely to eliminate the actions of what Steinach would have called the homologous sex hormone. One infers the same from the reintroduction of bilateral oophorectomy or x-ray castration of premenopausal women with inoperable breast cancer.[341] Parts of the fundamental concept of the antagonism of the male and female sex hormones, in short, were alive and well in some areas of clinical practice, even after the demolition of that concept in the biochemical laboratory.[342]

It was not just cancer per se that fascinated investigators but also what the endocrine treatment of cancer was revealing about that enigmatic field of chemical forces that generated and sustained the sexual body. As two American surgeons admitted, the confusing results of androgenic and estrogenic treatment in breast cancer led one

> to the disconcerting conclusion that by upsetting a patient's sex-hormone homeostasis,—whether by androgens, estrogens or castration,—one can profoundly alter a primary breast cancer or its more remote metastases. . . . By enhancing the subtle influences of hormones that are normally present within the body, cancer may yet be palliated, if, indeed, not conquered.[343]

The evolving endocrine approaches to cancer showed how numerous were the ways in which one could modify the endocrine framework of the sexual body. Breast cancer patients who did not respond to androgens, for instance, rarely responded better to the estrogens, and those refractory to the estrogens did not respond better to the androgens. Patients who responded well to one, however, usually also responded to the other.[344]

The comfortable, gender-polarized certainties of the 1920s were crumbling but only in some contexts. Even Hans Husslein, who explicitly rejected the older antagonistic paradigms of hormone function, slipped into traditional mode when discussing the actual clinical use of sex hormones. In explaining the effects of sex hormones in breast and prostate cancer, Husslein averred that the single most important element here was the inhibition by the crossed-sex (*andersgeschlechtliche*) hormone of the proliferation induced by the intrinsic (*körpereigenen*) hormone.[345] When "crossed-sex" treatment was advised—such as the use of androgens in females—it was usually because of the assumed antagonistic action of the opposite hormone and not necessarily because the androgens were expected to perform a normal, "physiological" function in the female organism. Some clinical investigators of the 1950s, to be sure, had entered a new, unpredictable universe where the hormonal basis of sex was as uncertain as in the laboratories of biochemical researchers, but the change was so patchy that one can talk, at best, only of a partial displacement of notions of hormonal antagonism. Much of the new therapeutic research, moreover, was still very empirical, guided more by clinical results and more-or-less ad hoc hypotheses rather than by the principles established by biochemists and physiologists.

The Hormone Therapy of the Future? Pituitary Gonadotropins in the Clinic

> Until now, hormone therapy was substitution therapy. Hormones not produced by the body—or produced insufficiently—were provided to the end organs. The identification of hormonotropic secretions of the pituitary creates the possibility of stimulating the hormone-producing endocrine glands themselves. . . . Hormonal substitution therapy had aimed at the sites of use; the future hormonotropic stimulation therapy, if I may call it that, will act on the sites of production.
> —BERNHARD ZONDEK, 1935[346]

> So far as the human subject was concerned, the newly discovered gonadotrophins proved to be almost entirely useless, a situation which is only now being corrected by the use of human pituitary gonadotrophins
> —ALAN PARKES, 1965[347]

Musing on the options available for treatment of pituitary disorders in 1912, Harvey Cushing had admitted that surgery was the mainstay. "It is conceivable," he had hoped nonetheless,

> that the day is not far distant when our present methods of dealing with hypophyseal enlargements, with scalpel, rongeur and curette—new as these measures actually are and brilliant as the results may often be—will seem utterly crude and antiquated, for it is quite probable that surgery will, in the end, come to play a less, rather than a more important, role in ductless gland maladies. This Utopia,

however, will be reached only when a sufficient understanding of the underlying aetiological agencies enables us to make more precocious diagnoses.[348]

In 1935, Bernhard Zondek felt that Cushing's utopia was almost within reach. Looking back on recent research on the pituitary hormones, it was clear to him that the pituitary was the motor of all endocrine function, rather than just *sexual* function. It was not just another endocrine gland but the master of them all. This fact was not simply of physiological importance; it had the greatest clinical potential. All hitherto existing endocrine treatments had aimed to replace a hormone or hormones. The results of such substitution therapy had never been spectacular, and many clinicians suspected that the natural products of the glands were superior to crude extracts or even to the purer, standardized products that were beginning to be introduced. Endocrine therapeutics, Zondek declared, would be revolutionized once clinicians could reactivate underperforming endocrine glands by enhancing the secretion of the appropriate trophic hormone by the pituitary. Even the purest, most rigorously standardized hormone could not hope to compete with the body's own products. This would be the hormone therapy of the future: the anterior pituitary held the keys to a new therapeutic kingdom.[349]

Zondek's hopes were widely shared, and there had already been many attempts to use pituitary hormones in the clinic.[350] Although Zondek's claim—supported by many researchers—that there were two distinct Prolans governing follicular and luteal growth faced significant challenge from some, the urine of pregnant women had been used to extract "anterior pituitary-like" hormones and these were already available commercially in roughly standardized forms.[351] Their effects, however, were uncertain, and the action on the ovary "is certainly not what would be expected from the pituitary hormones themselves."[352] Still, even these less than satisfactory products had produced better results than any available therapy in troublesome uterine bleeding.[353] In 1932, gynecologist Jean Paul Pratt observed in *Sex and Internal Secretions* that the anterior pituitary gonadotropic hormone—he was not entirely confident that the follicle-stimulating and luteinizing hormones had been definitively separated as yet—was of use in treating cases of severe uterine bleeding (especially a condition called metropathia hemorrhagica) and there were some reports of success in amenorrhea.[354] In 1935, Emil Novak referred to an "extensive and somewhat bewildering literature" that had developed around the clinical use of pituitary hormones, even though only two of the six or seven anterior pituitary hormones identified up to that point were available for clinical use—these being the growth hormone and "the gonadotropic hormone or hormones."[355]

The point of using gonadotropin was not just its efficacy, impressive as it had been so far, but the fact that an active corpus luteum hormone was not yet widely available. Where one needed a luteal effect—as in habitual abortions and, according to some researchers, dysmenorrhea—the only hope was to encourage the production of the luteal hormone by gonadotropic stimulation of the ovary, even though results remained entirely unpredictable.[356] Such experimental treatment, however, was frowned upon as too risky by many. In 1932, the *Lancet* pleaded for caution with gonadotropins:

> We are dealing here with substances whose potency is astonishing and whose biological action is imperfectly understood. We know to what chaos the ovaries of the lower animals are reduced by their administration even for a brief period: maturing follicles, blood follicles, atretic follicles, luteinising follicles, atretic corpora lutea and normal corpora lutea are produced in magnificent profusion with a corresponding reduction in the number of primary follicles available for subsequent normal maturation. . . . Until the action of the ovary-stimulating hormones is better understood and can be accurately controlled would it not be wiser to limit ovarian therapy to the use of the less volacanic ovarian hormones, alone or in combination?[357]

A year later, an American researcher, surveying many reports of gonadotropin treatment of "hypoovarian conditions" as manifested by amenorrhea, irregular or delayed menstruation, and hypomenorrhea, concluded that final judgment as to the true value of gonadotropic treatment should be delayed since the picture was still very confused. The reports he had reviewed were inconsistent, "some authors claiming a large percentage of cures, others almost total failures, and practically no one giving complete records of careful, well-controlled studies."[358]

Less predictably, gonadotropin treatment was also found useful in encouraging the descent of undescended testicles and, in an even more bizarre 1920s-like twist, in curing certain types of baldness and, less dependably, psoriasis and menopausal eczema.[359] The utility of the gonadotropins was not necessarily limited to therapy. As we have seen, the Aschheim-Zondek pregnancy reaction and its variants had enabled the reliable diagnosis of pregnancy. Even though physiologists continued to debate the actual origin of the gonadotropins in the urine of pregnant women, that uncertainty had "no bearing on the reliability of the pregnancy test."[360]

The pituitary gonadotropins do not, however, seem to have inaugurated a renewed quest for hormonal rejuvenation, although the conditions for such interest were not unpropitious. Max Reiss, an expert on endocrine influences on mental illness, wrote in 1940 that monkey-gland implants or the Steinach operation were now only of historical importance. "To-day we have at our disposal the gonadotrophic hormone of the pituitary anterior lobe which is often successful in stimulating testicular function. The results so far seen by the writer after treatment with this hormone are very promising."[361] Nevertheless, there was in fact very little interest in using pituitary extracts for rejuvenation. Most of the debilities and deficiencies targeted by Voronoff and Steinach were now being treated by the gonadal hormones. Zondek's vision of moving beyond hormone substitution into the brave new world of gonadotropic stimulation did not have much impact on those who specialized on the climacteric.

Eventually, the gonadotropins proved to be useful in the treatment of infertility, but that did not happen at all smoothly. In 1933, two Detroit investigators published a report of a thirty-three-year-old man who had been rendered infertile by an attack of mumps at the age of twenty-seven. Semen tests revealed complete aspermia, and the testicles felt atrophic on clinical examination. After treatment with the gonadotropin-like substance from pregnancy urine, "occasional nonmotile spermatozoa were found" in the semen and the testicles felt larger and firmer. There was also "marked subjective improvement, including an increase in capacity for both mental and physical work." After many weeks of interrupted treatment, the patient's semen finally began to swarm with active spermatozoa.[362] Not many general conclusions could be drawn, of course, from one case, and in two of his own cases Emil Novak had failed to bring about spermatogenesis by similar means.[363] Still, gonadotropin treatment became reasonably common in infertility, especially in female cases. The goal was to encourage luteinization of the ovaries and thereby enhance the level of the luteal hormone, which was expected to encourage the "nidation" of the fertilized ovum.[364]

The problem was to obtain pure, potent and standardized versions of the gonadotropins. Since these hormones are proteins, not steroids, they could not be chemically synthesized like the estrogens or the androgens. As Naomi Pfeffer has pointed out, "it was not until the arrival of biotechnology that synthetic gonadotrophins became available in the late 1980s. Until then, all gonadotrophins used in clinical practice were isolated from naturally occurring raw materials."[365] Gonadotropins extracted from animal pituitaries were used in human treatment, but by the end of the 1930s it was clear that these were species-specific hormones with no action on humans. And "by 1950, it had been established that, in response to injections of gonadotrophins of non-human animal origin, the human body produces antigonadotrophins ... [which] antagonize what pituitary gonadotrophins the patient herself produces and inactivate her own pituitary hormones, making her condition worse rather than better."[366] Yet another grand therapeutic dream, it seemed, had failed, and it would not return to life within the time-span of our story.

Compared with the bravura experimentation and visionary aspirations of the 1920s, the endocrine therapeutics of the mid-twentieth century can seem sober and almost boring. Although the story of mid-twentieth-century clinical endocrinology is not one of continuity with the 1920s, it is not characterized by any total break with the past either. The picture, instead, is one of technical sophistication and conceptual incoherence. All clinicians had moved far beyond the 1920s as far as the purity and potency of their therapeutic agents were concerned, but some old conceptual uncertainties had not really been resolved definitively. For example, not every clinician of the 1940s held true to the notion of sexual antagonism—but quite a few did. Some physicians used male hormones in women with more alacrity than female hormones in men. The new sex hormone products were easier to use and purer, but they were far from dramatic in their efficacy. The overall impression one gets from clinical reports on the sex hormones is one of great expectations and unpredictable returns. Speculations and old habits of thought were being jostled and only occasionally displaced by the new. The kind of Kuhnian paradigm shift that Nelly Oudshoorn has demonstrated among biochemists did not occur in the clinic until long after 1950.

The Gonads, the Brain, and the Neurohumoral Body

In the beginning was the nerve. The sex glands produced ova and sperm—but that was all they produced. All their other actions on the organism—which, for the ovary, were apparently numerous—were mediated by nervous connections. This complex system became simpler in the late nineteenth century, when the gonads came to be regarded as secretors of potent fluids with profound effects. The gonads were quickly joined by other organs, and the new discourse of "glands" led to thoroughgoing redefinitions of the body and its functions. The lore on the sex glands piled up with astonishing speed but remarkably little coherence. Inconsistencies, of course, were only to be expected. Glandular science was not a finished piece of specialized knowledge dropping into the laps of scientists but pieces of information acquired unpredictably and used in different ways in different contexts to make sense of different issues. All contemporaries—scientists, laypeople, popularizers, patients—knew that innumerable questions about glandular function remained to be resolved, but most of them shared the conviction that this new and as yet confusing science would resolve longstanding medical problems and eventually improve, enhance, and transform the human condition.

The nerves were scarcely mentioned in those utopian speculations, and the glands seemed virtually omnipotent. The nerves carried messages, to be sure, but they were no longer seen as much more important than one's postman. The really crucial business of regulating the organism had been made over to the endocrine glands and their secretions. Writing in 1962, Ernst and Berta Scharrer commented that over the previous, heroic decades of endocrine science,

> sight has sometimes been lost of the many instances of close interrelation of ner-
> vous and endocrine functions. It is only in recent years that the nervous control
> of organs of internal secretion has been studied. It has also become increasingly
> evident that the latter, through their hormones, act on the nervous system. In-
> deed, there are not many functions which are under neural or hormonal control
> exclusively; the great majority are under the overlapping authority of both the
> endocrine and the nervous systems.[1]

Of course, Edward Schäfer had said almost exactly the same at the dawn of what became endocrinology—his New Physiology was not *just* glandular physiology but an integrated physiology of the endocrine glands and the nervous system. Few, however, had paid much heed to that admonition. "Brown-Séquard, a contemporary of Berthold, never seems to have become interested in possible connections between the nervous system and the endocrine glands, although he was one of the pioneers in the investigation of both," commented the Scharrers.[2] Walter Langdon-Brown, who himself was more of a neurohumoral integrationist than most of his contemporaries, remarked in 1935 that "for a time after Starling developed his conception of the chemical control of the body by hormones, the pendulum swung so far in that direction that it seemed as if the nervous system would be deprived of its pride of place."[3]

Although hardly anybody cared to investigate them in detail, nobody dismissed the importance of the links between the central nervous system and the endocrines. "It is often thought that neuro-endocrinology is a recent innovation," observed Alan Parkes, "but, as it concerns reproduction, many of the basic facts have been realised for a long time."[4] Even in the gland-obsessed 1920s, many links were hypothesized between the hypothalamic nuclei and the pituitary. In the mid-1920s, Herbert Evans, then revolutionizing the study of the pituitary, reminded his readers that many investigators attributed the supposed consequences of pituitary malfunction to neurological lesions in the neighboring regions of the brain. Although recent experimental research had provided unequivocal evidence of the importance of pituitary secretions, Evans acknowledged that it was impossible "to overlook the newer conceptions of the adjacent area of the brain," or, to put it briefly, the hypothalamus.[5] Some of these concepts, he mused, could even undermine the autonomy of his own field: "Will," he asked, "the new nihilism effectually upset all notions of [the pituitary's] function?"[6]

So far, the great "pituitary-hypothalamus" question was framed largely in adversarial terms. Until the 1950s, neither neurologists nor endocrinologists regarded their specialities as two faces of a single intellectual entity. Evolution, Louis Berman had written at the commencement of the golden years of glandular research, had created two separate systems for communication within the body. One was the nervous system—younger in the phylogenetic sense, powered by electricity, and specializing on quick transmission and action. The other was the far older endocrine system, which sent slower, chemical messengers around the body.[7] It would take decades before the two would be brought together. Speaking in 1965, Alan Parkes lauded current efforts to combine the neural perspective of the nineteenth century with the glandular paradigm of the early twentieth as "a happy marriage."[8] The full story of that "happy marriage," which was far from consummated at the time of Parkes's speech, falls outside the limits of this study, but it would be worthwhile to outline some of the issues involved.

The Hypothalamus, the Pituitary and the Gonads: Feedback upon Feedback

The first and most complex issue was undoubtedly the question of feedback. In the earliest days of glandular research, the gonads were largely conceptualized as self-sufficient generators of masculine and feminine principles. Very soon, that simplistic scheme was challenged by more complicated visions of a sexual body that was constructed and maintained by (and always vulnerable to malfunction in) a whole network of glands. To hypothesize the glandular system as an "interlocking directorate" was, however, easier than to explicate the mechanism of their interaction and reciprocal influence. It was only with the emergence of concepts of feedback between the endocrine glands—in particular the pituitary and the gonads—that the hypotheses of polyglandular function gained intellectual muscle.

Even if we restrict ourselves to the feedback loops between the gonads, the other endocrine glands, and the central nervous system, we are faced with a complex set of episodes. One of the most influential models of the interrelationship of the endocrine glands was the Moore-Price hypothesis of feedback between the gonads and the pituitary. This was a major leap in conceptualization of the interrelationship of the endocrine glands and provided an elegant explanation of the endocrine function of the sex glands.

The anterior pituitary secreted gonadotropic substances that stimulated the endocrine cells of the gonads to pour forth their hormones. If there was too much gonadal hormone in circulation, then the pituitary would produce less gonadotropin, thus inhibiting the endocrine functions of the gonads. It was essentially what later researchers would call a negative feedback loop: *more* gonad hormones led to *less* pituitary gonadotropin. This feedback model, as we know, soon became far more complicated, and the emergence of the concept of a positive feedback, where *more* sex hormones leads to *more* gonadotropins, revealed that Moore and Price had identified but one mode of communication between the pituitary and the gonads.

As the links of the pituitary with the gonads—and then with the other endocrine glands—continued to be clarified, researchers also began to address the connection of the endocrine glands with the brain. Although a cogent, coherent model of neuroendocrine relations lay far in the future, anecdotal evidence of links between those two sectors had long been scattered in the medical literature.[9] Some suggestive evidence, for instance, came from the psychiatric domain. For a time, the Scharrers recorded, it had become "fashionable to treat the mentally ill with irradiation and surgical removal of endocrine organs or administration of hormones, although the causal relationships between psychiatric disorders and endocrine function were but little understood."[10] Despite this early enthusiasm, endocrine approaches to psychiatry did not prove very successful.[11] Neurological research, too, had contributed suggestive facts. Brain tumors had been found, for example, to be associated with atrophy of the gonads or amenorrhea.[12] In 1901, Viennese physician Alfred Fröhlich had published a highly influential report on what came to be known as the adiposogenital syndrome, the elements of which, we recall, were explained by Harvey Cushing as manifestations of pituitary tumors.

Other investigators, especially in Central Europe, disagreed: the simple mechanical pressure exerted by a tumor on a healthy pituitary could not, they argued, bring about the striking obesity and genital atrophy of Fröhlich's syndrome. Even though the condition might *seem* to be due to pituitary tumors or pituitary compression (and, therefore, to endocrine causes), the genital abnormalities, they argued, were actually caused by the damage of the brain tissue by the tumor. The adiposogenital syndrome, it was shown by two Viennese investigators in 1905, could result from brain tumors having no connection with the pituitary.[13] In 1912, Bernhard Aschner found that damage to the hypothalamic region produced genital atrophy in adult animals that was, in the long run, more severe than the genital atrophy following the removal of the pituitary. Aschner was so convinced of the neural governance of the genital sphere that in 1918 he even referred to a specific "genital center" (*Genitalzentrum*) in the hypothalamus.[14]

Neurosurgeon Cushing, for all his emphasis on the secretions of the pituitary, was eloquent on the importance of the central nervous system. Referring to the diencephalon, the region of the brain most closely linked with the pituitary, Cushing sang: "Here in this well concealed spot, almost to be covered by a thumb-nail, lies the very mainspring of primitive existence. . . . It proves to have direct connections with the first of the organs of internal secretion to become recognisably differentiated, and on which the very perpetuation of the species depends."[15] Walter Langdon-Brown, the physician who had famously called the pituitary the "leader of the endocrine orchestra," did not dispute the importance of the nervous system.[16]

For specialists in the study of internal secretion, however, the most important question in the debate concerned the governance of the pituitary itself. It was entirely possible that the pituitary was governed by the brain, but this was, at best, a hypothesis supported by some intriguing evidence. In his well-known textbook of physiology, Samson Wright had commented in 1936 that despite the image of the pituitary as the autocrat of the endocrine realm, it was a simple physiological axiom that "all the organs of the body work together in a co-ordinated and harmonious manner to maintain the constancy of the internal environment." The functions of the pituitary, therefore, were unlikely to be fully autonomous, but virtually nothing was known about its connections with organs other than the endocrine glands. "If we regard the anterior pituitary as the centre of the hormonal reflex arc," explained Wright, then we can say that we know a great deal about the efferent side of the arc, i.e., what the pituitary does to other organs, but comparatively little about the afferent side of the arc, i.e., the control of the pituitary itself."[17] So little was known about the governance of the pituitary that one could, if one wished, continue to assert the autonomy of the pituitary. Langdon-Brown, for all his stress on the importance of the nervous system, emphasized that it was the pituitary which was still the undisputed leader of the *endocrine* orchestra; the hypothalamus was connected to the endocrine sector but did not govern it.[18] Roughly around the same period, it was also becoming clear that some endocrine functions were influenced by the sympathetic nervous system, although, as Cushing confessed, it was "difficult to tell whether the sympathetic nervous system is the performer that plays on the ductless glands . . . or whether the secretion of the ductless glands twangs the strings of the nervous system. . . . The subject has become almost too complicated for ordinary understanding."[19]

We know now that the pituitary hormones are controlled by the hypothalamus, but the mechanism by which different hypothalamic releasing factors (or releasing hormones) stimulate the pituitary to release its respective trophic hormones was elucidated only very slowly.[20] One important early pioneer in research on the endocrine functions of the hypothalamus was the British physiologist Geoffrey Wingfield Harris (1913–71).[21] Harris began his research on the hormones of the *posterior* lobe of the pituitary at Cambridge toward the end of the Second World War. At that time, it was not even known that these hormones (vasopressin and oxytocin) were not synthesized in the pituitary itself.[22] It was gradually established that the posterior pituitary hormones were not products of glandular secretion at all but neurosecretory substances produced by nerve cells situated in the supraoptic and paraventricular nuclei of the hypothalamus, which "pass down the nerve fibres, through the infundibulum to the posterior pituitary where they 'distil' into the blood as the octapeptide hormones oxytocin and vasopressin."[23]

Naturally, then, investigators began to wonder whether the better-known hormones of the anterior pituitary (such as the gonadotropins) were also of neural provenance.[24] It was not a desultory question. The influence of the central nervous system on reproductive and sexual physiology was by now indubitable. The great reproductive biologist Francis Hugh Adam Marshall had shown (with E. B. Verney) in the mid-1930s that, in female rabbits, diffuse electrical stimulation of the head could lead to ovulation. Harris, in one of his earliest pieces of research, obtained similar results in rats; then, on Marshall's suggestion, he tried to apply "precisely localized electrical stimuli to different regions of the hypothalamus of oestrous female rabbits." This, too, led to ovulation, which, of course, was governed by the gonadotropic hormones of the anterior pituitary. It was clear, therefore, that stimulating certain regions of the hypothalamus stimulated the release of pituitary gonadotropins.[25] Simultaneously, Harris and others also demonstrated the ways in which the hypothalamus triggered the secretion of other hormones of the pituitary such as Adrenocorticotropic Hormone (ACTH), Thyroid Stimulating Hormone (TSH), and Growth Hormone.[26]

By the mid-1950s, it "became possible to plot a tentative map of localization of the hypothalamic areas influential, in some way, in regulating the secretion of anterior pituitary hormones." But how exactly was the hypothalamic influence on the pituitary mediated? A direct nervous stimulation was a possibility, but in spite of much imaginative experimentation, this could not be demonstrated. If not nervous stimulation, then the only possibility was humoral stimulation. In the 1930s, this was suggested by many (including Harris himself in his early papers cited above), but nobody could identify a vascular pathway that could carry the humoral stimulants from the hypothalamus to the pituitary. "At this time," Harris recalled decades later, "there was little generally acceptable information about a small localized vascular system present in the pituitary stalk called the hypophysial portal system." Once the improbability of nervous links between the hypothalamus and the anterior pituitary had become plain, researchers turned again to the all-but-forgotten network of blood vessels between the two. If the stalk of the pituitary—through which these blood vessels passed—were cut, then the experimental animals did not ovulate and their reproductive organs underwent atrophy. Another convincing piece of evidence was that a pituitary transplanted far from its actual anatomical site underwent a sharp diminution in function. "These findings stood in marked contrast to those observed on other transplanted endocrine glands, such as the testis, ovary, thyroid and adrenal cortex." The pituitary, then, was probably governed by blood-borne humoral substances produced in the hypothalamus, but these substances were not present in the general systemic circulation of the organism. They were to be found only in the portal system of blood vessels connecting the hypothalamus with the pituitary.[27]

Initially, scientists found it hard to conceive of hypothalamic nuclei secreting humoral substances directly into blood capillaries, and light microscopy could not reveal such minute links between the nerves and the capillaries.[28] So, researchers toyed with a complicated idea. The median eminence of the hypothalamus, they surmised, filtered out those hormones from the blood which were produced by the endocrine glands controlled by the pituitary. This filtered blood then reached the pituitary with "varying concentrations of these hormones and the secretion of the different [pituitary] trophic hormones varied accordingly."[29] The hypothalamic nuclei, in other words, governed the pituitary simply by removing certain substances from the blood rather than by secreting any special agent or agents into it. This was only one theory; there were many other and even more complicated possibilities. physician Douglas Hubble summarized in 1961:

> Let us assume that high levels of circulating oestrogen depress the stimulating centre, or excite the inhibitory centre in the hypothalamus. What then passes down the portal vessels to the anterior pituitary to depress the release of F.S.H.? Is it a higher concentration of oestrogen than is already reaching the anterior hypophysis through the systemic arteries? Is the humoral agent an oestrogen, with some changed protein binding? Is there some other hormone secreted by the hypothalamus, specific for one trophic hormone, or common for several ... which lies ready in the median eminence [of the hypothalamus] to play such a part? ... It

would have been pleasant to produce a less complex and better-defined account of the hypothalamic control of the anterior pituitary hormones. . . . When the final picture emerges, although clarification will have been achieved, simplification is hardly possible.[30]

Actually, however, the solution turned out to be quite simple and even elegant.

From the mid-1950s, investigators had been trying to make extracts of the hypothalamic tissue and injecting those extracts into test animals. The procedure would have been familiar to Brown-Séquard, but hypothalamic extracts were not as easy to obtain as the essence of dog testicles. "One major difficulty," remarked Harris, "has been the small amount of activity present in any one hypothalamus or median eminence. This meant extracting large numbers of brain 'fragments.' . . . Since it is necessary to start chemical extraction from some 50,000–500,000 hypothalami, in order to obtain a few milligrams of purified extract, the initial financial outlay is great."[31] The early extracts were anything but pure or potent, but an extract capable of stimulating the secretion of the adrenocorticotropic hormone was elaborated relatively swiftly.

No hypothalamic extract, however, seemed able to bring about the secretion of the gonadotropic hormones by the pituitary, and it would be 1960 before Harris's team and another group in the United States managed to produce a hypothalamic extract that could induce the pituitary to secrete the Luteinizing Hormone. Harris and his team also established that there were at least two centers in the hypothalamic region stimulating and depressing the secretion of the two gonadotropic hormones: Follicle Stimulating Hormone (FSH) and Luteinizing Hormone (LH).[32] It took almost another decade before scientists realized that the Luteinizing Hormone was released from the pituitary in pulses, reflecting the pulsatile secretion of the Luteinizing Hormone Releasing Hormone (LHRH) from the hypothalamus.[33]

Many new riddles were encountered and solved over subsequent decades, but in the course of that research it became clear that even the hypothalamus was unlikely to be the supreme controller of the endocrine sexual economy. Hypothalamic neurosecretory cells might reign over the pituitary, but they themselves were under the sway of the higher cerebral centers. Differences in levels of neurotransmitters such as dopamine could modulate the secretion of the hypothalamic releasing factors, thereby influencing the secretion of anterior pituitary hormones and, eventually, the secretion of the other endocrine glands. Recently, even the symptoms of menopause have come to be seen in a complex neurohumoral context. The "age-related changes in the central nervous system and the hypothalamo-pituitary unit initiate the menopausal transition," observed a research team in 1997. The aging process involved in menopause and andropause have all been linked, hypothetically, to "a central pacemaker in the hypothalamus or higher brain areas." The ovary as well as the brain are now being seen as "key pacemakers in menopause." Aging, in short, is more than a matter of the aging of a certain endocrine gland or certain glands: the brain may be the key to aging and, indeed, to any possibility of its reversal.[34]

The bottom line, as Medvei has put it recently, is that "the hypothalamus receives information from the general environment and other parts of the brain; it then translates this neural information into the language of the chemical messenger-interpreters (neurohormones), which act as release factors" and which, in turn, lead to the purely endocrine changes.[35] The endocrine glands, then, are not only connected with one another but plugged into the very core of the human organism. The genesis, maintenance and involution of the sexual body and sexuality are not simply hormonal phenomena, as Steinach's generation had imagined, but may well involve the highest functions known to man. Our historical journey, which commenced with the links between the gonads and the nervous system, has now taken us back, almost unwittingly, to the same interface, albeit at a far higher level. While nineteenth-century investigators were preoccupied with the nervous connections between ganglia, plexuses, and the gonads, we must contemplate—and strive to explain—an infinitely more complex neurochemical nexus involving the glands, the brain, and, if one may use an old, oft-despised but virtually indestructible word, the mind.

Notes

Abbreviations

AEM	*Archiv für Entwicklungsmechanik der Organismen*
AfG	*Archiv für Gynäkologie*
AgP	*Archiv für gesamte Physiologie* (see also *PAgP*)
BMJ	*British Medical Journal*
CfG	*Centralblatt/Zentralblatt für Gynäkologie*
DmW	*Deutsche medizinische Wochenschrift*
JAMA	*Journal of the American Medical Association*
KW	*Klinische Wochenschrift*
PAgP	*Pflügers Archiv für gesamte Physiologie* (see also *AgP*)
MmW	*Münchener medizinische Wochenschrift*
WkW	*Wiener klinische Wochenschrift*
WmedW	*Wiener medizinische Wochenschrift*

Introduction

1. On Galen's concept of sympathy and its later transformations, see Rudolph E. Siegel, *Galen's System of Physiology and Medicine* (Basel: Karger, 1968), 360–82.

2. See Edwin Clarke and L. S. Jacyna, *Nineteenth-Century Origins of Neuroscientific Concepts* (Berkeley and Los Angeles: University of California Press, 1987), 312–13.

3. Endocrine or ductless glands poured their secretions directly into the blood as it coursed through them (hence, their secretions were internal). The term "internal secretion" was coined in 1855 by Claude Bernard. The major examples of such glands, which we categorize as endocrine, are the thyroid, testicles, and adrenals. Secretions of *exocrine* glands (or glands with ducts, e.g., the salivary glands) were poured out through the ducts into specific organs or cavities, with no direct access

to the entire organism. See V. C. Medvei, *The History of Clinical Endocrinology* (Carnforth, Lancs.: Parthenon, 1993), 5.

4. Schäfer added, however, that the nervous system remained crucial in bioregulation—the glands acted in concert with or parallel to the nerves. See E. Sharpey-Schafer, "Endocrine Physiology," *Irish Journal of Medical Science*, 6th ser., no. 69 (September 1931): 483–505, at 484. In 1918, Schäfer changed his name to Edward Sharpey-Schafer. On his career, see Merriley Borell, "Setting the Standards for a New Science: Edward Schäfer and Endocrinology," *Medical History* 22 (1978): 282–90.

5. "Solidism" and "humoralism" are simple labels for complex (and internally diverse) orientations. For our purposes, we can define humoralism narrowly as the view that, within the body, actions at a distance are mediated by circulating fluids and solidism as the opposed conviction of distant action being mediated by solid structures, most notably the nerves. Unfortunately, there is no comprehensive historical analysis of these important concepts and their contexts. But see Russell C. Maulitz, "The Pathological Tradition," in W. F. Bynum and Roy Porter, eds., *Companion Encyclopaedia of the History of Medicine,* 2 vols. (London: Routledge, 1993), 1:169–89; and Robert J. Miciotto, "Carl Rokitansky: A Reassessment of the Hematohumoral Theory of Disease," *Bulletin of the History of Medicine* 52 (1978): 183–99.

6. Today, scientists would agree that many if not most or even all the sexual characters are dependent, at least up to a point, on hormones. This endocrine basis of sex, however, was neither "discovered" all at once, nor was it a matter only of the testicles in the male or the ovaries in the female. From the mid-nineteenth to the mid-twentieth century, hypotheses of sexual endocrinology evolved steadily and with them changed fundamental concepts of sex and indeed, of the nature of the body itself.

7. See Eugen Steinach, *Sex and Life: Forty Years of Biological and Medical Experiments* (New York: Viking, 1940), 5.

8. See Henry Smith Williams, *Your Glands and You* (New York: McBride, 1936), 128.

9. This phrase has been used by Lesley Hall in a broader sense. See L. Hall, "The Sexual Body," in Roger Cooter and John Pickstone, eds., *Companion to Medicine in the Twentieth Century* (London: Routledge, 2003), 261–75.

10. The genetic determination of sex, as I show later, was far from established even in the 1920s, and physiologists, in any case, argued that the genetic blueprint was just a permissive scheme—it determined possibilities, not outcomes. Our protagonists, in short, were concerned entirely with what we call the phenotype.

11. In this book, however, I often use expressions such as "male sex hormone" in an unproblematized manner. The use is purely for convenience and the lack of analysis simply because Nelly Oudshoorn has analyzed the history and presuppositions of this nomenclature at great depth. See Nelly Oudshoorn, "On the Making of Sex Hormones: Research Materials and the Production of Knowledge," *Social Studies of Science* 20 (1990): 5–33; and Oudshoorn, "On Measuring Sex Hormones: The Role of Biological Assays in Sexualizing Chemical Substances," *Bulletin of the History of Medicine* 64 (1990): 243–61.

12. On nineteenth-century biological convictions of the hazy morphological boundary between the sexes, see Frank J. Sulloway, *Freud, Biologist of the Mind: Beyond the Psychoanalytic Legend* (London: Fontana, 1979), esp. 158–60 and 292–96; Ornella Moscucci, *The Science of Woman: Gynaecology and Gender in England, 1800–1929* (Cambridge: Cambridge University Press, 1990), 17–22; and Chandak Sengoopta, *Otto Weininger: Sex, Science, and Self in Imperial Vienna* (Chicago: University of Chicago Press, 2000), esp. 44–46. For a different view, see Thomas Laqueur, *Making Sex: Body and Gender from the Greeks to Freud* (Cambridge, Mass.: Harvard University Press, 1991).

13. See Julia E. Rechter, "'The Glands of Destiny': A History of Popular, Medical, and Scientific Views of the Sex Hormones in 1920s America" (Ph.D. diss., University of California at Berkeley, 1997), 142.

14. Very recently, the hypothalamic releasing factors are being used in diagnostic tests related to depression, but the clinical utility of neuroendocrinology is still limited in general and virtually insignificant in areas concerned with sex.

15. Wilhelm Falta, *The Ductless Glandular Diseases,* trans. and ed. Milton K. Meyers (Philadelphia: Blakiston's, 1916), 3–4.

16. And here, I should emphasize that although I deal with much experimental research on the glandular functions of animals, the focus of the book is on humans. It is not the history of *experimental* endocrinology that I have tried to write, eminently worthy as that subject is for scholarly investigation. Rats and guinea pigs appear often enough in these pages but only when their physiologies were considered, at the time, to be of relevance to understanding human physiology or treating human disorders.

17. On these issues, see Adele E Clarke, *Disciplining Reproduction: Modernity, American Life Sciences, and the Problems of Sex* (Berkeley and Los Angeles: University of California Press, 1998).

18. See Nelly Oudshoorn, *Beyond the Natural Body: An Archeology of Sex Hormones* (London: Routledge, 1994), especially 65–108.

19. V. C. Medvei, *The History of Clinical Endocrinology* (Carnforth, Lancs.: Parthenon, 1993).

20. Diana Long Hall, "Biology, Sex Hormones and Sexism in the 1920s," *Philosophical Forum* 5, nos. 1–2 (1973/74): 81–96.

21. Anne Fausto-Sterling, *Sexing the Body: Gender Politics and the Construction of Sexuality* (New York: Basic Books, 2000).

22. See, for an earlier perspective, Robert T. Frank, *The Female Sex Hormone* (London: Baillière, Tindall and Cox, 1929), 127–28; and, for a later analysis, Egon Diczfalusy and Christian Lauritzen, *Oestrogene beim Menschen* (Berlin: Springer Verlag, 1961), 8–9.

23. See Kenneth Bowes, "The Oestrogens: Synthetic and Natural," *The Practitioner* 169 (1952): 243–52, at 243, 248. It was later found to cause genital anomalies and cancer. See Naomi Pfeffer, "Lessons from History: The Salutary Tale of Stilboestrol," in P. Alderson, ed., *Consent to Health Treatment and Research: Differing Perspectives* (London: Social Science Research Unit, University of London, 1992), 31–36.

Chapter One

1. E. A. Schäfer, "Address in Physiology on Internal Secretions," *Lancet,* 10 August 1895, 321–24, at 321; idem, "The Hormones Which are Contained in Animal Extracts: Their Physiological Effects," *Pharmaceutical Journal* 79 (1907): 670–74, at 670; and Merriley Borell, "Organotherapy and the Emergence of Reproductive Endocrinology," *Journal of the History of Biology* 18 (1985): 1–30, at 12–13.

2. R. Lawson Tait, *The Pathology and Treatment of Diseases of the Ovaries,* 4th ed. (Birmingham: Cornish, 1883), 3.

3. Nelly Oudshoorn, *Beyond the Natural Body: An Archeology of the Sex Hormones* (London: Routledge, 1994), 19.

4. Rudolf Virchow, "Der puerperale Zustand: Das Weib und die Zelle," in Virchow, *Gesammelte Abhandlungen zur wissenschaftlichen Medizin* (Frankfurt am Main: Meidinger, 1856), 735–79. On the contexts of the lecture, see Hans H. Simmer, "Zum Frauenbild Rudolf Virchows in den späten 1840er Jahren," *Medizinhistorisches Journal* 27 (1992): 292–319; and Simmer, "Der junge Rudolf Virchow und die Gesellschaft für Geburtshülfe in Berlin in den Jahren 1846–1848," *Sudhoffs Archiv* 77 (1993): 72–96. All translations in this work, unless otherwise attributed, are my own.

5. For Virchow's explanation of his unusual definition of the puerperium, see "Der puerperale Zustand," 735–36.

6. On this issue, see Stefanie Holle, "Die Widerlegung des Postulates von der Gleichzeitigkeit der Ovulation und Menstruation bei der Frau: Klinische und histologische Untersuchungen im frühe 20. Jahrhundert" (Inaugural diss., Medical Faculty, University of Erlangen-Nürnberg, 1984).

7. On Virchow's opposition to humoralism, see L. J. Rather, "Virchow's Review of Rokitansky's *Handbuch* in the *Preussische Medizinal-Zeitung,* Dec. 1846," *Clio Medica* 4 (1969): 127–40. My thanks to Harry Marks for this reference.

8. See Russell C. Maulitz, "The Pathological Tradition," in W. F. Bynum and Roy Porter, eds., *Companion Encyclopaedia of the History of Medicine,* 2 vols. (London: Routledge, 1993), 1:169–89.

9. Virchow, "Der puerperale Zustand," 739–40, 746.

10. Ibid., 747. For a critique of Virchow's conception of virilization following loss of ovarian function, see H. H. Simmer, "Rudolf Virchow und der Virilismus ovarioprivus," *Medizinhistorisches Journal* 28 (1993): 375–401.

11. Childbearing, of course, was *the* feminine function; the uterus and its transformations during menstruation, therefore, fascinated investigators up to the twentieth century. What changed in the nineteenth century, however, was that the uterus and its functions came increasingly to be seen as linked intimately with, and often secondary to, the functions of the ovaries. This shift has not been sufficiently explored by historians. But see Anne E. Walker, *The Menstrual Cycle* (London: Routledge, 1997), 34–35.

12. On Van Helmont, see Simmer, "Zum Frauenbild Rudolf Virchows," 308–10.

13. T. L. W. Bischoff, *Beweis der von der Begattung unabhängigen periodischen Reifung und Loslösung der Eier der Säugethiere und des Menschen als der ersten Bedingung*

ihrer Fortpflanzung (Gießen: Ricker, 1844), cited by H. H. Simmer, "Rudolf Virchow und der Virilismus ovarioprivus," 395.

14. Carroll Smith-Rosenberg and Charles Rosenberg, "The Female Animal: Medical and Biological Views of Woman and Her Role in Nineteenth-Century America," in Judith Walzer Leavitt, ed., *Women and Health in America: Historical Readings* (Madison: University of Wisconsin Press, 1984), 12–27, at 13.

15. See Joachim Radkau, *Das Zeitalter der Nervosität: Deutschland zwischen Bismarck und Hitler* (Munich: Propyläen, 2000), 44–45, and compare with the earlier (British) history of the concept in G. J. Barker-Benfield, *The Culture of Sensibility: Sex and Society in Eighteenth-Century Britain* (Chicago: University of Chicago Press, 1992).

16. The American surgeon Robert Battey had initially described the removal of such ovaries as "normal ovariotomy." He soon regretted that nomenclature and clarified that "the ovaries removed, and the tubes as well, have presented visible signs of disease—signs which are evident to the naked eye and palpable to the sense of touch." It was his own early ignorance of the histology and pathology of the ovaries that, he claimed, had led him to ignore cysts, sclerotic patches, or fibrous degenerations. See Battey's untitled contribution to T. Spencer Wells, Alfred Hegar, and Robert Battey, "Castration in Mental and Nervous Diseases: A Symposium," *American Journal of the Medical Sciences,* new ser., 92 (1886): 483–90, at 483–84.

17. Although nineteenth-century medical usage was not always consistent, in English, "oophorectomy" usually referred to the removal of ostensibly healthy or normal-sized ovaries, whereas the etymologically inaccurate term "ovariotomy" was generally reserved for the removal of cystic or otherwise enlarged ovaries. German practitioners almost invariably used the term "kastration" for bilateral oophorectomy. For the sake of clarity, I have followed the English convention throughout, even when referring to German research. For more on terminological issues, see Ornella Moscucci, *The Science of Woman: Gynaecology and Gender in England, 1800–1929* (Cambridge: Cambridge University Press, 1990), 157, 239, n. 1; and H. H. Simmer, "Oophorectomy for Breast Cancer Patients: Its Proposal, First Performance, and First Explanation as an Endocrine Ablation," *Clio Medica* 4 (1969): 227–49, at 240, n. 1. On the history of oophorectomy in general, see Moscucci, *Science of Woman,* 134–64, esp. 157–60; Lawrence D. Longo, "The Rise and Fall of Battey's Operation: A Fashion in Surgery," in Leavitt, ed., *Women and Health in America,* 270–84; Andrew Scull and Diane Favreau, "'A Chance to Cut is a Chance to Cure': Sexual Surgery for Psychosis in Three Nineteenth Century Societies," *Research in Law, Deviance and Social Control* 8 (1986): 3–39; H. H. Simmer, "Bilaterale Oophorektomie der Frau im späten 19. Jahrhundert: Zum methodologischen Wert der Kastration für die Entdeckung ovarieller Hormone," *Geburtshilfe und Frauenheilkunde* 43 (1983), Sonderheft, 54–59; and Thilo G. Funk, "Uterine Fibromyome und Blutungen als Indikation für eine bilaterale Oophorektomie im späten 19. Jahrhundert," *Medizinhistorisches Journal* 21 (1986): 159–71. On the use of the operation in "functional" nervous disorders in German-speaking lands, see Günter Burger, "Nerven- und Geisteskrankheiten als Indikationen für eine bilaterale Oophorektomie im späten 19. Jahrhundert" (Inaugural diss., Medical Faculty, University of Erlangen-Nürnberg, 1984), 24–41, 61–108, and 126–43.

18. Since the history of the introduction and dissemination of oophorectomy in Britain and America is treated in great detail by the general works cited previously, I shall concentrate on the German oophorectomy saga. Surgical anesthesia came into use in the 1840s and some surgeons (such as Spencer Wells himself) began to use quasi-aseptic techniques to prevent wound infections in the early 1860s. Antiseptic surgery began formally later in that decade with the use of carbolic acid by the Scottish surgeon Joseph Lister, who based his practice explicitly on the germ theory of Louis Pasteur. Surgery, once the craft of cutting quickly and closing, could now become more ambitious, and the excision of diseased ovaries (ovariotomy) was one of the first in the long series of radical and invasive operations introduced over the last decades of the nineteenth century. See Ulrich Tröhler, "Surgery (Modern)," in W. F. Bynum and Roy Porter, eds., *Companion Encyclopedia of the History of Medicine*, 2 vols. (London: Routledge, 1993), 2: 984–1028, esp. 985–94.

19. "Functional" diseases were those in which there was no evidence of any structural abnormality—they were, however, considered to be treatable by somatic methods because they were attributed to disorders of physiology (i.e., function).

20. I must emphasize here that although the proponents of oophorectomy all agreed that the ovary was the source of the menstrual impulse, they did not usually offer much information on the exact nature of this impulse and its precise mechanism of action. This ovarian theory was not the sole explanation of menstruation at the time. Many criticized the vogue for oophorectomy not because it was an invasive or dangerous operation but because it was founded in the ovarian theory of menstruation, which, they argued, was only speculative. See Longo, "Rise and Fall," 275.

21. See, for a classic general account, Elaine Showalter, *The Female Malady: Women, Madness and English Culture, 1830–1980* (London: Virago, 1987) and for the German context, Joachim Radkau, *Das Zeitalter der Nervosität;* and Günter Burger, "Nerven- und Geisteskrankheiten als Indikationen für eine bilaterale Oophorektomie."

22. For a concise overview of these well-known themes, see Naomi Pfeffer, *The Stork and the Syringe: A Political History of Reproductive Medicine* (Cambridge: Polity, 1993), 34–35.

23. See Longo, "Rise and Fall," 274.

24. Quoted in Scull and Favreau, "'A Chance to Cut,'" 15.

25. See William Goodell, *Lessons in Gynecology* (Philadelphia: Brinton, 1890), 394–95; and Regina Morantz and Sue Zschoche, "Professionalism, Feminism, and Gender Roles: A Comparative Study of Nineteenth Century Medical Therapeutics," *Journal of American History* 47 (1980): 568–88.

26. See Hegar's untitled contribution (trans. Montagu Handfield-Jones) to Spencer Wells, Hegar, and Battey, "Castration in Mental and Nervous Diseases," 471–83, at 480.

27. A. Hegar and R. Kaltenbach, *A Hand-Book of General and Operative Gynaecology,* trans. anon., vols. 6–7 of Egbert H. Grandin, ed., *Cyclopaedia of Obstetrics and Gynaecology,* 12 vols. (Edinburgh: Pentland, 1889), 6:303.

28. See Thilo Funk, "Uterine Fibromyome und Blutungen als Indikation für eine bilaterale Oophorektomie" for a convincing demonstration of the importance of bleeding fibromyomas and the little that could be done for them before the

introduction of oophorectomy or, for that matter, blood transfusion. As another champion of oophorectomy pointed out once, there was a social component to the popularity of oophorectomy in such cases: the operation was especially suitable, he argued, for those patients whose social status did not permit them the prolonged bed-rest and comprehensive care necessary for the conservative management of profuse uterine bleeding. See H. Menzel, "Beiträge zur Castration der Frauen, I: Castrationen bei Ovarialprolaps, Uterusfibrom, Retroflexio uteri mit Descensus ovariorum und Hysterie" *AfG* 26 (1885): 36–57, at 47.

29. See the overview of cases oophorectomized in Hegar's Freiburg clinic for nervous symptoms: G. Schmalfuss, "Zur Castration bei Neurosen," *AfG* 26 (1885): 1–35.

30. See, for instance, Hermann Fehling, "Zehn Castrationen: Ein Beitrag zur Frage nach dem Werthe der Castration," *AfG* 22 (1883): 441–55, at 455.

31. See H. Menzel, "Beiträge zur Castration der Frauen, I," 50–51.

32. See, for instance, Paul Flechsig, "Zur gynaekologischen Behandlung der Hysterie," *Neurologisches Centralblatt* 3 (1884): 433–39, 457–68, at 458.

33. See Thomas W. Laqueur, *Making Sex: Body and Gender from the Greeks to Freud* (Cambridge, Mass.: Harvard University Press, 1990), 177. In nineteenth-century medical usage, the word "neurosis" denoted disorders thought to be caused by physiological disturbances in the nervous system, i.e., disorders that did not necessarily produce any identifiable structural abnormality. This usage, of course, was to change radically after the rise of psychoanalysis. For the broader context, see George Frederick Drinka, *The Birth of Neurosis: Myth, Malady, and the Victorians* (New York: Simon and Schuster, 1984).

34. On this point, see Günter Burger, "Nerven- und Geisteskrankheiten" and W. F. Bynum, "The Nervous Patient in Eighteenth- and Nineteenth-Century Britain: The Psychiatric Origins of British Neurology," in W. F. Bynum, Roy Porter, and Michael Shepherd, eds., *The Anatomy of Madness: Essays in the History of Psychiatry,* 3 vols. (London: Tavistock, 1985–88), 1 (1985): 89–102.

35. See Hegar's contribution to the symposium on "Castration in Mental and Nervous Diseases," 477.

36. Hegar and Kaltenbach, *Hand-Book of General and Operative Gynaecology,* 6:307.

37. See Benedikt Stilling, *Physiologisch-pathologische und medicinisch-praktische Untersuchung über Spinal-Irritation* (Leipzig, 1840).

38. See Moritz Romberg, *Lehrbuch der Nervenkrankheiten des Menschen,* 2d ed., 2 vols. (Berlin: Duncker, 1851), 1, pt. 2: 198–223, esp. 210–11. On Hegar's use of the concept, see Burger, "Nerven- und Geisteskrankheiten als Indikationen," 7–9, 154–55. Generally, on ovarian theories of hysteria, see Ilza Veith, *Hysteria: The History of a Disease* (Chicago: University of Chicago Press, 1965), 173, 210, 232; and Mark Micale, "Hysteria Male/Hysteria Female: Reflections on Comparative Gender Construction in Nineteenth-Century France and Britain," in Marina Benjamin, ed., *Science and Sensibility: Gender and Scientific Enquiry, 1780–1945* (Oxford: Blackwell, 1991), 200–239, at 225–26.

39. The Munich gynecologist Joseph Amann (1832–1906) rejected Romberg's view that *all* cases of hysteria were reflex neuroses originating from the reproductive

tract. The essential—though not the sufficient—cause of hysteria was an inherited predisposition: the disease was precipitated when certain releasing conditions (*veranlassenden Momenten*) acted upon an already predisposed constitution. This dissent from the Romberg-Hegar perspective, however, was of little more than theoretical importance, for Amann identified diseases of the female reproductive tract as the most important as well as the most frequent releasing factors of hysteria. See Josef Amann, *Über den Einfluss der weiblichen Geschlechtskrankheiten auf das Nervensystem mit besonderer Berücksichtigung des Wesens und der Erscheinungen der Hysterie,* 2d ed. (Erlangen: Enke, 1874), iv, 79, 86.

40. See F. Jolly, "Hysterie und Hypochondrie," in A. Eulenburg, H. Nothnagel, J. Bauer, H. von Ziemssen, and F. Jolly, *Handbuch der Krankheiten des Nervensystems II,* which constitutes vol. 12, pt. 2 (1877) of H. von Ziemssen, ed., *Handbuch der speciellen Pathologie und Therapie,* 2d ed., 17 vols. in numerous parts (Leipzig: Vogel, 1876–85), 489–709, at 495, 501–2.

41. A. Rheinstaedter, "Über weibliche Nervosität, ihre Beziehungen zu den Krankheiten der Generationsorgane und ihre Allgemeinbehandlung," *Sammlung klinischer Vorträge,* no. 188, Gynäkologie no. 56 (1880), 1493–1510, at 1495.

42. A. Döderlein and B. Krönig, *Operative Gynäkologie* (Leipzig: Thieme, 1905), 594.

43. Ibid., 596–97.

44. Ibid., 595–96; and O. Binswanger, *Die Hysterie* (1904), constituting vol. 12, pt. 1, section 2 of Hermann Nothnagel et al., eds., *Specielle Pathologie und Therapie,* 24 vols. in numerous parts (Vienna: Hölder, 1894–1908), 59–60, 843–44, 945–46 and, on the degenerative basis of hysteria, 36–44. On degenerationism in late-nineteenth-century psychiatry and culture, see Daniel M. Pick, *Faces of Degeneration: A European Disorder, c.1848–c.1918* (Cambridge: Cambridge University Press, 1989).

45. See Hegar's contribution to "Castration in Mental and Nervous Diseases: A Symposium," 471. As late as in 1924, British surgeon Kenneth Walker asserted that in humans, "removal of the testes is more far-reaching in its results than is removal of the ovaries. Destruction of the testes early in life . . . leads to the production of a eunuch who as regards certain characters . . . approaches the female. Disease or congenital deficiency in the ovaries is, however, not productive in the human being of male characteristics, but of infantilism." Kenneth M. Walker, "The Internal Secretion of the Testis," *Lancet,* 5 January 1924, 16–21, at 17.

46. A. Hegar, "Die Castration der Frauen," in Richard Volkmann, ed., *Sammlung klinischer Vorträge,* nos. 136–38, Gynäkologie no. 42 (1878), 925–1068, at 926–31.

47. Even the well-known physical changes after menopause he was inclined to attribute at least partly to aging in general and the mental changes to experience. See ibid., 1002.

48. Ibid., 1001.

49. A. Hegar, "Über die Exstirpation normaler und nicht zu umfänglichen Tumoren degenerirter Eierstöcke," *CfG* 1 (1877): 297–306, 298–99, 305; and "Die Castration der Frauen," 1004–5.

50. Hegar and Kaltenbach, *Hand-Book of General and Operative Gynaecology,* 6:300. For a powerful challenge to Hegar's argument that oophorectomy did not reduce sex

drive, see Ludwig Glaevecke, "Körperliche und geistige Veränderungen im weiblichen Körper nach künstlichem Verluste der Ovarien einerseits und des Uterus andererseits," *AfG* 35 (1889): 1–88, 53–55.

51. F. Keppler, "Das Geschlechtsleben des Weibes nach der Kastration," *WmedW* 41 (1891): 1489–92 and 1523–26, at 1525.

52. See Ludwig Glaevecke, "Körperliche und geistige Veränderungen im weiblichen Körper," 31–32, 50.

53. See Jutta Blönnigen, "Die Osteomalazie als Indikation für eine bilaterale Oophorektomie im späten 19. u. frühen 20. Jahrhundert: Ergebnisse und Erklärungsversuche" (Inaugural diss., Medical Faculty, University of Erlangen-Nürnberg, 1980); and H. H. Simmer, "Oophorectomy for Breast Cancer Patients."

54. See R. Ted Steinbock, "Rickets and Osteomalacia," in Kenneth F. Kiple, ed., *The Cambridge World History of Human Disease* (Cambridge: Cambridge University Press, 1993), 978–80.

55. The fewer cases of osteomalacia in men were often considered to be fundamentally different. See H. Fehling, "Über Wesen und Behandlung der puerperalen Osteomalakie," *AfG* 39 (1891): 171–96, at 180–81.

56. On contemporary convictions of endemicity, see, for instance, W. Thorn, "Zur Kasuistik der Kastration bei Osteomalakie," *CfG* 15 (1891): 828–31.

57. See Adolf Strümpell, *Lehrbuch der speciellen Pathologie und Therapie der inneren Krankheiten*, 2 vols. in 3 (Leipzig: Vogel, 1883–84), 2, pt. 2 (1884): 169–72.

58. See Paul Zweifel, "Zur Discussion über Porro's Methode des Kaiserschnittes," *AfG* 17 (1881): 355–77; idem, "Ein Fall von Osteomalacie, modificirter Porro-Kaiserschnitt, geheilt," *CfG* 14 (1890): 25–29, esp. 28.

59. Jutta Blönnigen, "Die Osteomalazie als Indikation für eine bilaterale Oophorektomie," 10–12.

60. See P. Zweifel,"Hermann Johannes Karl Fehling," *CfG,* 49 (1925): 2866–74; and K. Franz, "Hermann Fehling," *AfG* 127 (1926): i–iv.

61. H. Fehling, "Über Kastration bei Osteomalacie," *Verhandlungen der Deutschen Gesellschaft für Gynäkologie* 2 (1888): 311–18.

62. See Blönnigen, "Die Osteomalazie als Indikation," 29–39.

63. A proportion did show some relatively minor (and quite inconsistent) abnormalities such as hypertrophy, atrophy, cysts, hyaline degeneration, and chronic oophoritis. See Blönnigen, "Die Osteomalazie als Indikation," 31–39.

64. See Blönnigen, "Die Osteomalazie als Indikation," 51.

65. H. Fehling, "Über Wesen und Behandlung der puerperalen Osteomalakie," *AfG* 39 (1891): 171–96; and idem, "Weitere Beiträge zur Lehre von der Osteomalakie," *AfG* 48 (1895): 472–98.

66. On support for the neural hypothesis, see H. Eisenhart, "Beiträge zur Aetiologie der puerperalen Osteomalacie," *Deutsches Archiv für klinische Medizin* 49 (1892): 156–205; Guillaume Rossier, "Anatomische Untersuchung der Ovarien in Fällen von Osteomalacie," *AfG* 48 (1895): 472–98.

67. F. von Winckel, "Über die Erfolge der Kastration bei Osteomalakie," *Sammlung klinischer Vorträge,* new ser., no. 71, Gynäkologie no. 28 (1893): 657–82, at 673–74.

68. See Blönnigen, "Die Osteomalazie als Indikation," 60–61. Fehling ("Weitere Beiträge zur Lehre von der Osteomalakie," 484–85) pointed out that patients of osteomalacia had never improved after being chloroformed for purposes other than oophorectomy or after taking chloral as a sleeping draught. Oophorectomy even under anesthesia with ether, on the other hand, had led to striking results.

69. Hermann Senator, "Zur Kenntniss der Osteomalacie und der Organotherapie," *Berliner klinische Wochenschrift* 34 (1897): 109–12, 143–44; and Wilhelm Latzko and Julius Schnitzler, "Ein Beitrag zur Organotherapie bei Osteomalacie," *DmW* 23 (1897): 587–92.

70. For a survey of the various endocrine explanations of the condition in the 1930s, see Hermann Zondek, *The Diseases of the Endocrine Glands,* 3d ed., trans. Carl Prausnitz (London: Edward Arnold, 1935), 356–59.

71. See, for instance, Döderlein and Krönig, *Operative Gynäkologie,* 540–42.

72. M. Hofmeier, "Zur Frage der Behandlung der Osteomalacie durch Kastration," *CfG* 15 (1891): 225–28, at 224, 227. Osteomalacia is now known to be caused mostly by the deficiency of Vitamin D, and occasionally by other factors such as phosphate deficiency or renal disorders. Endocrine abnormalities are no longer considered to be of any causal importance. See http://www.nlm.nih.gov/medlineplus/ency/article/000376.htm (accessed 8 July 2005). Why removing the ovaries was effective in osteomalacia, however, is a question that cannot, as Jutta Blönnigen emphasizes, be solved even with the aid of today's knowledge of endocrinology, and no consensus was ever reached on the issue in the earlier decades of the century. Blönnigen, "Die Osteomalazie als Indikation," 84.

73. G. T. Beatson, "On the Treatment of Inoperable Cases of Carcinoma of the Mamma: Suggestions for a New Method of Treatment with Illustrative Cases," *Lancet,* 11 July 1896, 104–7; ibid., 18 July 1896, 162–65. On his life and career, see the obituary by G. H. Edington in the *BMJ,* 25 February 1933, 344–45. For examples of recent attempts by clinician-historians to portray Beatson as a pioneer of "hormonal cancer therapeutics," see Tim Gulliford and Richard J. Epstein, "Endocrine Treatment of Cancer," *Journal of the Royal Society of Medicine* 89 (1996): 448–53, quoted phrase at 448; and Robin Leake, "100 Years of the Endocrine Battle against Breast Cancer," *Lancet* 347 (1996): 1780–81.

74. Beatson, "On the Treatment of Inoperable Cases of Carcinoma," 106.

75. Ibid.

76. Ibid.

77. In the late nineteenth century, many theories of cancer argued that it was a bacterial or parasitic disease. On these theories, see Jacob Wolff, *The Science of Cancerous Disease from Earliest Times to the Present,* trans. Barbara Ayoub (Canton, Mass.: Science History Publications, 1989), 431–590.

78. Beatson, "On the Treatment of Inoperable Cases of Carcinoma," 107. In a later article, he stated that thyroid "powerfully affects the metabolism generally of the body cells, raising their tone and improving their vigour, while it acts favourably on the lymphatic system, lessening the chances of dissemination by it." Beatson, "The Treatment of Cancer of the Breast by Oöphorectomy and Thyroid Extract," *BMJ,* 19 October 1901, 1145–48, at 1147. The sequelae of thyroidectomy were well

documented by the 1880s, and were being treated with thyroid extracts by 1891: this has often been seen as "the first generally recognized success of organotherapy." V. C. Medvei, *The History of Clinical Endocrinology* (Carnforth, Lancs.: Parthenon, 1993), 160–62.

79. Beatson, "On the Treatment of Inoperable Cases of Carcinoma," 163.

80. Beatson, "The Treatment of Cancer of the Breast by Oöphorectomy and Thyroid Extract," 1146.

81. Beatson, "On the Treatment of Inoperable Cases of Carcinoma," 163.

82. Ibid., 164. Beatson thought that the testicles probably had a similar influence on cancer in men: "I am making inquires as to the existence of cancer amongst eunuchs," he wrote, "for if my view is correct they should not suffer from it" (ibid.).

83. George Neil Stewart, *A Manual of Physiology with Practical Exercises* (London: Baillière, Tindall and Cox, 1895). See Simmer, "Oophorectomy for Breast Cancer Patients," 232–33.

84. G. T. Beatson, "Remarks on the Etiology of Carcinoma: Has It a Physiological Function in the Body?," *BMJ*, 29 April 1905, 921–25, at 924.

85. Stanley Boyd, "On Oöphorectomy in the Treatment of Cancer," *BMJ*, 2 October 1897, 890–96. Boyd did not, however, use thyroid extract until the oophorectomy had obviously failed to bring about any improvement. He later explained that "when endeavouring to ascertain the effect of one mode of treatment based upon highly theoretical considerations it is surely unwise to combine it with another method resting *on still more shadowy grounds*." S. Boyd, "On Oöphorectomy in Cancer of the Breast," *BMJ*, 20 October 1900, 1161–67, at 1166, emphasis added.

86. Boyd, "On Oophorectomy in the Treatment of Cancer," 895.

87. Ibid., 896.

88. On this issue, see Timothy H O'Neill, "The Invisible Man? Problematising Gender and Male Medicine in Britain and America, 1800–1950." Ph.D. diss., University of Manchester, 2003.

89. Stanley Boyd, "On Oöphorectomy in Cancer of the Breast."

90. Ibid., 1166.

91. Ibid.

92. One such abnormality was prostatic enlargement, in which castration was recommended by some surgeons from the 1890s. See, for example, C. Mansell Moulin, "On the Treatment of Enlargement of the Prostate by Removal of the Testes," *BMJ*, 3 November 1894, 976. On the contexts, see Timothy O'Neill, "The Invisible Man?," 129–30.

93. According to Claudia Schirren ("Andrology: Origin and Development of a Special Discipline in Medicine," *Andrologia* 17 [1985]: 118–25), the term "andrology" was coined in 1951 by gynecologist Harald Siebke; there are now societies of andrologists and at least one journal devoted to the subject. Timothy O'Neill, however, credits the term to the surgeon Edred Corner and remarks that it was a mere re-labeling of the field of genito-urinary medicine and that andrology never involved the kind of "theorisation" of gender that occurred in gynecology. See O'Neill, "The Invisible Man?," 222–26; and E. M. Corner, *Male Diseases in General Practice: An Introduction to Andrology* (London: Oxford University Press, 1910).

94. The interest in women's bodies, high as it already was, was heightened further by the emergence of feminist activism at the end of the century. The New Woman appeared to symbolize the end of civilization, and the "science of woman" became central to debates on the rights and wrongs of feminist demands, especially in Central Europe. See Chandak Sengoopta, *Otto Weininger: Sex, Science and Self in Imperial Vienna* (Chicago: University of Chicago Press, 2000) and the numerous sources cited therein. None of this should be taken to imply that a dedicated profession of gynecology emerged everywhere without difficulty. In late-Victorian and Edwardian England, as we shall see later, it took a lot of effort and politicking to establish it on an equal basis to general medicine and general surgery. For further details, see Moscucci, *The Science of Woman.*

95. Aristotle, *History of Animals,* 5.14, 545a, 21, trans. D'Arcy W. Thompson, in Jonathan Barnes, ed., *The Complete Works of Aristotle: The Revised Oxford Translation,* 2 vols. (Princeton: Princeton University Press, 1984), 1:861.

96. W. Yarrell, "On the Influence of the Sexual Organ in Modifying External Character," *Journal of the Proceedings of the Linnean Society (Zoology)* 1 (1857): 76–82, at 81. On Yarrell, see Thomas R. Forbes, "William Yarrell, British Naturalist," *Proceedings of the American Philosophical Society* 106 (1962): 505–15.

97. C. Darwin, *The Variation of Animals and Plants under Domestication,* 2d ed., 2 vols. (New York: Appleton, 1897), 2:26–27.

98. Ibid., 2:27.

99. A. Weismann, *The Germ-Plasm: A Theory of Heredity,* trans. W. Newton Parker and Harriet Rönnfeldt (New York: Scribner's, 1893), 358–59.

100. The history of biological and medical concepts of male castration remains to be written. For a recent idiosyncratic attempt to show the importance of castration in cultural history and sexual psychology from Augustine to Freud, see Gary Taylor, *Castration: An Abbreviated History of Western Manhood* (New York: Routledge, 2000).

101. C. Barker Jørgensen, *John Hunter, A. A. Berthold, and the Origins of Endocrinology* (Odense: Odense University Press, 1971).

102. Ibid., 16. Although the phenomenon of graft rejection was widely known and feared, there was, as yet, no universally accepted immunological theory to explain rejection and guide the matching of donor and recipient.

103. Ibid., 18.

104. Ibid., 24. See also H. Ribbert, "Über die compensatorische Hypertrophie der Geschlechtsdrüsen," *Archiv der pathologische Anatomie und Physiologie* 120 (1890): 247–72, especially 271; S. Samuel, "Das Gewebswachsthum bei Störungen der Innervation," *Archiv der pathologische Anatomie und Physiologie* 113 (1888): 272–314, esp. 313–14; and idem, "Trophoneurosen," in Albert Eulenburg, ed., *Real-Encyclopädie der gesammten Heilkunde,* 22 vols. (Vienna: Urban & Schwarzenberg, 1885–90), 20 (1890): 188–263, esp. 223–24.

Chapter Two

1. H. M. Evans, "Present Position of Our Knowledge of Anterior Pituitary Function," *JAMA* 101 (1933): 425–32, at 425. Evans specifically mentioned

Brown-Séquard and the controversial physiologist of the 1920s, Eugen Steinach (1861–1944). For more on these figures, see below.

2. Hans Lisser, "The Endocrine Society—The First Forty Years (1917–1957)," *Endocrinology* 80 (1967): 5–28, at 7.

3. A. A. Berthold, "The Transplantation of Testes," trans. D. P. Quiring, *Bulletin of the History of Medicine* 16 (1944): 399–401.

4. Ibid., 399–400.

5. Ibid., 401.

6. A. A. Berthold, "Über die Transplantation der Hoden," *Nachrichten von der Georg August Universität und der königliche Gesellschaft der Wissenschaften zu Göttingen*, 19 February 1849, 1–6, especially 1, 6.

7. A. Foges, "Zur Lehre von den secundären Geschlechtscharakteren," *AgP* 93 (1903): 39–58, at 39.

8. Herman H. Rubin, *The Glands of Life: The Story of the Mysterious Ductless Glands* (New York: Bellaire, 1935), 9.

9. E. Steinach, "Zur Geschichte des männlichen Sexualhormones und seiner Wirkungen am Säugetier und beim Menschen," *WkW* 49 (1936): 161–72, 196–205, at 199. Steinach was actually reacting against the absence of references to his own experimental work in the reports of the biochemically oriented endocrinologists of the 1930s. In spite of this explicit self-interest, however, his question on Berthold was valid and historically astute.

10. "Observation of man and of animals of widely different species shows, beyond any manner of doubt," declared Biedl, "that castration does not impart characteristics peculiar to the opposite sex, and that transformation into the heterosexual type [i.e. feminization in males or masculinization in females] is never observed." Artur Biedl, *The Internal Secretory Organs: Their Physiology and Pathology*, trans. Linda Forster (London: John Bale, Sons and Danielsson, 1913), 378. This was a translation of the first edition of Biedl's *Innere Sekretion: Ihre physiologischen Grundlagen und ihre Bedeutung für die Pathologie* (Berlin: Urban and Schwarzenberg, 1910). On the significance of this work, see L. Feher and W. Kaiser, "Die Anfänge der modernen Endokrinologie: Pro memoria Artur Biedl (1869–1933)," *NTM* 7 (1990): 99–108.

11. W. Yarrell, "On the Influence of the Sexual Organ in Modifying External Character," *Journal of the Proceedings of the Linnean Society (Zoology)* 1 (1857): 76–82, at 81. On Yarrell, see Thomas R. Forbes, "William Yarrell, British Naturalist," *Proceedings of the American Philosophical Society* 106 (1962): 505–15. See also C. Darwin, *The Variation of Animals and Plants under Domestication*, 2d ed., 2 vols (New York: Appleton, 1897), 2:26–27; and A. Weismann, *The Germ-Plasm: A Theory of Heredity*, trans. W. Newton Parker and Harriet Rönnfeldt (New York: Scribner's, 1893), 358–59.

12. Foges had been preceded by Hugo Sellheim, "Zur Lehre von den sekundären Geschlechtscharakteren," *Beiträge zur Geburtshilfe und Gynäkologie* 1 (1898): 229–55. Sellheim's views, however, were based on even fewer experiments than those of Foges.

13. See F. Goltz and A. Freusberg, "Über den Einfluß des Nervensystems auf die Vorgänge während der Schwangerschaft und des Gebärakts," *AgP* 9 (1874): 552–65.

Goltz—student of Hermann von Helmholtz and teacher of Jacques Loeb—was far better known in his time for his experimental investigation of reflex phenomena in frogs and for his critical stance on the theory of cerebral localization. For details, see J. R. Ewald, "Friedrich Goltz," *AgP* 94 (1903): 1–64, and Karl E. Rothschuh's biographical sketch in *Dictionary of Scientific Biography*, 5:462–64. See also Philip J. Pauly, "The Political Structure of the Brain: Cerebral Localization in Bismarckian Germany," *International Journal of Neuroscience* 21 (1983): 145–50.

14. See H. H. Simmer, "Bilaterale Oophorektomie der Frau im späten 19. Jahrhundert: Zum methodologischen Wert der Kastration für die Entdeckung ovarieller Hormone," *Geburtshilfe und Frauenheilkunde* 43 (1983), Sonderheft, 54–59, at 57.

15. See A. Foges, "Zur Lehre von den secundären Geschlechtscharakteren." The young and idiosyncratic philosopher Otto Weininger, in his 1903 study of masculinity and femininity, appreciated the biological significance of Goltz's experiment more deeply than any scientist or medical man that I know of. On Weininger's use of Goltz's experiment and other early reports on the origin and nature of sex, see Chandak Sengoopta, *Otto Weininger: Sex, Science, and Self in Imperial Vienna* (Chicago: University of Chicago Press, 2000), 74–76.

16. See Pauly, "Political Structure of the Brain."

17. See Humphry Davy Rolleston, *The Endocrine Organs in Health and Disease with an Historical Review* (London: Oxford University Press, 1936), 28–29, 319–21.

18. Lisser, "The Endocrine Society: The First Forty Years," 7.

19. On the spermatic economy, see Ben Barker-Benfield, "The Spermatic Economy: A Nineteenth Century View of Sexuality," *Feminist Studies* 1 (1972): 45–74; and H. Tristram Engelhardt Jr, "The Disease of Masturbation: Values and the Concept of Disease," in Judith W. Leavitt and Ronald L. Numbers, eds., *Sickness and Health in America: Readings in the History of Medicine and Public Health* (Madison: University of Wisconsin Press, 1985), 13–21.

20. Merriley Borell, "Brown-Séquard's Organotherapy and its Appearance in America at the End of the Nineteenth Century," *Bulletin of the History of Medicine* 50 (1976): 309–20, at 310; idem, "Organotherapy and the Emergence of Reproductive Endocrinology," *Journal of the History of Biology* 18 (1985): 1–30; and idem, "Origins of the Hormone Concept: Internal Secretions and Physiological Research, 1889–1905" (Ph.D. diss., Yale University, 1976).

21. Benjamin Harrow, *Glands in Health and Disease* (New York: Dutton, 1922), 185–86.

22. Ibid., 186–87.

23. Spencer G. Strauss, "Endocrine Gland Extracts: Their Manufacture and Use," *New York Medical Journal* 113 (1921): 395–97, 468–69.

24. Even Benjamin Harrow was compelled to concede that "it cannot be said that we have accomplished much with glandular extracts beyond their use in thyroid disease.... With a persistence worthy of some admiration, glandular advocates, failing to get results with any one extract, tried each one of the others in turn. Still without result, they adopted pluriglandular treatment.... It cannot be said that these methods yielded any better results." Harrow, *Glands in Health and Disease*, 186–87.

25. Paul de Kruif, *The Male Hormone* (Garden City, N.Y.: Garden City Publishing Co., 1945), 59.

26. W. S. Bainbridge, "Transplantation of Human Ovaries: Present Status and Future Possibilities," *American Journal of Obstetrics and Gynecology* 5 (1923): 493–98, 493–94.

27. Bainbridge was neither the first gynecologist nor the only one to express similar views. For another example, see W. R. Nicholson, "A Review of the Literature of Ovarian Transplantation," *University of Pennsylvania Medical Bulletin* 14 (1901–2): 401–7, at 401.

28. See Merriley Borell, "Brown-Séquard's Organotherapy," and idem, "Organotherapy, British Physiology, and Discovery of the Internal Secretions."

29. George W. Corner, "The Early History of the Oestrogenic Hormones," *Journal of Endocrinology* 31 (1964–65): iii–xvii, at vi.

30. H. H. Simmer, "Organotherapie mit Ovarialpräparaten in der Mitte der neunziger Jahre des 19. Jahrhunderts: Medizinische und pharmazeutische Probleme," in Erika Hickel and Gerald Schröder, eds., *Neue Beiträge zur Arzneimittelgeschichte* (Stuttgart: Wissenschaftliche Verlagsgesellschaft, 1982), 229–64.

31. Bernhard Zondek, *Die Hormone des Ovariums und des Hypophysenvorderlappens: Untersuchungen zur Biologie und Klinik der weiblichen Genitalfunktion* (Berlin: Springer, 1931), v.

32. Corner, "Early History," vi.

33. Erna Lesky, *The Vienna Medical School of the 19th Century*, trans. L. Williams and I. S. Levij (Baltimore: Johns Hopkins University Press, 1976), 427, 487.

34. See R. Chrobak, "Über Einverleibung von Eierstocksgewebe," *CfG* 20 (1896): 521–24.

35. Ibid., 522. Chrobak referred to an earlier report by F. Mainzer, "Vorschlag zur Behandlung der Ausfallserscheinungen nach Castration," *DmW* 22 (1896): 188. Mainzer had reported on the successful treatment of post-oophorectomy symptoms in one patient with ovarian extracts and had proffered the same analogy with thyroid extracts. The idea of treating these symptoms with ovarian substance had not been Mainzer's own but had been suggested to him by the gynecologist Leopold Landau. See Simmer, "Organotherapie mit Ovarialpräparaten." The analogy between the thyroid and the ovary was also used, albeit less frequently, to justify the use of ovarian extracts in women suffering from hypothyroidism. See René Moreau, *De l'opothérapie ovarienne dans la maladie de Basedow chez la femme* (Paris: Impr. des Théses de la Faculté de Médecine de Paris, 1899). On the use of thyroid therapy in the late nineteenth century, see V. C. Medvei, *The History of Clinical Endocrinology* (Carnforth, Lancs.: Parthenon, 1993), 160–62.

36. He reported having administered powdered ovarian substance to eight patients in all, with no controls and sometimes no follow-up. Of those, three patients reported a significant diminution of symptoms—of the others, one had dropped out and the rest had only very recently been put on the treatment. The powder had not caused any negative symptoms in any of the patients. Chrobak admitted that his uncontrolled experiments did not permit any conclusions about the efficacy of the treatment or its mechanism of action. Chrobak, "Über Einverleibung," 523–24.

37. See Simmer, "Organotherapie mit Ovarialpräparaten."

38. American endocrinologist Roy Hoskins would later state: "As commercially prepared [in the late nineteenth and early twentieth centuries], the ovarian preparations amounted merely to organic débris quite devoid of hormone content." R. G. Hoskins, *The Tides of Life: The Endocrine Glands in Bodily Adjustment* (London: Kegan Paul, Trench, Trubner, 1933), 204–5.

39. See Hans H. Simmer, "Innere Sekretion der Ovarien als Ursache der Menstruation: Halbans Falsifikation der Pflügerschen Hypothese," in Kurt Ganzinger, Manfred Skopec and Helmut Wyklicky, eds., *Festschrift für Erna Lesky zum 70. Geburtstag* (Vienna: Hollinek, 1981), 123–48, at 124.

40. Corner, "Early History," vi.

41. Chrobak and Knauer were apparently unaware that ovarian transplantations were being attempted at the time on humans by the American surgeon Robert T. Morris (1857–1945), who, too, was interested in allaying the sequelae of oophorectomy and had been influenced by reports of successful thyroid grafting. By 1901, Morris had performed twelve ovarian transplants and in 1906, reported that one of his transplantees had had a baby. For further information, see R. T. Morris, *Lectures on Appendicitis and Notes on Other Subjects* (New York and London: G. P. Putnam, 1895), 156–59; idem, "The Ovarian Graft," *New York Medical Journal* 62 (1895): 436–37; Hans H. Simmer, "Robert Tuttle Morris (1857–1945): A Pioneer in Ovarian Transplants," *Obstetrics and Gynecology* 35 (1970): 314–28; and J. D. Biggers, "In Vitro Fertilization and Embryo Transfer in Historical Perspective," in Alan Trounson and Carl Wood, eds., *In Vitro Fertilization and Embryo Transfer* (Edinburgh and New York: Churchill Livingstone, 1984), 3–15, on 5–6 (my thanks to Dr. Biggers for alerting me to his article). Another early report on ovarian transplantation in humans also came from the U.S.: James Glass, "An Experiment in Transplantation of the Entire Human Ovary," *Medical News* 74 (1899): 523–25. Glass (1854–1931) was well acquainted with the work of the Viennese Knauer but not with that of his fellow American Morris. Knauer was apparently unaware of the American reports until 1900. Compare E. Knauer, "Einige Versuche über Ovarientransplantation bei Kaninchen: Vorläufige Mitteilung" *CfG* 20 (1896): 524–28, 528; and idem, "Die Ovarientransplantation: Experimentelle Studie," *AfG* 60 (1900): 322–75, 323–24.

42. Again, the example of the thyroid was important. Many medical scientists, beginning with Moritz Schiff in 1884, had reported that thyroid grafting was feasible and effective in removing symptoms of thyroid deficiency, especially those following thyroidectomy. See V. C. Medvei, *History of Clinical Endocrinology,* 136, 160–62.

43. The other gynecologists who noticed the occurrence of premature menopause after oophorectomy were equally silent on the possibility that the ovary produced an internal secretion affecting the female reproductive organs. See Simmer, "Organotherapie mit Ovarialpräparate," 243–45.

44. E. Knauer, "Einige Versuche über Ovarientransplantation bei Kaninchen: Vorläufige Mitteilung"; idem, "Über Ovarientransplantation," *WkW* 12 (1899): 1219–22; and idem, "Die Ovarientransplantation: Experimentelle Studie." For an attempt to refute Knauer's claims, see E. Arendt, "Demonstration und Bemerkungen zur Ovarientransplantation," *CfG* 22 (1898): 1116–17; and for a rebuttal, Knauer, "Zu Dr. Arendt's 'Demonstration und Bemerkungen zur Ovarientransplantation' auf der

70. Versammlung deutscher Naturforscher und Ärzte zu Düsseldorf," ibid., 1257–60. On Knauer, see Hermann Knaus, "Emil Knauer, Graz," *AfG* 159 (1935): 429–31.

45. E. Knauer, "Zur Ovarientransplantation (Geburt am normalen Ende der Schwangerschaft nach Ovarientransplantation beim Kaninchen)," *CfG* 22 (1898): 201–3.

46. Knauer, "Die Ovarientransplantation: Experimentelle Studie," 340.

47. Knauer, "Über Ovarientransplantation," 1220.

48. Ibid., 1222.

49. On Halban's career and research, see Robert Köhler, "Josef Halban," *CfG* 61 (1937): 1458–66; H. H. Simmer, "Josef Halban (1870–1937): Pionier der Endokrinologie der Fortpflanzung," *WmedW* 121 (1971): 549–52; and Simmer, "Innere Sekretion der Ovarien als Ursache der Menstruation."

50. J. Halban, discussion on E. Knauer, "Über Ovarientransplantation," *WkW* 12 (1899): 1243–44. For Knauer's acceptance of the endocrine explanation, see E. Knauer, "Die Ovarientransplantation," 354.

51. See Norbert Pecher, "Halbans Lehre von der protektiven Wirkung der Sexualhormone: Eine frühe Konzeption über den Wirkungsmechanismus der Hormone" (Inaugural diss., Medical Faculty, University of Erlangen-Nürnberg, 1985).

52. In 1905, Moritz Nussbaum offered a rather strained compromise by suggesting that although sexual development could not occur in the absence of the internal secretions of the gonads, those secretions did not act directly on the end-organs but, rather, exerted their influence through the nerves. M. Nussbaum, "Innere Sekretion und Nerveneinfluß," *Ergebnisse der Anatomie und Entwickwlungsgeschichte* 15 (1905): 39–89, at 78–80.

53. Alexander Lipschütz, *The Internal Secretions of the Sex Glands: The Problem of the "Puberty Gland"* (Cambridge: Heffer, 1924), 92.

54. See Corner, "Early History of the Oestrogenic Hormones," xi; and A. J. Carlson, "Physiology of the Mammalian Ovaries" *JAMA,* 83 (1924): 1920–23.

55. W. R. Nicholson, "A Review of the Literature of Ovarian Transplantation."

56. Many of them were by biologists who seemed intent on demonstrating the mere possibility of ovarian transplantation and finding ever more sophisticated techniques for the operation. See, for example, Woldemar Grigorieff, "Die Schwangerschaft bei der Transplantation der Eierstöcke," *CfG* 21 (1897): 663–68; Hugo Ribbert, "Über Transplantation von Ovarium, Hoden und Mamma," *AEM* 7 (1898): 688–708; and G. L. Basso, "Über Ovarientransplantation," *AfG* 77 (1906): 51–62. For a review of these experiments, see W. E. Castle and John C. Phillips, *On Germinal Transplantation in Vertebrates* (Washington, D.C.: The Carnegie Institution, 1911), 2–6; and Knud Sand, "Transplantation der Keimdrüsen bei Wirbeltieren," in A. Bethe, G. von Bergmann, G. Embden, and A. Ellinger, eds., *Handbuch der normalen und pathologischen Physiologie*, 18 vols. in 25 (Berlin, 1925–32), 14, pt. 1 (1926): 251–92, at 274–92.

57. See Franklin H. Martin, "Ovarian Transplantation," *Surgery, Gynecology and Obstetrics* 35 (1922): 573–85, at 573, 582.

58. See, for instance, the negative opinion of F. Unterberger, "Hat die Ovarientransplantation praktische Bedeutung?," *DmW* 44 (1918): 903–4. Some

researchers accepted that tissue antagonisms between donor and recipient were the likely cause of graft failures. See Franklin H. Martin, "Ovarian Transplantation: A Review of the Literature and Bibliography up to and including the Earlier Months of 1915," *Surgery, Gynecology and Obstetrics* 21 (1915): 568–78, at 572.

59. W. Blair-Bell, "The Relation of the Internal Secretions to the Female Characteristics and Functions in Health and Disease," *Proceedings of the Royal Society of Medicine* 7 (1913), pt. 2, Obstetrical and Gynaecological Section, 47–100, on 66–67. In his comments on Blair-Bell's lecture, gynecologist Amand Routh added that since the entire endocrine system was closely interlinked "in a biochemical chain... we must be even more careful to avoid double oöphorectomy than hitherto, for such an operation had not only to be considered from the ovarian point of view, but from that of the functions of all the other endocrine glands, for their metabolism might be seriously affected by the exclusion of the ovarian link" (ibid., 97).

60. See obituary in *Nature* 114 (1924): 904.

61. Hans H. Simmer, "The First Experiments to Demonstrate an Endocrine Function of the Corpus Luteum: On the Occasion of the 100. Birthday of Ludwig Fraenkel (1870–1951)," *Sudhoffs Archiv* 55 (1971): 392–417.

62. A. Prenant, "La valeur morphologique du corps jaune, son action physiologique et thérapeutique possible," *Revue générale des sciences pures et appliquées* 9 (1898): 646–50.

63. See H. H. Simmer, "On the History of Hormonal Contraception, I: Ludwig Haberlandt (1885–1932) and His Concept of 'Hormonal Sterilization,'" *Contraception* 1 (1970): 3–27, at 5–9.

64. See Simmer, "The First Experiments to Demonstrate an Endocrine Function of the Corpus Luteum." Fraenkel had been trained in gynecology by, among others, Alfred Hegar (ibid., 403).

65. L. Fraenkel, "Die Funktion des Corpus luteum," *AfG* 68 (1903): 438–545. For a briefer presentation of his conclusions together with comments by other scientists of the time, see L. Fraenkel, "Weitere Mitteilungen über die Funktion des Corpus luteum," *CfG* 28 (1904): 621–36, 657–68.

66. Fraenkel, "Weitere Mitteilungen," 624.

67. It was Vilhelm Magnus's little-known research that indicated that other portions of the ovary might produce their own, different internal secretions. See Simmer, "The First Experiments to Demonstrate an Endocrine Function of the Corpus Luteum, Part II: Ludwig Fraenkel versus Vilhelm Magnus" *Sudhoffs Archiv* 56 (1972): 76–99, at 80.

68. Fraenkel, "Die Funktion des Corpus luteum," 439. Fraenkel's arguments were far from incontestable. Josef Halban, for instance, pointed out that since the uterine mucosa began to proliferate before the corpus luteum had taken shape, it was likely that the proliferation was induced by an unknown internal secretion of the fertilized ovum. Halban's chief Friedrich Schauta as well as other researchers felt that Fraenkel's method of extirpating the corpora lutea may well have damaged the ovarian tissue—his experimental results, therefore, could not indubitably be attributed to the absence of the corpora lutea alone. For Halban's comments, see Fraenkel, "Weitere Mitteilungen," 628–32, and for Schauta's, ibid., 660–61.

69. L. Fraenkel, "Die Function des Corpus luteum," 439.

70. Ibid., 489–91.

71. Ideally, remarked Fraenkel, one would use human corpora lutea but that was prevented by "ethical and other difficulties." He did not elaborate on these difficulties, but physiologically significant corpora lutea would, of course, be present only in the ovaries of pregnant women, and his own work had shown that the removal of the corpora lutea during pregnancy would lead to abortion. See ibid., 496. Non-pregnant women also developed corpora lutea after release of the ovum from the follicle, but this corpus luteum lasted only up to the commencement of menstrual bleeding.

72. Ibid., 492. Fraenkel provided the address of a chemist from whom Lutein could be obtained at the rate of 4.50 Marks for 100 tablets. See ibid., footnote 1.

73. Ibid., 496.

74. Ibid., 497; "Weitere Mitteilungen," 622.

75. "Die Function des Corpus luteum," 498–99.

76. "Weitere Mitteilungen," 627–28. The Beard-Prenant hypothesis, of course, had claimed that it was the corpus luteum that prevented ovulation, and the experimental work of Leo Loeb had shown in 1909 that the corpus luteum was one of the important agencies for preventing ovulation, which was confirmed by other researchers. Hence, it was actually more logical to use the corpus luteum (or, indeed, the ovary as a whole) to *prevent* pregnancy, and this was attempted successfully in the 1920s by the Innsbruck physiologist Ludwig Haberlandt. Haberlandt transplanted ovaries of pregnant rabbits into nonpregnant animals, which resulted frequently in temporary sterilization. For further information, see H. H. Simmer, "On the History of Hormonal Contraception, I," and idem, "On the History of Hormonal Contraception, II: Ottfried Otto Fellner (1873–19??) and Estrogens as Antifertility Hormones," *Contraception* 3, no. 1 (1971): 1–20.

77. By the early 1930s, however, the fame of the corpus luteum had spread far and wide. In Aldous Huxley's 1932 novel, *Brave New World,* pregnancy was forbidden, babies being generated artificially. But women, although subject to rigid Malthusian drills to prevent pregnancy, were required to have periodic Pregnancy Substitutes beginning from the age of twenty one. Syrup of Corpus Luteum was one of the essential components of that program. See Aldous Huxley, *Brave New World* (1932; London: Flamingo, 1994), 33.

78. Artur Biedl, *The Internal Secretory Organs,* 398–410.

79. A. Louise McIlroy, comment in the discussion following W. Blair-Bell, "The Relation of the Internal Secretions to the Female Characteristics and Functions in Health and Disease," 77.

80. Simmer, "The First Experiments to Demonstrate an Endocrine Function of the Corpus Luteum, Part II," 82.

81. "Glands," however, were not alone in assisting British gynecologists in their struggle for professional respectability. Earlier, "nerves" had served similar ends. For a discussion, see Chandak Sengoopta, "'A Mob of Incoherent Symptoms'?: Neurasthenia in British Medical Discourse, 1860–1920," in Marijke Gijswijt-Hofstra and Roy Porter, eds., *Cultures of Neurasthenia from Beard to the First World War* (Amsterdam: Rodopi, 2001), 97–115, esp. 100–102.

82. V. B. Green-Armytage, "The Rise of Surgical Gynaecology," in J. M. Munro-Kerr, R. W. Johnstone, and Miles H. Phillips, eds., *Historical Review of British Obstetrics and Gynaecology, 1800–1950* (Edinburgh: Livingstone, 1954), 357–69, at 363.

83. For biographical accounts, see John Peel, *William Blair-Bell: Father and Founder* (London: Royal College of Obstetricians and Gynaecologists, 1986); and D'Arcy Power and W. R. Le Fanu, *Lives of the Fellows of the Royal College of Surgeons of England, 1930–1951* (London: Royal College of Surgeons of England, 1953), 85–87.

84. Blair-Bell saw gynecology as "a science limited no longer . . . to the 'region below the belt,' but embracing many aspects of medicine and surgery, to both of which, indeed, it has itself largely contributed." W. Blair-Bell, letter to *BMJ*, 23 March 1929, 572, quoted in William Fletcher Shaw, *Twenty-Five Years: The Story of the Royal College of Obstetricians and Gynaecologists, 1929–1954* (London: Churchill, 1954), 31–33.

85. In a historical introduction written for the last edition of his textbook of gynecology, Bell observed: "Many of us look forward to that day when knowledge of the whole of the Genital System in Woman—its morphology, physiology and pathology—will be included in the word 'gynaecology,' and when 'obstetrics' will form a natural subdivision only of the inclusive subject." William Blair-Bell, *The Principles of Gynaecology: A Text-Book for Students and Practitioners*, 4th ed. (London: Baillière, Tindall and Cox, 1934), 1.

86. Chris Lawrence has remarked that "the practice of gynaecology was not always viewed with approval by the Victorian medical élite who valued generalism as a gentlemanly ideal over specialism." Christopher Lawrence, *Medicine in the Making of Modern Britain, 1700–1920* (London: Routledge, 1994), 60. On the opposition of Victorian professionals and intellectuals to specialism in *any* sphere, see Rosemary Jann, *The Art and Science of Victorian History* (Columbus: Ohio State University Press, 1985), 230; and Stefan Collini, *Public Moralists: Political Thought and Intellectual Life in Britain, 1850–1930* (Cambridge: Cambridge University Press, 1991), 32–33.

87. The generalist-specialist debate was fundamental to nineteenth-century British medicine and although no longer as heated in Blair-Bell's time as it had been some decades earlier, many general physicians and surgeons still considered specialists (of whom gynecologists were a particularly controversial group) to be ignorant of all bodily parts or functions save the one they claimed to know. See M. Jeanne Peterson, *The Medical Profession in Mid-Victorian London* (Berkeley and Los Angeles: University of California Press, 1978), 273–75, 277–79; William F. Bynum, *Science and the Practice of Medicine in the Nineteenth Century* (Cambridge: Cambridge University Press, 1994), 191–96; and Ornella Moscucci, *The Science of Woman*, 57–59, 72–81.

88. T. Clifford Allbutt, "On Local and Constitutional Treatment in Uterine Diseases," *BMJ*, 26 September 1885, 589–90, at 589.

89. Quoted by Peterson, *Medical Profession in Mid-Victorian London*, 129.

90. William Blair-Bell, *The Principles of Gynaecology*, 4th ed., 1.

91. W. Blair-Bell, "The Relation of the Internal Secretions to the Female Characteristics and Functions in Health and Disease," *Proceedings of the Royal Society of Medicine* 7 (1913), pt. 2, Obstetrical and Gynaecological Section, 47–100, at 47.

92. Blair-Bell, *The Sex Complex* (London: Baillière, Tindall and Cox, 1916), vii (emphasis in the original), 5.

93. He remained an agnostic about the number or histological source of the ovarian secretion(s), topics that were beginning to interest some British physiologists such as Francis Marshall. Blair-Bell contented himself with the observation that "the internal secretion or secretions of the ovaries have never been isolated; indeed it is still a matter of dispute as to how and where it is or they are produced" (ibid., 23). With regard to the corpus luteum, he remarked: "Although it is well known that the corpus luteum in the ovary of the pregnant female is considerably larger than in the non-pregnant, it is not certain that this hyperplasia has any more importance than an epiphenomenon" (ibid.)

94. Blair-Bell, *Sex Complex*, 30–35, 46–47.

95. Ibid., 35. "Any influence," Bell added, "the ovary may have over the general metabolism is ... related to and dependent on its primary reproductive functions. I do not believe that this organ influences the metabolism except in so far as this special function is concerned" (ibid., 97).

96. Ibid., 37.

97. Ibid., 39–41.

98. Ibid., 48–55.

99. See W. Blair-Bell, "The Pituitary Body and the Therapeutic Value of the Infundibular Extract in Shock, Uterine Atony, and Intestinal Paresis," *BMJ*, 4 December 1909, 1609–13; and idem, *The Pituitary: A Study of the Morphology, Physiology, Pathology, and Surgical Treatment of the Pituitary, together with an account of the Therapeutical Uses of the Extracts made from this Organ* (London: Baillière, Tindall and Cox, 1919).

100. W. Blair-Bell and Pantland Hick, "Observations on the Physiology of the Female Genital Organs," *BMJ*, 27 February 1909, 517–22; 6 March 1909, 592–97; 13 March 1909, 655–58; 20 March 1909, 716–18; 27 March 1909, 777–83. The uterine effect of posterior pituitary extract had previously been noted by Henry H. Dale, who had supplied Blair-Bell with the extract (ibid., 779).

101. It was realized only subsequently that the extract was separated into pressor and oxytocic fractions. See J. M. Munro Kerr, "Labour," in Munro Kerr, Johnstone, and Phillips eds., *Historical Review of British Obstetrics and Gynaecology*, 97–98; idem, "The Haemorrhages," ibid., 104–14; and Medvei, *History of Clinical Endocrinology*, 262–64.

102. He added, however: "I do not use the word 'invade' in any derogatory sense, for we warmly welcome their aid, which has so long been withheld, but rather because the whole scientific basis of our subject—embryological, morphological, psychological, and physiological, including the biochemical and hormonal, and pathological—has been established by practising obstetricians and gynaecologists here and in other countries." W. Blair-Bell, "Lloyd Roberts Lecture on the Present and the Future of the Science and Art of Obstetrics and Gynaecology," *BMJ*, 9 January 1932, 45–50, at 45.

103. Ibid.

104. Ibid., 49.

105. W. Blair-Bell, "The Relation of the Internal Secretions to the Female Characteristics and Functions in Health and Disease," 48, 99.

106. W. Blair-Bell, "Lloyd Roberts Lecture," 49.

107. Moscucci, *The Science of Woman,* 206.

108. Roy Porter and Lesley Hall, *The Facts of Life: The Creation of Sexual Knowledge in Britain, 1650–1950* (New Haven, Conn.: Yale University Press, 1995), 173.

109. Blair-Bell, *Sex Complex,* 106, 109.

110. W. Blair-Bell, "Hermaphroditism," *Liverpool Medico-Chirurgical Journal* 35 (1915): 272–92, at 289.

111. W. Blair-Bell, "The Correlation of Function: With Special Reference to the Organs of Internal Secretion and the Reproductive System," *BMJ,* 12 June 1920, 787–91, at 789.

112. W. Blair-Bell, "Disorders of Function," in Thomas Watts Eden and Cuthbert Lockyer, eds., *The New System of Gynaecology,* 3 vols. (London: Macmillan, 1917), 1:287–415, at 373.

113. Ibid., 401.

114. Blair-Bell, *Sex Complex,* 108.

115. W. Blair-Bell, "Lloyd Roberts Lecture," 45.

116. Ibid.

117. C. Lawrence, *Medicine in the Making of Modern Britain,* 60–61.

118. As Louis Berman, an American biochemist and popularizer of endocrine science, put it: "The reproductive mechanism of woman has rendered her whole internal secretion system, and so her nervous system, all her organs, her mind, definitely and sharply more tidal in their currents, more zigzag in their phases, more angular in their ups and downs of function, and so less predictable, reliable and dependable." L. Berman, *The Glands Regulating Personality: A Study of the Glands of Internal Secretion in relation to the Types of Human Nature* (New York: Macmillan, 1922), 154. Despite the wide popularity of Berman's book in the 1920s, there is little information to be found about the author himself. I have not been able to consult Christer Nordlund, "Hormoner och visioner i mellankrigstid: Louis Bermans idéer om möjligheten att förädla mänskligheten," *Lychnos* (2004). An English summary of the article is available at the journal's website under the title "Hormones and Visions in 1930s America: Louis Berman's Ideas on New Creations in Human Beings" (http://www.vethist.idehist.uu.se/lychnos/summaries2004.html#nordlund; accessed on 11 July 2005).

119. One should not, however, overestimate its depth—even in 1935, endocrinologist Hermann Zondek remarked that "knowledge of the male sex hormone is less advanced than that of the female substances." Hermann Zondek, *The Diseases of the Endocrine Glands,* 3d ed., trans. Carl Prausnitz (London: Edward Arnold, 1935), 69.

120. F. Leydig, "Zur Anatomie der männlichen Geschlechtsorgane und Analdrüsen der Säugethiere," *Zeitschrift für wissenschaftliche Zoologie* 2 (1850): 1–57.

121. J. Tandler and S. Grosz, "Untersuchungen an Skopzen," *WkW* 21 (1908): 277–82, at 281–82.

122. Bouin and Ancel published extensively on the subject between 1903 and 1926. See the following examples: Bouin and Ancel, "Recherches sur les cellules

interstitielles du testicule des mammifères," *Archives de zoologie expérimentale et générale,* 4th ser., 1 (1903): 437–523; and Ancel and Bouin, "Les cellules séminales ont-elles une action sur les caractères sexuels? Discussion et nouvelles recherches," *Comptes rendus hebdomadaires des séances et mémoires de la Societé de Biologie* (Paris) 75 (1923): 175–78. On their work, see Marc Klein, "Ancel, Paul Albert," *Dictionary of Scientific Biography* 1 (1970): 152–53; Klein, "Bouin, Pol André," ibid., 2 (1970): 344–46; and Klein, "Sur les interférences des sciences fondamentales et de la clinique dans l'essor de l'endocrinologie sexuelle," *Clio Medica* 8 (1973): 31–52, at 40–41.

123. Steinach's work awaits comprehensive exploration. See, however, Harry Benjamin, "Eugen Steinach, 1861–1944: A Life of Research," *Scientific Monthly* 61 (1945): 427–42; Marc Klein, "L'Oeuvre de Steinach dans l'histoire de la biologie de la reproduction," in Erna Lesky, ed., *Wien und die Weltmedizin* (Vienna: Böhlaus, 1974), 204–13; and Heiko Stoff, "Vermännlichung und Verweiblichung: Wissenschaftliche und utopische Experimente im frühen 20. Jahrhundert," in Ursula Pasero and Friederike Braun, eds., *Wahrnehmung und Herstellung von Geschlecht: Perceiving and Performing Gender* (Opladen and Wiesbaden: Westdeutscher Verlag, 1999), 47–62. Meanwhile, Steinach's self-serving but factually detailed autobiography, *Sex and Life: Forty Years of Biological and Medical Experiments* (New York: Viking, 1940) remains indispensable for an overview of his long, complex, and controversial career and should be supplemented with the perceptive discussions in Diana Long Hall, "Biology, Sex Hormones and Sexism in the 1920s," *Philosophical Forum* 5, nos. 1–2 (1973/74): 81–96, and Anne Fausto-Sterling, *Sexing the Body: Gender Politics and the Construction of Sexuality* (New York: Basic Books, 2000).

124. Walther Koerting, *Die Deutsche Universität in Prag: Die letzten hundert Jahre ihrer medizinischen Fakultät* (Munich: Bayerische Landesärztekammer, 1968), 118–21.

125. See Harry Benjamin, "Eugen Steinach, 1861–1944."

126. J. R. Tarchanoff, "Zur Physiologie des Geschlechtsapparates des Frosches," *AgP* 14 (1887): 331–51.

127. Steinach, *Sex and Life,* 21–22.

128. In both groups of castrated rats, the seminal vesicles and the prostate were markedly reduced in size. This last observation would provide Steinach with a physiological index of masculinity. Although the seminal vesicles had nothing to do with sexual excitability, its dimensions were valuable in assessing the physiological sexual status of an animal. A senile animal or a castrate would always have shrunken vesicles.

129. E. Steinach, "Untersuchungen zur vergleichenden Physiologie der männlichen Geschlechtsorgane insbesondere der accessorischen Geschlechtsdrüsen," *AgP* 56 (1894): 304–38, at 337–38.

130. There is no institutional history of the Vivarium. There are brief descriptions in Arthur Koestler, *The Case of the Midwife Toad* (New York: Random House, 1971), 21–23, 122; Wilhelm Kühnelt, "Zoologische Forschung im Bereich der Wiener Universität," *Archiv der Geschichte der Naturwissenschaften* 14/15 (1985): 663–79, at 668–70; Karl Przibram, "Hans Przibram," *Neue Österreichische Biographie ab 1815: Große Österreicher* 13 (1959): 184–91; D'Arcy W. Thompson, "Dr. Hans Przibram," *Nature* 155 (1945): 782; and Richard Meister, *Geschichte der Akademie der Wissenschaften in Wien 1847–1947* (Vienna: Holzhausen, 1947), 151, 344. Of the

238 · Notes to Page 58

scientists working at the Vivarium, only Paul Kammerer (1880–1926) has received some historical attention. Apart from Koestler's polemical biography (*The Case of the Midwife Toad*, above), see Albrecht Hirschmüller, "Paul Kammerer und die Vererbung erworbener Eigenschaften," *Medizinhistorisches Journal* 26 (1991): 26–77.

131. On the dispute among turn-of-the-century biologists over observation and experimentation, see Hans Querner, "Beobachtung oder Experiment? Die Methodenfrage in der Biologie um 1900," *Verhandlungen der Deutschen Zoologischen Gesellschaft* 68 (1975): 4–12.

132. Roux had declared in his programmatic writings that the goal of the "new" science of developmental mechanics (*Entwickelungsmechanik*) was to determine "the causes of organic forms, and hence . . . the causes of the origin, maintenance, and involution [Rückbildung] of these forms." To analyze development, the normal sequence had to be distorted by "isolating, transposing, destroying, weakening, stimulating, false union, passive deformation, changing the diet and the functional size of the parts of eggs, embryos, or more developed organisms, by the application of unaccustomed agencies like light, heat, electricity, and by the withdrawal of customary influences." W. Roux, "The Problems, Methods, and Scope of Developmental Mechanics: An Introduction to the 'Archiv für Entwickelungsmechanik der Organismen,'" trans. W. M. Wheeler, in Jane Maienschein, ed., *Defining Biology: Lectures from the 1890s* (Cambridge, Mass.: Harvard University Press, 1986), 105–48, at 107, 125 (emphases removed). On Roux and his work, see E. S. Russell, *Form and Function: A Contribution to the History of Animal Morphology* (London: John Murray, 1916; reprint, Chicago: University of Chicago Press, 1982), 314–34; and Hans Querner, "Die Entwicklungsmechanik Wilhelm Roux und ihre Bedeutung in seiner Zeit," in Gunter Mann and Rolf Winau, eds., *Medizin, Naturwissenschaft, Technik und das Zweite Kaiserreich* (Göttingen: Vandenhoeck and Ruprecht, 1977), 189–200. On the influence of Roux on endocrinology, see Wolfram Kaiser and Janos Kenéz, "Morphologische Detailforschung in der Frühgeschichte der modernen Endokrinologie," *Anatomischer Anzeiger* 136 (1974): 193–206, 334–48, at 337.

133. W. Roux, "Zur Orientirung über einige Probleme der embryonalen Entwickelung" (1885), in Roux, *Gesammelte Abhandlungen über Entwickelungsmechanik der Organismen,* 2 vols. (Leipzig: Engelmann, 1895), 2:154–55. The translation is a modified version of the one in Frederick Churchill, "Roux, Wilhelm," *Dictionary of Scientific Biography,* 11 (1975): 570–75, 572.

134. Of the scientists working at the Vivarium, Roux himself listed Przibram, Steinach, Paul Kammerer, Theodor Koppányi, and Franz Megušar as fellow-practitioners of Entwickelungsmechanik. See W. Roux, "Wilhelm Roux in Halle," in L. R. Grote, ed., *Die Medizin der Gegenwart in Selbstdarstellungen* (Leipzig: Meiner, 1923), 141–206, at 173. Many important publications of the Vivarium group—including several of Steinach's most important papers—were published in the *Archiv für Entwickelungsmechanik der Organismen,* which was founded in 1895 and edited until his death in 1924 by Roux himself. Frederick Churchill has remarked, however, that Roux's list of "followers" contains the names of all the leading embryologists and cytologists of the period and should be treated with some reserve. See Churchill, "Roux, Wilhelm," *Dictionary of Scientific Biography,* 11: 574.

135. For examples of transplantations attempted at the Zoological Section of the Vivarium, see Hans Przibram, *Tierpfropfung: Die Transplantation der Körperabschnitte,*

Organe und Keime (Braunschweig: Vieweg, 1926) and the fourteen papers from the Section in *Archiv für mikroskopische Anatomie und Entwickelungsmechanik* 99 (1923), no. 1. See also Eugen Korschelt, *Regeneration and Transplantation,* trans. Sabine Lichtner Ayed, ed. Bruce M. Carlson, 2 vols. in 3 (orig. pub. 1927–31; Canton, Mass.: Science History Publications, 1990), 2, pt. 1: 352–59.

136. E. Steinach, "Geschlechtstrieb und echt sekundäre Geschlechtsmerkmale als Folge der innersekretorischen Funktion der Keimdrüse," *Zentralblatt für Physiologie* 24 (1910): 551–66.

137. Ibid., 552–53.

138. Ibid., 555–58. Similar results were obtained by other investigators. See W. Harms, "Hoden- und Ovarialinjektionen bei Rana fusca-Kastraten," *PAgP* 133 (1910): 27–44; and A. Biedl, "Die Keimdrüsenextrakte," in A. Bethe, G. von Bergmann, G. Embden, and A. Ellinger, eds., *Handbuch der normalen und pathologischen Physiologie,* 18 vols in 25 (Berlin: Springer, 1925–32), 14 (*Fortpflanzung, Entwicklung und Wachsthum*), pt. 1 (1926): 357–426, at 372–73.

139. Steinach, "Geschlechtstrieb und echt sekundäre Geschlechtsmerkmale," 558–59. Steinach, however, provided no details regarding the preparation of the brain extracts, let alone regarding their standardization.

140. Ibid., 559–60.

141. At the end of his life, he would record a glowing tribute to laboratory rats as "my favourite test animals," while claiming, not without justification, to have introduced them into experimental biology. He was delighted to note that "in almost all biological experimentation, particularly in hormone research, the rat has since made a place of honour for itself because its cooperation, though involuntary, has contributed to countless and important successes." Steinach, *Sex and Life,* 16. For the broader context of the use of rats in Steinach's research as well as in modern experimental biology, see Cheryl A. Logan, "'Are Norway Rats . . . Things?': Diversity versus Generality in the Use of Albino Rats in Experiments on Development and Sexuality," *Journal of the History of Biology* 34 (2001): 287–314, esp. 297, where Logan agrees that Steinach "perhaps more than any other individual" established the rat's physiological utility.

142. Steinach, "Geschlechtstrieb und echt sekundäre Geschlechtsmerkmale," 560.

143. His favored location for the grafts was the internal surface of the lateral abdominal muscles. He usually transplanted both testicles in their entirety. See ibid.

144. Ibid., 562, 564.

145. Ibid., 565–66. On the persistence of sexuality in human eunuchs and castrates, see Julius Tandler and Siegfried Grosz, "Untersuchungen an Skopzen," *Wiener klinische Wochenschrift* 21 (1908): 277–82, at 279; and Alexander Lipschütz, *Internal Secretions of the Sex Glands,* 17–18. Eunuchs, however, were not usually regarded as adequately sexual, as we shall see in the next chapter. As Sir Arthur Keith put it, "in the eunuch we seem to be dealing with a new species of mankind." A. Keith, *The Engines of the Human Body,* 3d ed. (London: Williams and Norgate, 1926), 229. A comprehensive historical study of the representation of the eunuch in medicine and broader cultural realms remains to be written, but for some interesting arguments, see Gary Taylor, *Castration: An Abbreviated History of Western Manhood* (New York: Routledge, 2000).

146. Steinach, "Willkürliche Umwandlung von Säugetier-Männchen in Tiere mit ausgeprägt weiblichen Geschlechtscharakteren und weiblicher Psyche," *PAgP* 144 (1912): 71–108, at 74.

147. The contention that vasectomy caused degeneration of the germinal cells was denied by Chicago biologist Carl Moore, who was one of Steinach's most tenacious antagonists. On the basis of his own animal experiments, Moore demonstrated that vasectomy did not lead to the germinal degeneration that Steinach (or, for that matter, Ancel and Bouin) had claimed. It was the *position* of the testicle in the scrotum and the scrotal *temperature* that determined the status of the germinal element. See Carl R. Moore, "The Behavior of the Testis in Transplantation, Experimental Cryptorchidism, Vasectomy, Scrotal Insulation, and Heat Application," *Endocrinology* 8 (1924): 493–508.

148. Steinach, "Geschlechtstrieb und echt sekundäre Geschlechtsmerkmale," 564. See also E. Steinach, "Willkürliche Umwandlung," 73. This hyperproliferation, Steinach felt, was responsible for the hypersexuality of some of the graftees.

149. Steinach, "Willkürliche Umwandlung," 74.

150. Ibid., 75. In the 1920s, after the international sensation over Steinach's operation for rejuvenation (discussed in the next chapter), this term would become tediously familiar to readers of German-language medical journals. Steinach's contention that the puberty gland produced the masculinizing secretions of the testicle was strongly disputed by Central European biologists and medical scientists. The leitmotif of the German criticism was that no experimental procedure could completely destroy the germinal tissue while leaving the interstitial cells undamaged. Even if that were possible, the critics argued, the endocrine function could well be due to survival of the third cellular element of the testicle: the supportive cells of Sertoli. See the following examples drawn from a voluminous and repetitious literature: Hermann Stieve, "Entwickelung, Bau und Bedeutung der Keimdrüsenzwischenzellen," *Ergebnisse der Anatomie und Entwicklungsgeschichte* 23 (1921): 1–249; Carl Benda, "Bemerkungen zur normalen und pathologischen Histologie der Zwischenzellen des Menschen und der Säugetiere," *Archiv für Frauenkunde und Eugenetik* 7 (1921): 30–40; Carl Sternberg, "Über Vorkommen und Bedeutung der Zwischenzellen," *Beiträge zur pathologischen Anatomie und allgemeinen Pathologie* 69 (1921): 262–80; A. Schmincke and B. Romeis, "Anatomische Befunde bei einem männlichen Scheinzwitter und die Steinachsche Hypothese über Hermaphroditismus," *AEM* 47 (1921): 221–38; and Benno Romeis, "Geschlechtszellen oder Zwischenzellen? Kritisches Referat über die Ergebnisse der einschlägigen Arbeiten des letzten Jahres," 3 pts., *KW* 1 (1922): 960–64, 1005–10, 1064–67. Steinach, Bouin, and Ancel did, however, have their supporters on the "interstitial cell question." Among them were Artur Biedl, Jürgen W. Harms, Julius Tandler, Albert Kuntz, Knud Sand, and Alexander Lipschütz. See Klein, "Sur les interférences," 42; K. Sand, "L'Hermaphrodisme expérimental," 485; idem, "Degenerative Changes in the Seminal Epithelium and Associated Hyperplasia of the Interstitial Tissue in the Mammalian Testis," *Endocrinology* 5 (1921): 190–204; and E. Steinach, "Zur Geschichte des männlichen Sexualhormons und seiner Wirkungen am Säugetier und beim Menschen," 2 pts, *WkW* 49 (1936): 161–72, 196–205, at 168–69. Steinach himself thought that since the idea of rejuvenation was repugnant to many scientists, but since the results of the Steinach Operation were

unassailable, the critics attempted to demolish the theory on which the operation was founded. Later, Carl Benda and Benno Romeis both admitted that Steinach had been right about the interstitial origin of the testicular hormones. See B. Romeis, "Über ein beinahe acht Jahre altes Hodentransplantat mit erhaltener inkretorischer Funktion," *KW* 12 (1933): 1640–42, 1642; idem, "Über weitere Fälle von langjährigen Hodentransplantationen mit nachgewiesener inkretorischer Funktion," *Anatomischer Anzeiger* 94 (1943): 401–16; and C. Benda, "Diskussion" following E. Steinach, "Antagonistische Wirkungen der Keimdrüsenhormone," in Max Marcuse, ed., *Verhandlungen des I. Internationalen Kongresses für Sexualforschung, Berlin,* 5 vols. (Berlin: A. Marcus and E. Weber, 1927–28), 1 (1927): 222–23.

151. The ovaries were grafted either on the peritoneal surface of the abdominal muscles or subcutaneously in the outer layers of those muscles. See Steinach, "Willkürliche Umwandlung," 78–81.

152. The rats were between three and four weeks old and the guinea pigs between two and three weeks.

153. Ibid., 81–82.

154. These findings, Steinach asserted, invalidated Josef Halban's claim that the breasts, from conception onward, were *either* male *or* female (ibid., 87–91). For Halban's argument, see his study, "Die Entstehung der Geschlechtscharaktere: Eine Studie über den formativen Einfluß der Keimdrüse," *AfG* 70 (1903): 205–308, at 66.

155. So-called secondary pseudohermaphroditism, where *either* testes *or* ovaries co-existed with sexual characters of *both* sexes, was explainable by an incomplete differentiation of the puberty gland. In such cases, the testis probably contained female puberty gland cells. This almost casual comment foreshadowed Steinach's later involvement in the "treatment" of homosexuality. See Steinach, "Willkürliche Umwandlung," 84–86.

156. Ibid., 103–4. Many of these findings were confirmed by other researchers. For references, see Lipschütz, *Internal Secretions of the Sex Glands,* 297–99, 304–6. Lipschütz himself remained chary of drawing sweeping (and anthropomorphic) conclusions about the behavior of small animals (see ibid., 298, 365). For another critique of Steinach's purely anthropomorphic interpretations of the sexual behavior of laboratory animals, see Heinrich Poll, "Die biologischen Grundlagen der Verjüngungsversuche von Steinach," *Medizinische Klinik* 16 (1920): 917–20, at 920.

157. E. Steinach, "Feminierung von Männchen und Maskulierung von Weibchen," *Zentralblatt für Physiologie* 27 (1913): 717–23, at 720–21. Steinach's findings on lactation by feminized males were replicated by a number of investigators. For a discussion of their results and references, see Lipschütz, *The Internal Secretions of the Sex Glands,* 291–94.

158. The ovaries of an intact infant female could be similarly affected by x-ray destruction of the generative cells. The results were the same as in feminized males: full development of the breasts and lactation. In female subjects, there was also massive growth of the uterus. See Steinach, "Feminierung von Männchen," 721–22. For more details on the irradiation experiments, see Steinach and Guido Holzknecht, "Erhöhte Wirkungen der inneren Sekretion bei Hypertrophie der Pubertätsdrüsen," *AEM* 42 (1917): 490–507.

159. Steinach, "Feminierug von Männchen," 722–23.

160. He now corrected his earlier view that the female puberty gland retained germinal elements. The follicles, he now said, were retained only for about a month after the grafting. After this period, the interstitial cells were the only surviving elements, as in testicular grafts. See Steinach, "Pubertätsdrüsen und Zwitterbildung," *AEM* 42 (1917): 307–32, at 312–18. Steinach cited the major papers of Ludwig Fraenkel and also discussed the experiments of Edmund Herrmann (1875–1930) and Marianne Stein with extracts of the corpus luteum. Herrmann and Stein had found that the extract of the corpus luteum stimulated the development of the breasts but, when injected in males, stunted testicular development and inhibited spermatogenesis. See Steinach "Pubertätsdrüsen und Zwitterbildung," 319; E. Herrmann and M. Stein, "Über den Einfluß eines Hormones des Corpus luteum auf die Entwicklung männlicher Geschlechtsmerkmale," *WkW* 29 (1916): 177–78; Herrmann and Stein, "Über die Wirkung eines Hormones des Corpus luteum auf männliche und weibliche Keimdrüsen," ibid., 778–82. During the discussion on Herrmann and Stein's first report, Otfried Otto Fellner pointed out that a substance with the same properties was also present in the testicle. In a secretory sense, therefore, the male and female sex glands were hermaphroditic and only *predominantly* male or female. See Herrmann and Stein, "Über den Einfluß," 177; and O. O. Fellner, "Über die Wirkung des Placentar- und Hodenlipoids auf die männlichen und weiblichen Sexualorgane," *PAgP* 189 (1921): 199–214, at 210–12. On Fellner, see H. H. Simmer, "On the History of Hormonal Contraception, II: Otfried Otto Fellner (1873–19??) and Estrogens as Antifertility Hormones," *Contraception* 3 (1971): 1–21.

161. Steinach, "Pubertätsdrüsen und Zwitterbildung," 320–23.

162. Ibid., 323–25. A physiologist in Aldous Huxley's 1923 novel *Antic Hay* explained the situation pithily, if not entirely accurately. Referring to a cock grafted with an ovary, he observed: "When he's with hens . . . he thinks he's a cock. When he's with a cock, he's convinced he's a pullet." Aldous Huxley, *Antic Hay* (1923; Harmondsworth: Penguin, 1948), 252. Similar experimental "hermaphrodites" were created independently by Knud Sand (b. 1887) in Copenhagen. See K. Sand, "Experimenteller Hermaphroditismus: Vorläufige Mitteilung," *PAgP* 173 (1919): 1–7; and idem, "L'Hermaphrodisme expérimental," *Journal de physiologie et de pathologie générale* 20 (1922): 472–87. Sand's animals were constantly bisexual—and not, like Steinach's animals, cyclically feminine and masculine. Steinach doubted the validity of Sand's observation. See Steinach, "Künstliche und natürliche Zwitterdrüsen," 20–21.

163. Citing the identical view of embryologist Curt Herbst (1866–1946), Steinach asserted that Herbst's theoretical argument had now been proved experimentally. See Steinach, "Willkürliche Umwandlung," 105; and C. Herbst, *Formative Reize in der tierischen Ontogenese: Ein Beitrag zum Verständnis der tierischen Embryonalentwicklung* (Leipzig: Georgi, 1901), 75–76. For other conceptions similar to Herbst's, see Max Kassowitz, *Allgemeine Biologie*, 4 vols. (Vienna: Perles, 1899–1906), 2 (1899): 67; and Hans Kurella, "Zum biologischen Verständniss der somatischen und psychischen Bisexualität," *Zentralblatt für Nervenheilkunde und Psychiatrie*, new ser., 7 (1896): 234–41, at 237–38.

164. Steinach, "Pubertätsdrüsen und Zwitterbildung," 319–20. On a later occasion, Steinach described the antagonism of puberty glands as the "Battle of the

Gonads" (*Kampf der Gonaden*). See Steinach, "Künstliche und natürliche Zwitterdrüsen," 13. These phrases were reminiscent of Wilhelm Roux's celebrated concept of the "Battle of the Parts' in the development of an organism. See W. Roux, *Der Kampf der Theile im Organismus: Ein Beitrag zur Vervollständigung der mechanischen Zweckmässigkeitslehre* (Leipzig: W. Engelmann, 1881). See also Steinach's discussion of Roux's concept in his *Sex and Life*, 131.

165. In double transplantations of ovaries and testicles, prior castration of the experimental animal reduced the antagonism but did not remove it completely. See E. Steinach, "Pubertätsdrüsen und Zwitterbildung," 309–12. In 1921, Steinach's American critic Carl Moore reported successful transplantations of testicles in female rats with intact ovaries and dismissed Steinach's notion of antagonism. See C. R. Moore, "On the Physiological Properties of the Gonads as Controllers of Somatic and Psychical Characteristics, III: Artificial Hermaphroditism in Rats," *Journal of Experimental Zoology* 33 (1921): 129–71, at 164–69. Conceding that his usage of the word "antagonism" had been careless, Steinach now restricted its use to the actions of male and female hormones on the sexual characters. The rejection of grafts, he now stressed, was not necessarily due to hormonal antagonism alone. See E. Steinach and H. Kun, "Antagonistische Wirkungen der Keimdrüsen-Hormone," *Biologia Generalis* 2 (1926): 815–34, at 817–20. This refined notion, too, was rejected by Moore on the evidence of injection experiments with glandular extracts. A potent ovarian extract, for instance, produced no feminizing effects on the sex characters of castrated male rats. See C. R. Moore, "A Critique of Sex Hormone Antagonism," in A. W. Greenwood, ed., *Proceedings of the Second International Congress for Sex Research, London, 1930* (London: Oliver and Boyd, 1931), 293–303. Out of the controversy between Moore and Steinach evolved the feedback theory of the regulation of sex-glands by the pituitary, on which see chapter 4 below.

166. The results of his laboratory experiments, Steinach added, were similar to those experiments of nature, in which the puberty gland was incompletely differentiated. In a male individual with such a gland, the male cells might predominate for a long period, resulting in masculine development. Later, due to disease or functional reasons, the male cells might lose their precarious predominance, leading to the activation of the dormant female cells. The masculine traits of the individual would now be inhibited, and he would begin to show some feminine sex characters. The earlier the female cells were switched on, the greater would be the degree of feminization. In adult life, the changes might well be slight: a man, for instance, might simply develop a fullness of one breast. A female in a comparable situation might develop some facial hair. See Steinach, "Willkürliche Umwandlung," 104.

167. Steinach, "Willkürliche Umwandlung," 106.

168. In order to study sexual development, it was necessary, of course, to select sexually immature animals; glandular manipulations might lead to visible changes in the sexually mature adult, too—and Steinach's later work would be founded in that fact—but only in prepubertal animals could the ensuing changes be attributed clearly and unequivocally to the experimental procedure.

169. Steinach's challenge to Halban, the Viennese physician Wilhelm Falta agreed in 1916, was a serious one: "If Steinach's experiments were to receive full corroboration, were someone in converse manner successful in bringing to previously

castrated females the male sexual characters by implantation of the testicles, we would have to agree that they would stand in contradiction to Halban's view." Wilhelm Falta, *The Ductless Glandular Diseases,* trans. and ed. Milton K. Meyers (Philadelphia: Blakiston's, 1916), 389.

170. See Artur Biedl, *Innere Sekretion: Ihre physiologischen Grundlagen und ihre Bedeutung für die Pathologie,* 2d ed., 2 vols. (Berlin: Urban and Schwarzenberg, 1913), 2:199–343; and Francis H. A. Marshall, *The Physiology of Reproduction,* 2d ed. (London: Longmans, Green, 1922), 320–92. For a recent endorsement of Steinach's historical importance, see Anne Fausto-Sterling, *Sexing the Body,* 158–63.

171. Benjamin Harrow, *Glands in Health and Disease,* 101.

172. For numerous examples, see Lipschütz, *The Internal Secretions of the Sex Glands.*

173. Steinach was nominated for the prize in 1921 by J. H. Zaaijer (Leiden), in 1922 by Y. Sakaki (Fukuoka), in 1927 by L. Haberlandt (Innsbruck), in 1930 by H. H. Meyer, S. Klein, and E. Pick (Vienna), in 1934 by A. Durig (Vienna), and in 1938 by J. Bock et al. (Copenhagen). My thanks to the Nobel Committee for supplying me with this information.

174. See Lipschütz, *The Internal Secretions of the Sex Glands,* 298, 365; and Heinrich Poll, "Die biologischen Grundlagen der Verjüngungsversuche von Steinach," *Medizinische Klinik* 16 (1920): 917–20, at 920.

175. The relevant section is as follows: "[Meine] fem. masc. hermaph.-Versuche sind keine Geschlechts*bestimmungen,* sondern nur Geschlechts*beeinflussungen* im Sinne der zünftigen Biologen. (Weil das Geschlecht durch die Chromosomen *voraus*bestimmt sei: in praxi nutzt diese 'Bestimmung' aber nichts, da wir das Geschlecht oder wenigstens die Geschlechtsmerkmale *umstimmen* können.") Steinach's letter to Harry Benjamin dated 28 February 1923, in Eugen Steinach-Harry Benjamin Correspondence, The New York Academy of Medicine Historical Collections. (All letters in this collection are filed by date and henceforth, will be cited either as "ES to HB" or "HB to ES," followed by the date.) My thanks to the Academy for making this collection available to me. On the collection's history, see Ernest Harms, "Forty-four Years of Correspondence between Eugen Steinach and Harry Benjamin," *Bulletin of the New York Academy of Medicine* 45 (1969): 761–66; and on the career of Benjamin, see the contributions to "Memorial for Harry Benjamin," *Archives of Sexual Behavior* 17, no. 1 (1988), 3–31.

176. The sexual status of the early embryo was a subject of great interest to biologists during the nineteenth and the early twentieth centuries. Most investigators agreed that in its earliest stages, the embryo of sexually differentiated species was sexually indistinguishable. The interpretation of this apparent indeterminacy was the subject of debate. Some influential figures—such as Karl Friedrich Burdach (1776–1847) and Gabriel Gustav Valentin (1810–33)—argued that the indeterminacy was only apparent. The embryo's sex was determined at conception but simply remained indistinct during the early part of gestation. Heinrich Rathke (1793–1860), Johannes Müller (1801–58), and Karl Ernst von Baer (1792–1876) held, on the contrary, that the embryo *developed* into a male or a female from an initially undifferentiated state. The Berlin anatomist Wilhelm Waldeyer (1836–1921) argued in 1870 that the embryo was initially neutral only with regard to the external genitals. It

possessed precursors of both male and female internal genitals and sex glands and was, therefore, hermaphroditic until the development of the ovaries or the testes. For a concise overview of these theories, see W. Nagel, "Über die Entwickelung des Urogenitalsystems des Menschen," *Archiv für mikroskopische Anatomie* 34 (1889): 269–384, at 299–304. Waldeyer's view, often with the caveat that the hermaphroditism was only "potential" or even without using the word "hermaphroditism" but accepting the presence of the precursors of the genitals of *both* sexes, was widely accepted by the turn of the century. See Patrick Geddes and J. Arthur Thomson, *The Evolution of Sex* (New York: Scribner's, [1889]), 32–33, 78–79; and Robert Müller, *Sexualbiologie* (Berlin: L. Marcus, 1907), 177; Artur Biedl, *The Internal Secretory Organs: Their Physiology and Pathology*, 358, 363; and Oscar Hertwig, *Lehrbuch der Entwicklungsgeschichte des Menschen und der Wirbeltiere*, 2d ed. (Jena: G. Fischer, 1888), 303. On the concept of "potential hermaphroditism," see Paul Kammerer, "Steinachs Forschungen über Entwicklung, Beherrschung und Wandlung der Pubertät," *Ergebnisse der inneren Medizin und Kinderheilkunde* 17 (1919): 295–398, 378–80. The evidence of phylogeny appeared to be in agreement. See Charles Darwin, *The Descent of Man, and Selection in Relation to Sex*, 2 vols. in 1 (London: John Murray, 1871; rpt., Princeton, N.J.: Princeton University Press, 1981), 1:213; and Ernst Haeckel, "Gonochorismus und Hermaphrodismus: Ein Beitrag zur Lehre von den Geschlechts-Umwandlungen (Metaptosen)," *Jahrbuch für sexuelle Zwischenstufen* 13 (1913): 259–87, at 287. In early-twentieth-century Central Europe, the assumption of initial embryonic hermaphroditism faced severe criticism from Michael von Lenhossék, *Das Problem der geschlechtsbestimmenden Ursachen* (Jena: G. Fischer, 1903) and from Josef Halban, "Die Entstehung der Geschlechtscharaktere" (1903), 259–77. Both argued that sex was determined *ab ovo*. The sex-glands played an important protective role with regard to the sexual characters but did not, in any sense, determine sex.

177. Steinach, "Feminierung von Männchen," 723. Steinach's hypothesis was supported by Alexander Lipschütz, "Die Gestaltung der Geschlechtsmerkmale durch die Pubertätsdrüsen," *AEM* 44 (1918): 396–410, at 401–3. The notion of the "asexual embryo" was strongly criticized by Paul Kammerer, who was otherwise a supporter of Steinach. The embryo, said Kammerer, was potentially hermaphroditic; to say it was "asexual" was to imply that the sex characters were formed by sex-specific puberty gland secretions from "absolutely undifferentiated plasm, and therefore, so to speak, out of nothing." Kammerer, "Steinachs Forschungen," 382. None of this implied, of course, that the puberty gland played no role in adult life. Adult male rats, after castration, lost their libido, and their seminal vesicles regressed to an infantile state. Testicular implantation returned the animal to its normal condition. See E. Steinach, "Künstliche und natürliche Zwitterdrüsen und ihre analogen Wirkungen: Drei Mitteilungen," *AEM* 46 (1920): 12–37, at 15–16.

178. Not every researcher considered the gonads to be the sole endocrine determinants of sex. William Blair-Bell, for example, argued in 1915 that it was the entire endocrine system that came into play once the essential direction of sex had been set by non-endocrine factors. See W. Blair-Bell, "Hermaphroditism," 289.

179. Richard Goldschmidt, "Die biologischen Grundlagen der konträren Sexualität und des Hermaphroditismus bein Menschen," *Archiv für Rassen- und Gesellschafts-Biologie* 12 (1916–18): 1–14, at 13.

180. Julian Huxley, "Sex Biology and Sex Psychology" (1922), in Huxley, *Essays of a Biologist* (Harmondsworth: Penguin, 1939), 111–42, at 116–17. Huxley was profoundly impressed by Steinach's masculinization and feminization operations. "Steinach and others have taken new-born male guinea-pigs and have removed their testes and grafted ovaries in their place," he wrote. "The result has been an animal almost completely feminized both as regards body and mind. . . . The reverse operation, the masculinization of females, was equally successful, the animals growing large and showing all the instincts of a normal male and none of those of a normal female" (ibid., 119–20). Note Huxley's emphasis on the suppression of the characters of the other sex—a specifically Steinachian motif.

181. R. G. Hoskins, *The Tides of Life: The Endocrine Glands in Bodily Adjustment* (London: Kegan Paul, Trench, Trubner, 1933), 223.

182. Knud Sand, "Die Physiologie des Hodens," in Max Hirsch, ed., *Handbuch der inneren Sekretion,* 3 vols. in 5 (Leipzig: Kabbitzsch, 1928–32), 2, pt. 2 (1933): 2017–2268, at 2073–74, 2109.

183. See Emil Witschi and William F. Mengert, "Endocrine Studies on Human Hermaphrodites and Their Bearing on the Interpretation of Homosexuality," *Journal of Clinical Endocrinology* 2 (1942): 279–86, at 286.

184. L. R. Broster, *Endocrine Man: A Study in the Surgery of Sex* (London: William Heinemann Medical Books, 1944), 100.

Chapter Three

1. Aldous Huxley, *Brave New World* (1932; London: Flamingo, 1994), 48–49. On seeing an old person in a savage reservation, one of Huxley's characters wonders why the old people of the civilized world look so youthful. The explanation: "We keep their internal secretions artificially balanced at a youthful equilibrium. . . . Youth almost unimpaired till sixty, and then, crack! the end" (99).

2. On the general background, see V. C. Medvei, *History of Clinical Endocrinology* (Carnforth, Lancs.: Parthenon, 1993), and specifically on insulin, Michael Bliss, *The Discovery of Insulin* (Chicago: University of Chicago Press, 1982).

3. David Hamilton, *The Monkey Gland Affair* (London: Chatto and Windus, 1986), 69.

4. There were many theories of the nature of mongolism in the early twentieth century—some held that it was a congenital condition caused by endocrine dysfunction. For a brief overview, see K. Codell Carter, "Early Conjectures that Down Syndrome is Caused by Chromosomal Nondisjunction," *Bulletin of the History of Medicine* 76 (2002): 528–63, at 539–40. For an example of contemporary material, see R. M. Clark, "The Mongol: A New Explanation," *Journal of Mental Science* 74 (1928): 265–80, 739–47; 75 (1929): 261–62; 79 (1933): 328–35. The treatment of the condition with preparations of thymus, gonad, thyroid, or pituitary was often attempted: the well-known London physician Francis Crookshank had reported beneficial effects of such agents in mongolism from 1912. See F. G. Crookshank, *The Mongol in Our Midst: A Study of Man and his Three Faces,* 3d ed. (London: Kegan Paul, Trench, Trubner, 1931), 309.

5. Louis Berman, *The Personal Equation* (New York: Century, 1925), 286–87.

6. William J. A. Bailey, quoted in Charles Evans Morris, *Modern Rejuvenation Methods* (New York: Scientific Medical Publishing Co., 1926), 5–6.

7. Max Goldzieher, Introduction to Florence Mateer, *Glands and Efficient Behavior* (New York: Appleton-Century, 1935), vii. Mateer emphasized, however, that "no hormone or any other treatment so far dreamed of can actually improve a person's intelligence.... With the added hormones the patient finds it less and less difficult to overcome handicaps which have caused retardation, slowness, inertia or disinterest" (57, 71).

8. W. Cimbal, "Die Bedeutung der endokrinen Vorgänge für die Psychosen und Neurosen," in Max Hirsch, ed., *Handbuch der inneren Sekretion*, 3 vols. in 5 (Leipzig: Kabitzsch, 1928–33), 3, pt. 2 (1933), 1183–1284, at 1200.

9. See Max G. Schlapp and Edward H. Smith, *The New Criminology: A Consideration of the Chemical Causation of Abnormal Behavior* (New York: Boni and Liveright, 1928), 28. Schlapp was Professor of Neuropathology at New York Post-Graduate Medical School and Smith was the author of works such as *Famous Poison Mysteries* (New York: Dial Press, 1927).

10. Charles Evans Morris, *Modern Rejuvenation Methods* (New York: Scientific Medical Publishing Co., 1926), 8.

11. Schlapp and Smith, *The New Criminology*, 160, 209. In 1935, Herman Rubin put it more pithily: it was the organic constitution (especially the endocrine profile) that drove a man to crime. Social factors, including alcoholism and addictions, were simply "provocative causes." Herman H. Rubin, *The Glands of Life: The Story of the Mysterious Ductless Glands* (New York: Bellaire, 1935), 130.

12. Dorothy L. Sayers, *The Unpleasantness at the Bellona Club* (1921; London: New English Library, 1983), 154.

13. Agatha Christie, *The Murder at the Vicarage* (1930; London: Agatha Christie Ltd/Planet Three, [2001]), 99.

14. L. Berman, *The Glands Regulating Personality: A Study of the Glands of Internal Secretion in Relation to the Types of Human Nature* (New York: Macmillan, 1922), 167.

15. Ibid., 202–3.

16. Ibid., 243. It was a great pity, observed physician Herman Rubin, that Darwin had not lived to see the dawn of gland treatment. "If he could only have had a mixture of adrenal, gonad, and perhaps a little thyroid extract, he might have been a much better physical man." Rubin, *The Glands of Life*, 95.

17. Berman, *Glands Regulating Personality*, 234.

18. Ibid., 177–78, 183.

19. Morris, *Modern Rejuvenation Methods*, 8.

20. A. W. Rowe, *The Differential Diagnosis of Endocrine Disorders* (Baltimore: Williams and Wilkins, 1932), 110–11.

21. Morris, *Modern Rejuvenation Methods*, 80, 82.

22. Note, however, that in America at least, popular interest in psychoanalysis took forms somewhat similar to the interest in glands. See Nathan G. Hale Jr, "From Berggasse XIX to Central Park West: The Americanization of Psychoanalysis,

1919–1940," *Journal of the History of the Behavioral Sciences* 14 (1978): 299–315, esp. 303.

23. Samuel Wyllis Bandler, *The Endocrines* (Philadelphia: Saunders, 1921), iv.

24. Berman, *Glands Regulating Personality,* 21.

25. Bandler, *The Endocrines,* 312–13.

26. Berman, *Glands Regulating Personality,* 228–29.

27. Ibid., 23–24, 228–29, 283–85, 22. Reviewing Berman's book for the British newspaper *The Observer,* the well-known literary journalist John Collings Squire observed, "it is a strange idea of a God. I suppose it doesn't much matter. The man who wants to Get Omniscient Quick is no new type. . . . The frontiers of psychology and physiology are infested by hosts of these ill-balanced persons who get hold of a little truth and turn it into an idol." J. C. Squire, "Glands," review of Louis Berman, *The Glands Regulating Personality, The Observer,* 26 February 1922, 4.

28. "The Present Position of Organotherapy," a discussion led by Swale Vincent at the Section on Therapeutics and Pharmacology of the Royal Society of Medicine, *Lancet,* 20 January 1923, 130–32, at 131. On Vincent and generally on the different approaches to endocrine research in Britain in the 1920s, see Diana Long Hall, "The Critic and the Advocate: Contrasting British Views on the State of Endocrinology in the Early 1920s," *Journal of the History of Biology* 9 (1976): 269–85.

29. For another instance of such a critical approach, see A. J. Clark, "The Experimental Basis of Endocrine Therapy," *BMJ,* 14 July 1923, 51–53. Initial successes (especially with thyroid treatment) had, according to Clark, Professor of Pharmacology at University College London, "suggested boundless possibilities in the use of organ extracts. . . . There has been every temptation for therapeutic practice to outrun scientific fact, and in addition commercial enterprise has certainly done its full share in assisting the development of endocrine therapy. The result is that endocrine therapy is coming to bear a suspicious resemblance to mediaeval magic" (51).

30. "The Present Position of Organotherapy," *Lancet,* 20 January 1923, 132; and W. Langdon-Brown, *The Endocrines in General Medicine* (London: Constable, 1927), 131.

31. E. H. Starling, "Hormones," *Nature,* 1 December 1923, 795–98, at 795. On Starling, see Diana Long Hall, "The Critic and the Advocate."

32. Artur Biedl, "Organotherapy," *Harvey Lectures* 19 (1923–24): 27–38, 32–33.

33. Van Buren Thorne, "The Craze for Rejuvenation," review of Benjamin Harrow, *Glands and Health and Disease, New York Times,* 4 June 1922, Section 3, 18.

34. Swale Vincent, "The Arris and Gale Lecture on a Critical Examination of Current Views on Internal Secretion," *The Lancet,* 12 August 1922, 313–20, at 313.

35. Quoted by Van Buren Thorne, "The Craze for Rejuvenation."

36. Berman, *Glands Regulating Personality,* 130–31.

37. The widespread fascination with endocrine glands did not necessarily die out in subsequent decades. As late as in 1939, George Orwell noted that apart from death rays, rockets, and Martians, British weekly magazines for boys carried "far-off rumours of psychotherapy and ductless glands." George Orwell, "Boys' Weeklies" [written in 1939], in Orwell, *Essays* (London: Penguin, 2000), 78–100, at 92.

38. E. Steinach and R. Lichtenstern, "Umstimmung der Homosexualität durch Austausch der Pubertätsdrüsen," *Mm W* 65 (1918): 145–48; E. Steinach, "Histologische Beschaffenheit der Keimdrüse bei homosexuellen Männern," *Archiv für Entwickelungsmechanik* 46 (1920): 29–37. For a contextual analysis, see Chandak Sengoopta, "Glandular Politics: Experimental Biology, Clinical Medicine, and Homosexual Emancipation in Fin-de-Siècle Central Europe," *Isis* 89 (1998): 445–73; and Heiko Stoff, "Degenerierte Nervenkörper und regenerierte Hormonkörper: Eine kurze Geschichte der Verbesserung des Menschen zu Beginn des 20. Jahrhunderts," *Historische Anthropologie: Kultur, Gesellschaft, Alltag* 11 (2003): 225–39.

39. E. Steinach, "Künstliche und natürliche Zwitterdrüsen und ihre analogen Wirkungen: Drei Mitteilungen," *Archiv für Entwickelungsmechanik* 46 (1920): 12–37, 25; E. Steinach, "Pubertätsdrüsen und Zwitterbildung," *Archiv für Entwickelungsmechanik* 42 (1916): 307–32, at 328–30.

40. M. Foucault, *The History of Sexuality,* vol. 1: *An Introduction,* trans. Robert Hurley (New York: Vintage Books, 1978), 43.

41. On turn-of-the-century homosexual emancipation movements, see John Lauritsen and David Thorstad, *The Early Homosexual Rights Movement (1864–1935)* (New York: Times Change Press, 1974); James D. Steakley, *The Homosexual Emancipation Movement in Germany* (New York: Arno, 1975); Hans-Georg Stümke and Rudi Finkler, *Rosa Winkel, Rosa Listen: Homosexuelle und "Gesundes Volksempfinden" von Auschwitz bis heute* (Reinbek bei Hamburg: Rowohlt, 1981), 16–66; and John C. Fout, "Sexual Politics in Wilhelmine Germany: The Male Gender Crisis, Moral Purity, and Homophobia," *Journal of the History of Sexuality* 2 (1992): 388–421. On the history of biomedical research on homosexuality, see Rainer Herrn, "On the History of Biological Theories of Homosexuality," in John De Cecco and David Allen Parker, eds., *Sex, Cells, and Same-Sex Desire: The Biology of Sexual Preference* (New York: Haworth Press, 1995), 31–56; Rüdiger Lautmann, ed., *Homosexualität: Handbuch der Theorie- und Forschungsgeschichte* (Frankfurt am Main: Campus, 1993) and the essays in Vernon A. Rosario, ed., *Science and Homosexualities* (London: Routledge, 1997). What I, for the sake of convenience and relative euphony, call "homosexuality" was indeed referred to as "Homosexualität" by many German physicians of the early twentieth century, including the protagonists mentioned here. In their usage, however, the word did not simply indicate the phenomenon of same-sex attraction but also the theoretical conviction that homosexual desire was an indication of psychological and biological gender transposition: the male homosexual was psychosexually female or at least feminized. In contemporary English texts, such as those of Havelock Ellis, the designation "sexual inversion" reflected the conceptual reality more clearly, as did the older German term "konträre Sexualempfindung" ("contrary sexual feeling"). For a succinct discussion of these terminological and conceptual complexities, see Havelock Ellis, *Studies in the Psychology of Sex,* 2 vols., vol. 1, pt. 4: *Sexual Inversion* (New York: Random House, 1936), 310–17.

42. On Hirschfeld, see Manfred Herzer, *Magnus Hirschfeld: Leben und Werk eines jüdischen, schwulen und sozialistischen Sexologen* (Frankfurt: Campus, 1992), and James D. Steakley, *"Per scientiam ad justitiam:* Magnus Hirschfeld and the Sexual Politics of Innate Homosexuality," in Rosario, ed., *Science and Homosexualities,* 133–54.

43. On Hirschfeld's positivistic approach to sexual issues, see Ralf Seidel, "Sexologie als positive Wissenschaft und sozialer Anspruch: Zur Sexualmorphologie von Magnus Hirschfeld" (Inaugural diss., University of Munich, 1969).

44. Ellis, *Studies,* vol. 1, pt. 4, 73.

45. See Steakley, *Homosexual Emancipation Movement;* Lauritsen and Thorstad, *The Early Homosexual Rights Movement;* Fout, "Sexual Politics"; and Herzer, *Magnus Hirschfeld.* On the history of Paragraph 175, see Stümke and Finkler, *Rosa Winkel,* 39–48.

46. See the "Petition an die gesetzgebenden Körperschaften des deutschen Reiches behufs Abänderung des §175 des R.-Str.-G. B. und die sich daran anschliessenden Reichstags-Verhandlungen," composed by Hirschfeld and signed by the members and numerous supporters of the Committee, in *Jahrbuch für sexuelle Zwischenstufen* 1 (1899): 239–66.

47. On medical theories of homosexuality, see Frank J. Sulloway, *Freud, Biologist of the Mind: Beyond the Psychoanalytic Legend* (New York: Basic Books, 1979), 277–319, and David F. Greenberg, *The Construction of Homosexuality* (Chicago: University of Chicago Press, 1988), 397–433. On degeneration in general, see Annemarie Wettley, "Zur Problemgeschichte der 'Dégénérescence,'" *Sudhoffs Archiv* 43 (1959): 193–212; J. Edward Chamberlin and Sander L. Gilman, eds., *Degeneration: The Dark Side of Progress* (New York: Columbia University Press, 1985); Annemarie Wettley and Werner Leibbrand, *Von der "Psychopathia sexualis" zur Sexualwissenschaft* (Stuttgart: Enke, 1959), 45–55; Daniel Pick, *Faces of Degeneration: A European Disorder, c.1848–c.1918* (Cambridge: Cambridge University Press, 1989); and Rafael Huertas, "Madness and Degeneration, I: From 'Fallen Angel' to Mentally Ill," *History of Psychiatry* 2 (1992): 391–441. See also Françoise Castel, "Dégénérescence et structures: Réflexions méthodologiques à propos de l'oeuvre de Magnan," *Annales médico-psychologiques* 125 (1967): 521–36.

48. Although degeneration itself was believed to be hereditary, the specific disorders it *caused* were not necessarily so. The offspring of homosexuals, for instance, were degenerates but not necessarily homosexual. Anybody with a degenerate nervous system, on the other hand, could *acquire* homosexuality, especially if seduced in adolescence. See Richard von Krafft-Ebing, "Über gewisse Anomalien des Geschlechtstriebs und die klinisch-forensische Verwerthung derselben als eines wahrscheinlich functionellen Degenerationszeichens des centralen Nerven-Systems," *Archiv für Psychiatrie und Nervenkrankheiten* 7 (1877): 291–312, at 305–12; and R. v. Krafft-Ebing, *Psychopathia Sexualis: A Medico-Forensic Study,* 12th ed., trans. anon. (New York: Pioneer, 1939), 245–46. On the medical, juridical, and cultural importance of Krafft-Ebing's work, see Harry Oosterhuis, *Stepchildren of Nature: Krafft-Ebing, Psychiatry and the Making of Sexual Identity* (Chicago: University of Chicago Press, 2000). Individual theorists differed on the importance they accorded to heredity or environment in the genesis of homosexuality, but even the staunchest environmentalist accepted that a generalized *predisposition* to perversion (as opposed to the specific perversion itself) could well be inherited. See, for examples of such views, Alfred Binet, "Le Fétichisme dans l'Amour," *Revue philosophique* 24 (1887): 143–67, 252–74, at 153, 164–67; Albert von Schrenck-Notzing, *Die Suggestions-Therapie bei krankhaften Erscheinungen des Geschlechtssinnes* (Stuttgart: Enke,

1892), 150, 157–59, 193; Emil Kraepelin, *Psychiatrie: Ein kurzes Lehrbuch fur Studierende und Ärzte,* 8th ed., 4 vols. (Leipzig: Barth, 1909–15), 4:1952–60; and Kraepelin, "Geschlechtliche Verirrungen und Volksvermehrung," *MmW* 65 (1918): 117–20. Conversely, physicians who claimed that homosexuality was entirely innate did not deny that homosexual behavior could be learned or resorted to in exceptional circumstances, as, for instance, in prisons or boarding schools. Such sodomitic acts, however, did not constitute a "perversion." For the Berlin sexologist Iwan Bloch (1872–1922), isolated homosexual acts signified "pseudohomosexuality," whereas true homosexuality was inborn and wholly integral to the personality.

Magnus Hirschfeld accepted and popularized this distinction, basing his whole political crusade on the conviction that homosexuality was an innate condition that grew out of and in turn molded one's very being. See I. Bloch, *Das Sexualleben unserer Zeit in seinen Beziehungen zur modernen Kultur* (Berlin: Marcus, 1907), 590–91; M. Hirschfeld, *Die Homosexualität des Mannes und des Weibes,* 2d ed. (Berlin: Marcus, 1920), 187, 193–94; and Hirschfeld, "Die Ursachen und Wesen des Uranismus," *Jahrbuch für sexuelle Zwischenstufen* 5 (1903): 1–193, at 5. Hirschfeld also constructed a separate category of "bisexuality." While a pseudohomosexual was simply capable of being sexually potent with members of his own sex, the true bisexual possessed an inner sexual drive directed toward both sexes. See Hirschfeld, *Die Homosexualität,* 199–200.

49. Medical sexologists found it easier to jettison the idea of disease than that of anomaly. Havelock Ellis, who rejected all degenerationist explanations of homosexuality, argued, quoting no less an authority than Rudolf Virchow (1821–1902), that homosexuality was pathological since any deviation from the norm was pathological, without necessarily being a disease. See H. Ellis, *Studies,* vol. 1, pt. 4, 321; and A. Moll, "Die Behandlung der Homosexualität," *Jahrbuch für sexuelle Zwischenstufen* 2 (1900): 1–29, at 6.

50. See Frank Sulloway, *Freud,* 277–319; and Stephen Jay Gould, *Ontogeny and Phylogeny* (Cambridge: Harvard University Press, 1977).

51. See Sulloway, *Freud,* 290–96.

52. J. G. Kiernan, "Sexual Perversion and the Whitechapel Murders," *Medical Standard* 4 (1888): 129–30, at 130. See also G. F. Lydston, "A Lecture on Sexual Perversion, Satyriasis and Nymphomania," in his *Addresses and Essays,* 2d ed. (Louisville, Ky: Renz and Henry, 1892), 243–64, at 247.

53. See K. H. Ulrichs, *Memnon,* 184; *Formatrix,* 62–78, both in Ulrichs, *Forschungen über das Rätsel der mannmännliche Liebe* (Leipzig: Spohr, 1898; reprint, New York: Arno, 1975); and Ulrichs, "Vier Briefe von Karl Heinrich Ulrichs (Numa Numantius) an seine Verwandten," *Jahrbuch für sexuelle Zwischenstufen* 1 (1899): 36–70, at 64–69. On Ulrichs, see Hubert Kennedy, *Ulrichs: The Life and Work of Karl Heinrich Ulrichs, Pioneer of the Modern Gay Movement* (Boston: Alyson, 1988). European physicians had always accepted this formulation without necessarily endorsing Ulrichs's theory that male homosexuality was brought about not by vice or disease but by the body developing in a masculine direction and the soul in a feminine one. See, for instance, Richard von Krafft-Ebing, *Psychopathia Sexualis,* 1st ed. (Stuttgart: Enke, 1886), 58.

54. Hirschfeld, "Die Ursachen und Wesen des Uranismus," 79–86.

55. A different kind of hermaphroditism had, of course, been implicit in previous theories of homosexuality. The hermaphroditism they implied was constituted by the incongruity between the brain and the genitals: the genitalia belonged to a distinct sex and the brain to another. Borrowing the terminology of the philosopher Eduard von Hartmann (1842–1906), sexologist Albert Moll had described the homosexual as a "body-mind hermaphrodite" (Leibseelenzwitter). See A. Moll, *Untersuchungen über die Libido sexualis* (Berlin; Fischer's medicinische Buchhandlung, 1898), 477; and E. v. Hartmann, *Ausgewählte Werke*, 4 vols., vol. 4: *Philosophie des Schönen* (Leipzig: Haacke, 1887), 237–38. It was the coexistence of male genitalia and a female psychosexual personality in a single organism that constituted the hermaphroditic phenomenon. Actual physical hermaphroditism was considered to be separate from homosexuality except in some severe and exceedingly rare cases of the latter. See Alice Dreger, *Hermaphrodites and the Medical Invention of Sex* (Cambridge, Mass.: Harvard University Press, 1998), esp. 133–34.

56. These signs, he emphasized, were not necessarily flamboyant. Women with flowing beards, for instance, were well known in the medical literature but they were rarely homosexual in orientation; the same applied to men with well-developed breasts. Male homosexuals, for example, often manifested periodic, menstruation-like phenomena such as nosebleeds, bleeding from the mouth or the anus, or migraine, backaches, and depression. In 463 cases examined by Hirschfeld, the Adam's apple, the characteristic sign of laryngeal maturity appearing in males at puberty, was undeveloped in 128 homosexuals, poorly developed in 219, and normal only in 116. While the shoulders of average men were wider than their hips and vice versa in average women, the hips of homosexual males tended to be wider than or as wide as their shoulders, their hands smaller and their handshakes limper than that of average males. Psychologically, they showed a far greater impressionability and lability of mood and disposition than a "complete man" (*Vollmann*). See Hirschfeld, *Die Homosexualität*, 130–31, 133–34, 137–40, 141–45, 161. Even this incomplete list should establish that Hirschfeld did not question traditional medical conceptions of male and female sexual characters and never bothered to establish exactly what he meant by the "average male" or the "complete man." If homosexuality represented a deviation from the norm, then it was the deviation that Hirschfeld was concerned with. The concept of the norm he neither defined nor problematized.

57. The principle is enunciated unchanged in virtually everything that Hirschfeld ever published, but one of the most detailed expositions is M. Hirschfeld, "Die objektive Diagnose der Homosexualität," *Jahrbuch für sexuelle Zwischenstufen* 1 (1899): 4–35. The essence of this idea was, of course, ancient, and Ulrichs, not Hirschfeld, had been the first to apply it to homosexuality. Hirschfeld, however, breathed new life into the hypothesis, popularizing it with activists, who welcomed it because of its message and meaning, while its rigorously "scientific" tone and language facilitated its quick dissemination in (although not necessarily universal acceptance by) medical circles. While using the concept of sexual intermediacy to claim the biological kinship of homosexuals and "normal" people, Hirschfeld did not, however, use it to blur the boundaries of concepts of "normal" or "full" masculinity and femininity. See Sulloway, *Freud*, 158–60, 292–96; and Gert Hekma, "'A Female Soul in a Male Body': Sexual Inversion as Gender Inversion in Nineteenth-Century Sexology," in Gilbert Herdt, ed.,

Third Sex, Third Gender: Beyond Sexual Dimorphism in Culture and History (New York: Zone, 1994), 213–39.

58. Elsewhere, he revealed his nineteenth-century roots even more clearly by arguing that although homosexuality had nothing to do with degeneration directly, it probably worked as a "prophylactic" against degeneration. When a family began to slide toward a degenerative sequence, Nature brought about the birth of a homosexual, which, by stopping reproduction, prevented the transmission of the degenerative taint. A "cure" for homosexuality, if found, might, therefore, harm the species by allowing homosexuals to reproduce. These fundamental inconsistencies came into harsh focus when one strand of Hirschfeld's thought (that male homosexuals were pathologically feminized) compelled him to support the "treatment" of a condition that he claimed was a mere variety of nature and which, according to still another element of Hirschfeld's thought, should be left untreated for eugenic reasons. See *Die Homosexualität*, 398; and Herzer, *Magnus Hirschfeld*, 98–99.

59. The earliest glandular explanation of homosexuality that I know of was by Otto Weininger, who recorded it in a 1901 draft and never published it in its full form. Weininger had reasoned that male homosexuality could be "cured" if the subject's weak masculinity could be supplemented by testicular extracts. He is known to have attempted experiments, possibly on himself. For more details, see Chandak Sengoopta, "Science, Sexuality, and Gender in the Fin de Siècle: Otto Weininger as Baedeker," *History of Science* 30 (1992): 249–79, at 266–67. For Hirschfeld's suggestion that masculine and feminine sexual desire were engendered, respectively, by hypothetical chemical substances, which he named Andrin and Gynäcin, see M. Hirschfeld, *Naturgesetze der Liebe* (Berlin: Pulvermacher, 1912), 179, 182. In 1914, Hirschfeld had suggested that male homosexuality might be caused by a deficiency of Andrin. See Hirschfeld, *Die Homosexualität*, 416.

60. M. Hirschfeld, "Die Untersuchungen und Forschungen von Professor E. Steinach über künstliche Vermännlichung, Verweiblichung und Hermaphrodisierung," *Vierteljahresberichte des Wissenschaftlich-humanitären Komitees/Jahrbuch für sexuelle Zwischenstufen* 17 (1917): 3–21.

61. Steinach, "Pubertätsdrüsen und Zwitterbildung," 326–27.

62. In 1918, declaring that Steinach's experiments were astonishing feats of research, Hirschfeld warned that the possibility of curing homosexuality, however, did not entail its justifiability. See M. Hirschfeld, *Sexualpathologie*, 3 vols., vol. 2: *Sexuelle Zwischenstufen* (Bonn: Marcus & Weber, 1917–20), 218.

63. As noted earlier (see chapter 1), scientists had not yet developed any consensus on the immunological causes of graft rejection. Transplantation of tissue, therefore, was an ad hoc procedure and the rejection of grafts a common but ill-understood phenomenon. See Michael F. A. Woodruff, *The Transplantation of Tissues and Organs* (Springfield, Ill.: Charles Thomas, 1960), 67–69.

64. See Goldschmidt, "Die biologischen Grundlagen der konträren Sexualität und des Hermaphroditismus beim Menschen," 14; and Kenneth Walker, *Male Disorders of Sex* (London: Cape, 1930), 76. Goldschmidt's priority was not acknowledged by Steinach or Lichtenstern.

65. See Hermann Rohleder, *Moderne Behandlung der Homosexualität und Impotenz durch Hodeneinpflanzung* (Berlin: Fischer's medizinische Buchhandlung, 1917), 7–8.

Patients who refused transplantation should, Rohleder suggested, be offered testicular irradiation with x-rays to reduce the libido (23–24). Rohleder had earlier tried to treat homosexuality in men as well as women with gonadal extracts to no avail (11–13). By the time the work was printed, Steinach and Lichtenstern had published their first report of testicular transplantation in homosexuality and Rohleder added a note at the end recording the vindication of Steinach's earlier beliefs and his own endorsement of it (31). Rohleder also called for systematic postmortem examinations of testicles, to determine whether, as predicted by Steinach's experiments, the testes of bisexuals and homosexuals were less differentiated histologically than those of heterosexuals (4). This, incidentally, was before Steinach himself announced that he had found ovarian cells in the testicles of human homosexuals, on which see below. Testicular implantation, Rohleder also suggested, might be useful in cases of impotence due to deficiency of internal secretions but hoped that it would not be used in senile men, which would reprise the spectacle (*Schauspiel*) of the elderly Brown-Séquard with his testicular extracts (26, 30–31). See also Hermann Rohleder, "Heilung von Homosexualität und Impotenz durch Hodeneinpflanzung," *DmW* 43 (1917): 1509–10.

66. See R. G. Hoskins, "Studies on Vigor, IV. The Effect of Testicle Grafts on Spontaneous Activity," *Endocrinology* 9 (1925): 277–96 for a brief but exhaustive review with references to earlier work. Of the earliest reports of human testicular transplantation, one by Victor Lespinasse of Chicago had significant impact, if subsequent citations are anything to go by. See Victor D. Lespinasse, "Transplantation of the Testicle," *JAMA* 61 (22 November 1913): 1869–70. Lespinasse reported only on one case, that of a thirty-eight-year-old man, one of whose testicles had been lost in accident and the other during a hernia operation. The patient's chief reason for consulting Lespinasse was impotence. "A testicle from a normal man was easily obtained. In fact," confided Lespinasse, without revealing details of the source, "I was surprised at the number of testicles that are available for transplantation purposes. . . . On the fourth day after the operation the patient had a strong erection accompanied by marked sexual desire. He insisted on leaving the hospital to satisfy this desire." Lespinasse lost track of the patient after two years, but for that period, the patient remained potent as well as lusty. Lespinasse did not refer to internal secretions but emphasized the importance of the Leydig cells to the maintenance of male sexual characters and potency.

67. G. Frank Lydston, "Sex Gland Implantation: Additional Cases and Conclusions to Date," *JAMA* 66 (1916): 1540–43, at 1540. On testicular transplantation in effeminate men, see idem, "Two Remarkable Cases of Testicle Implantation," *New York Medical Journal* 113 (1921): 232–33; and idem, "Further Observations on Sex Gland Implantation," *JAMA* 72 (1919): 396–98.

68. Sometimes, they could be spectacular, as in the case of the "paretic dement" who was receiving antisyphilitic treatment without any improvement. Lydston, "in the hope of retarding the progress of the paresis," implanted two testicles taken from the cadaver of a teenage boy. Subsequently, the patient's condition improved remarkably and "sexual activity was so increased that he was rather inclined to complain of it." The patient died, possibly of a cerebral thrombosis, about fourteen months after the operation; the post-implantation improvements were sustained for the entire period. See Lydston, "Further Observations on Sex Gland Implantation," 397. The testicular

grafts, Lydston emphasized, often disappeared, but that did not prevent their therapeutic benefits from continuing (398).

69. R. Lichtenstern, "Mit Erfolg ausgeführte Hodentransplantation beim Menschen," *MmW* 63 (1916): 673–75, esp. 674. On Lichtenstern's earlier stint at Steinach's laboratory, see R Lichtenstern, "Bisherige Erfolge der Hodentransplantation beim Menschen," *Jahreskurse für ärztliche Fortbildung* 11, no. 4 (1920): 8–11, at 9. For references to many contemporary reports of successful testicular transplantations (for various reasons unrelated to homosexuality), see R. Lichtenstern, *Die Überpflanzung der männlichen Keimdrüse* (Vienna: Springer, 1924).

70. E. Steinach and R. Lichtenstern, "Umstimmung der Homosexualität durch Austausch der Pubertätsdrüsen," 147–48.

71. E. Steinach, "Histologische Beschaffenheit der Keimdrüse bei homosexuellen Männern," *Archiv für Entwickelungsmechanik* 46 (1920): 29–37, at 31–34. Steinach had earlier claimed to have found male interstitial cells in the ovaries of female "homosexual" goats: see Steinach, "Künstliche und natürliche Zwitterdrüsen," 26–28.

72. Julian Huxley, "Sex Biology and Sex Psychology" [1922], in Huxley, *Essays of a Biologist* (Harmondsworth: Penguin, 1939), 111–42, at 123.

73. See H. C. Rogge, "Die Bedeutung der Steinachschen Forschungen für die Frage der Pseudohomosexualität," in A. Weil, ed., *Sexualreform und Sexualwissenschaft: Vorträge gehalten auf der I. Internationalen Tagung für Sexualreform auf sexualwissenschaftlicher Grundlage in Berlin* (Stuttgart: Püttmann, 1922), 56–61.

74. See, for instance, R. Gaupp, "Das Problem der Homosexualität," *Klinische Wochenschrift* 1 (1922): 1033–38; and A. Moll, *Behandlung der Homosexualität: Biochemisch oder psychisch?* (Bonn: Marcus & Weber, 1921), 15–16, 20–21.

75. That one physician was Hermann Rohleder. Ideally, female homosexuality should, he said, be treated with ovarian transplantation, but, since that was a difficult operation, all that seemed to be feasible was x-ray treatment of the ovaries to lower the libido. See Hermann Rohleder, *Moderne Behandlung der Homosexualität und Impotenz durch Hodeneinpflanzung* (Berlin: Fischer's medizinische Buchhandlung, 1917), 25. The great clinical sexologists such as Krafft-Ebing, Havelock Ellis, and Magnus Hirschfeld always considered female homosexuality to be a manifestation of masculinization (i.e., the obverse of male homosexuality) but none of them, to my knowledge, considered a glandular solution for it. On sexological theories of female homosexuality, see Katharina Rowold, "A Male Mind in a Female Body: Sexology, Homosexuality and the Woman Question in Germany, 1869–1914," in Kurt Bayertz and Roy Porter, eds., *From Physico-Theology to Bio-Technology: Essays in the Social and Cultural History of Biosciences* (Amsterdam: Rodopi, 1998), 153–79.

76. Freud added, however, that impressive as it was, Steinach's treatment was relevant only to those cases presenting with "a very patent physical 'hermaphroditism.'" Freud, "The Psychogenesis of a Case of Homosexuality in a Woman" (1920), in James Strachey et al., eds., *The Standard Edition of the Complete Psychological Works of Sigmund Freud*, 24 vols (London: Hogarth Press, 1955), 18:145–76, at 171. "It would," Freud added, "be unjustifiable to assert that these interesting experiments put the theory of inversion on a new basis" and hasty to expect them to offer a universal means of "curing homosexuality." S. Freud, *Three Essays on the Theory of Sexuality* (1905), *Standard Edition*, 7:123–245, at 147. This comment

was added in 1920. Paul Roazen records that Freud warned a student of his that "the blind giant, the hormone man, will do a lot of damage if the dwarf psychologist does not take him out of the China shop." P. Roazen, *Freud and His Followers* (London: Penguin, 1979), 151. Personally as well as intellectually, Steinach and Freud were on good terms. See J. H. W. van Ophuijsen, "A New Phase in Clinical Psychiatry, Part I and Introduction. Endocrinologic Orientation to Psychiatric Disorders," *Journal of Clinical and Experimental Psychopathology* 12 (1951): 1–4, at 1; and Harry Benjamin, "Reminiscences," *Journal of Sex Research* 6, no. 1 (1970): 3–9, at 7.

77. George Sylvester Viereck, *Glimpses of the Great* (New York: Macaulay, 1930), 264–66. Those who did not approve of psychoanalysis, however, could easily co-opt Steinach against Freud. Julian Huxley remarked, for instance, that "if some abnormal individuals can be cured by implantation, and others are abnormal owing to an early failure of activation . . . the Freudian is robbed of some of his most cherished examples"; it was essential that "the quality of gonad secretion and the balance of all the endocrines . . . be taken into account far more than is done by the average psycho-analyst." Julian Huxley, "Sex Biology and Sex Psychology" [1922], in Huxley, *Essays of a Biologist,* 125.

78. See R. Lichtenstern, "Bisherige Erfolge der Hodentransplantation beim Menschen"; Richard Mühsam, "Der Einfluß der Kastration auf Sexualneurotiker," *DmW* 47 (1921): 155–56; and E. Kreuter, "Über Hodenimplantation beim Menschen," *Zentralblatt für Chirurgie* 46 (1919): 954–56.

79. R. Mühsam, "Über die Beeinflussung des Geschlechtslebens durch freie Hodenüberpflanzung," *DmW* 46 (1920): 823–25.

80. R. Mühsam, "Chirurgische Eingriffe bei Anomalien des Sexuallebens," *Die Therapie der Gegenwart* 67 (1926): 451–55, at 451.

81. A surgeon in Erlangen, for instance, transplanted a testicle from a homosexual in a heterosexual, who had been bilaterally castrated for undisclosed reasons. The subject failed to develop homosexual leanings and the grafted testis was histologically normal. See E. Kreuter, "Hodentransplantation und Homosexualität," *Zentralblatt für Chirurgie* 49 (1922): 538–40.

82. Max Thorek, *A Surgeon's World: An Autobiography* (Philadelphia: Lippincott, 1943), 198, 200.

83. For representative opinions, see F. Scheunig, "Zur Frage von Steinachs F-Zellen," *AfG* 116 (1923): 660–83; Benno Slotopolsky and Hans R. Schinz, "Histologische Hodenbefunde bei Sexualverbrechern," *Virchows Archiv für pathologische Anatomie und Physiologie* 257 (1925): 294–355; and K. Sand and H. Okkels, "L'Histopathologie du testicule humain chez des individus a sexualité anormale," *Comptes rendus hebdomadaires des séances et mémoires de la Société de Biologie, Paris* 123 (1936): 339–44. For a comprehensive, well-informed overview, see Benno Slotopolsky, "Über Sexualoperationen, ihre biologischen Grundlagen und ihre praktischen Ergebnisse," *KW* 7 (1928): 675–81.

84. M. Hirschfeld, "Operative Behandlung der Homosexualität," *Vierteljahresberichte des Wissenschaftlich-humanitären Comitées/Jahrbuch für sexuelle Zwischenstufen* 17 (1917): 189–90.

85. M. Hirschfeld, "Hodenbefunde bei intersexuellen Varianten," *Archiv für Frauenkunde und Eugenetik* 7 (1921): 173–74, at 174.

86. M. Hirschfeld, *Die Homosexualität*, xiv. A Hirschfeld associate attempted to save Steinach's hypothesis by arguing that other endocrine glands besides the gonads (such as the adrenals) could be involved in causing homosexuality. See Walter Grossmann, "Endokrine und psychische Mechanismen in der Ätiologie der Sexualinversion," *Zeitschrift für die gesamte Neurologie und Psychiatrie (Originalien)* 62 (1920): 309–32, at 319–23. For a different attempt to retrieve something of Steinach's hypothesis, see L. Berman, *The Personal Equation*, 252.

87. See M. Hirschfeld, "Untersuchungen," 15, 18.

88. See M. Hirschfeld, "Ist die Homosexualität körperlich oder seelisch bedingt?," *MmW* 65 (1918): 298–99. This was a rebuttal of Kraepelin's assertion that the German nation was being weakened by the "spread" of homosexuality. See Emil Kraepelin, "Geschlechtliche Verirrungen und Volksvermehrung," MmW 65 (1918): 117–20.

89. See Kurt Blum, "Homosexualität und Pubertätsdrüse," *Zentralblatt für die gesamte Neurologie und Psychiatrie* 31 (1923): 161–68, at 167.

90. M. Hirschfeld, *Geschlechtskunde*, 5 vols., vol. 3 (Stuttgart: Püttmann, 1930), 25, 537.

91. Steinach, *Sex and Life*, 91. By the time Steinach wrote his autobiography in the late 1930s, his own experimental work (unrelated to homosexuality) had led him to the conviction that even at the hormonal level, *all* human beings were male as well as female to varying degrees. See the discussion in the next chapter of his later experiments on luteinization of the ovary. Back in 1918, however, when he had introduced his transplantation treatment for homosexuality, his views on universal sexual intermediacy were actually far less expansive.

92. R. G. Hoskins, *The Tides of Life: The Endocrine Glands in Bodily Adjustment* (London: Kegan Paul, Trench, Trubner, 1933), 342, 345.

93. Steinach, *Sex and Life*, 22–23. Such endocrine analogies of senility were not unique to Steinach; nor were they confined to the sex glands alone. The signs of hypothyroidism—dry skin, loss of hair, diminished energy—were often considered to be analogous to senility, and there was a proposal to prevent the ravages of old age by administering thyroid extracts prophylactically to women from the age of thirty-five and to men from the age of forty. For a brief review of this topic, see Humphry Rolleston, *Medical Aspects of Old Age* (London, 1932), 79–83.

94. In a dismissive account of Steinach's work in his otherwise illuminating book, *The Monkey Gland Affair*, David Hamilton erroneously claimed that Steinach believed that his operation exerted its effects by preventing sperm from leaving the testis. Nothing could be farther from Steinach's conception. See Hamilton, *The Monkey Gland Affair*, 45.

95. The breed of rats used in the experiment rarely lived beyond 30 months and began to show signs of senility between 18 and 23 months. See E. Steinach, *Verjüngung durch experimentelle Neubelebung der alternden Pubertätsdrüse* (Berlin, 1920), 15. For a powerful refutation of the claim that vasectomy caused irreversible germinal atrophy, see Carl R. Moore, "The Behavior of the Testis in Transplantation, Experimental Cryptorchidism, Vasectomy, Scrotal Insulation, and Heat Application," *Endocrinology* 8 (1924): 493–508.

96. Steinach, *Verjüngung*, 30. For another contemporary report on sexual hyperexcitation after vasectomy in dogs, see Albert Kuntz, "Degenerative Changes in

the Seminal Epithelium and Associated Hyperplasia of the Interstitial Tissue in the Mammalian Testis," *Endocrinology* 5 (1921): 190–204, at 203.

97. The major signs of aging in laboratory rats, according to Steinach, were loss of weight, loss of fur, and diminution in potency and libido. Internally, the seminal vesicles remained the best guides: the more senile the animal, the more shriveled, inconspicuous, and castrate-like the vesicles. Microscopically, senile testicular tissue showed extensive degeneration of the germinal as well as the interstitial cells. Signs of senility usually appeared between 18 and 23 months. The contrast between young and old was stark in rats and it was impossible, claimed Steinach, to deceive oneself about a true rejuvenation of these animals. See Steinach, *Verjüngung*, 14–19.

98. Eugen Steinach, "Untersuchungen über die Jugend und über das Alter" (1912), appendix to *Verjüngung*, 61–63.

99. Steinach's implicit equation of aging in laboratory rats with human aging was rarely questioned; but for two rare exceptions, see Kurt Mendel, "Zur Beurteilung der Steinachschen Verjüngungsoperation," *DmW* 47 (1921): 986–89; and Ernst Payr, "Über die Steinach'sche Verjüngungsoperation," *Zentralblatt für Chirurgie* 47 (1920): 1130–39, at 1133–34. Generally, on rejuvenation in the early twentieth century, see D. Schultheiss, J. Denil and U. Jonas, "Rejuvenation in the Early Twentieth Century," *Andrologia* 29 (1997): 351–55.

100. See Robert Lichtenstern, "Die Erfolge der Altersbekämpfung beim Manne nach Steinach," *Berliner klinische Wochenschrift* 57 (1920): 989–95, at 990. It was not so much the vasectomy as the ligation that, according to Steinach, produced the beneficial effects of the operation by causing "back pressure" on the germinal portions of the testis and, soon, their atrophy. The operation was hence often called "vasoligature." See H. Benjamin, "The Steinach Operation: Report of 22 Cases with Endocrine Interpretation," *Endocrinology* 6 (1922): 776–86, at 776–77.

101. See Robert Lichtenstern, "Die Erfolge der Altersbekämpfung beim Manne nach Steinach," 990.

102. Anatomist Benno Romeis pointed out that even the simple relief of hydrocele often restored energy, vitality, and sexual desire and potency. The vasectomy may not, therefore, have made any contribution to the patient's apparent revitalization. See Benno Romeis, "Steinachs Verjüngungsversuche," *MmW* 67 (1920): 1020–21, at 1021. See also Benno Slotopolsky, "Über Sexualoperationen," 681.

103. Robert Lichtenstern, "Die Erfolge der Altersbekämpfung," 990.

104. Ibid., 992. The stimulation of hair growth was found to be a common result of the operation. Harry Benjamin explained it as being due to the reactivation of the adrenal glands (known to be involved in hair distribution) by the re-energized gonads. See H. Benjamin, "The Effects of Vasectomy (Steinach Operation)," *American Medicine* 28 (1922): 435–43, at 438.

105. Steinach, *Verjüngung*, 54–55. Since the operation of vasectomy had been performed for many years to relieve prostatic hypertrophy, some critics wondered why the rejunevative potential of vasectomy had not been noticed before Steinach. See, for instance, Benno Romeis, "Steinachs Verjüngungsversuche," 1021; and Carl R. Moore, "The Regulation of Production and the Function of the Male Sex Hormone," *JAMA* 97 (1931): 518–22, at 519. German physiologist Alfons Pütter cited older reports

claiming that far from rejuvenation, vasectomy in cases of prostatic hypertrophy had often led to grave consequences, including physical and mental decline and psychoses. See A. Pütter, "Der Nachweis der Verjüngung," *Die Naturwissenschaften* 8 (1920): 948–54, at 953. Surgeon Kenneth Walker in London remarked that the earlier vasectomists had not observed the rejuvenation of their patients simply because they had never taken sufficient care to prevent damage to the blood vessels and the nerves. See Kenneth M. Walker and J. Lumsden Cook, "Steinach's Rejuvenation Operation," *Lancet,* 2 February 1924, 223–26, at 224. See also Viktor Blum, "Verjüngung und Verjüngungsoperationen," *WmedW* 86 (1936): 989–94, at 991.

106. For reviews and excerpts of case reports, see Norman Haire, *Rejuvenation: The Work of Steinach, Voronoff, and Others* (London: Allen and Unwin, 1924); Peter Schmidt, *The Theory and Practice of the Steinach Operation with a Report on One Hundred Cases* (London: Heinemann, 1924); and Peter Schmidt, *The Conquest of Old Age: Methods to Effect Rejuvenation and to Increase Functional Activity* (New York: Dutton, 1931). See also Knud Sand, "Vasoligature (Epididymectomy) Employed ad mod. Steinach with a View to Restitution in Cases of Senium and Other States (Impotency, Depression): Operation on Man," *Acta Chirurgica Scandinavica* 55 (1923): 386–426; and Erwin Horner, "Über das Problem der Steinachschen Vasoligatur und ihre Erfolge in den ersten 10 Jahren," *Medizinische Klinik* 27 (1931): 1096–98, at 1096.

107. "Gland Treatment Spreads in America," *New York Times,* 8 April 1923, sec. 9, 2, cols. 6–7.

108. Judge John de B. Limley, "Science Can Make You Grow Younger: Dr. Lorenz Illustrates from His Own Experience How a Simple Operation Renews Youthful Energy," *Liberty,* 13 March 1926, 19–22, at 22.

109. See Harry Benjamin, "Preliminary Communication Regarding Steinach's Method of Rejuvenation," *New York Medical Journal* 114 (1921): 687–92, at 688.

110. Steinach, *Sex and Life,* 170–71.

111. On this usage, see Norman Haire, *Rejuvenation: The Work of Steinach, Voronoff, and Others,* 7; and on Haire, see Ivan Crozier, "Becoming a Sexologist: Norman Haire, the 1929 London World League for Sexual Reform Congress, and Organizing Medical Knowledge about Sex in Interwar England," *History of Science* 39 (2001): 299–329, at 299. See also Julia Rechter, "'Glands of Destiny': A History of Popular, Medical and Scientific Views of the Sex Hormones in 1920s America," Ph.D. diss, University of California at Berkeley, 1997, 184.

112. R. G. Hoskins, "Gland Research Aims to Prolong Active Life," in "Science Promises an Amazing Future: Leaders of Research in Twelve Fields Forecast Some of the Tremendous Advances Yet to be Made in Man's Conquest of Nature," *New York Times,* 20 January 1924, section 8, 3.

113. Limley, "Science Can Make You Grow Younger," 19, 21.

114. Tissue culture experiments, associated most famously with Alexis Carrel, also suggested that tissues could survive for much longer when cultivated in an artificial environment where their wastes could be washed away. See G. Stanley Hall, *Senescence: The Last Half of Life* (New York: Appleton, 1922), 285–95.

115. Steinach, *Verjüngung,* 41.

116. Horner, "Über das Problem der Steinachschen Vasoligatur," 1097; and Lichtenstern, "Die Erfolge der Altersbekämpfung beim Manne," 994.

117. See Peter Schmidt, *Conquest of Old Age,* 27–28; and E. Steinach, "Zur Geschichte des männlichen Sexualhormons und seiner Wirkungen am Säugetier und beim Menschen," *WkW* 49 (1936): 161–72, 198–205, at 168.

118. See Manfred Fraenkel, *Die Verjüngung der Frau zugleich ein Beitrag zum Problem der Krebsheilung* (Bern: Bircher, 1924), 13–14; Horner, "Über das Problem der Steinachschen Vasoligatur," 1097. Paul Kammerer suggested that the revitalized gonad exerted a stimulatory effect on the other ductless glands: it was the entire, revivified endocrine system that produced the signs of rejuvenation. See Paul Kammerer, *Rejuvenation and the Prolongation of Human Efficiency: Experiences with the Steinach-Operation on Man and Animals* (London: Methuen, 1924), 219–26. Soon, Steinach himself endorsed this claim. See E. Steinach and H. Kun, "Die entwicklungsmechanische Bedeutung der Hypophysis als Aktivator der Keimdrüseninkretion: Versuche an infantilen, eunuchoiden und senilen Männchen," *Medizinische Klinik* 24 (1928): 524–29, at 524. Steinach's critics had long pointed out that it was physiologically wrong to focus on the sex glands in isolation from the endocrine system in explaining as complex a phenomenon as aging. See, for instance, Heinrich Poll, "Die biologischen Grundlagen der Verjüngungsversuche von Steinach," *Medizinische Klinik* 16 (1920): 917–20, at 917.

119. A Prague physiologist reported, for instance, that the Steinach operation inhibited the progressive "thickening" of cellular fluids that occurred with aging. Vladimir Ruzicka, "Die Protoplasmahysteresis und das Verjüngungsproblem," *DmW* 48 (1922), 931–32.

120. Steinach, *Sex and Life,* 24–25.

121. See Stephen Lock, "'O That I Were Young Again': Yeats and the Steinach Operation" *BMJ* 287 (1983): 1964–68; and Diana Wyndham, "Versemaking and Lovemaking—W. B. Yeats' 'Strange Second Puberty': Norman Haire and the Steinach Rejuvenation Operation," *Journal of the History of the Behavioral Sciences* 39 (2003): 25–50.

122. G. S. Viereck, *Glimpses of the Great* (New York: Macaulay, 1930), 38. The belief that the Steinach operation had an ameliorative effect on cancer was quite common. See, for example, Kammerer, *Rejuvenation and the Prolongation of Human Efficiency,* 106–8; and Benjamin, "Preliminary Communication Regarding Steinach's Method of Rejuvenation," 691–92. Surgeon Erwin Horner later deprecated this early belief, blaming it entirely on credulous newspapermen. See Horner, "Über das Problem der Steinachschen Vasoligatur," 1097.

123. "Don't talk about it as long as I am alive," Freud warned Benjamin. See H. Benjamin, "Reminiscences," 7.

124. Ernest Jones, *Sigmund Freud: Life and Work,* vol. 3, *The Last Phase, 1919–1939,* (London: Hogarth Press, 1957), 104.

125. Limley, "Science Can Make You Grow Younger," 22.

126. Noel Coward, *Private Lives* (1930), in Coward, *Plays: Two* (London: Methuen, 1979), 1–90, at 50–51. Coward, of course, combined bits and pieces from

various experiments reported by Steinach and others. I am indebted to Natsu Hattori and the late Roy Porter for this reference.

127. Norman Haire, *Rejuvenation,* 126–29. Acknowledging that many of the reported improvements were subjective, Harry Benjamin asserted that "if we hear over and over again that certain subjective symptoms have been ameliorated in a very identical way," he asserted, "we are justified in attributing the effects to the cause [i.e., the operation]." H. Benjamin, "New Clinical Aspects of the Steinach Operation," *Medical Journal and Record* 122 (1925): 452–57, 515–18, 592–94, at 452.

128. One Mr. Alfred Wilson of London, whose operation in Vienna had been supervised by Steinach himself, died suddenly the day before he was scheduled to deliver a lecture at the Albert Hall on "How I was made Twenty Years Younger." This cast a shadow over the Steinach operation, although its champions attributed Wilson's death to his irresponsible attempt to "live like a young man in the twenties." See "Youth Renewed at 70 by Thyroid [*sic*] Operation, Dies on Day of Lecture on His Rejuvenation," *New York Times,* 13 May 1921, 19; Haire, *Rejuvenation,* 8–9; and Benjamin "Preliminary Communication Regarding Steinach's Method," 691. For a fictional treatment of the same theme, see Mikhail Bulgakov, *Heart of A Dog,* trans. Michael Glenny (London: Harvill, 1968), 22–23. Incidentally, from the report on the inquest on Wilson, we get an idea of the kind of fee charged by Steinach for the operation: 700 pounds. See "'Rejuvenated' Man's Death: A £700 Operation," *The Times* (London), 14 May 1921, 7c. Fees, however, varied widely. For impressionistic comments on the range in America, see Limley, "Science Can Make You Grow Younger," 21.

129. Horner, "Über das Problem der Steinachschen Vasoligatur," 1097. Whether the operation was really useful in impotence remained controversial even among its enthusiasts. "Anyone who would operate too freely in cases of impotence," Benjamin remarked, "will surely experience great disappointment." Nothing if not imaginative, Benjamin used the disappointing results to hit back at those critics who sneered at the operation as a way of re-erotizing dotards: "After such conclusions it appears rather preposterous to still regard the Steinach operation as a 'sexual operation.'" See Benjamin, "New Clinical Aspects of the Steinach Operation," 518.

130. Erwin Horner, "Über das Problem der Steinachschen Vasoligatur," 1097. Horner also speculated that the operation might well reverse male homosexuality but regretted that no studies had been conducted of its efficacy in homosexuals (ibid.). He seemed unaware of Steinach's earlier attempts to cure homosexuality with testicular transplants. "Eunuchoid" is a term that will recur in this book and was used to denote post-pubertal males who were not hermaphrodites but presented with varying degrees of physical "feminization."

131. Ibid., 1098.

132. H. Benjamin, "New Clinical Aspects of the Steinach Operation," 594; see also Haire, *Rejuvenation,* 8–9; Horner, "Über das Problem der Steinachschen Vasoligatur," 1098.

133. Schmidt, *Conquest of Old Age,* 233.

134. Steinach, *Verjüngung,* 47, 50–51, 58–60. For a very different approach to the evaluation of such experiments, see David Hamilton, *The Monkey Gland Affair,* 44.

135. For a medical report on the rejuvenation of aging women by *human* ovarian transplants (taken from patients undergoing oophorectomy for gynaecological reasons), see F. Sippel, "Die Ovarientransplantation bei herabgesetzter und fehlender Genitalfunktion," *AfG* 118 (1923): 445–89, esp. 477–81.

136. Bulgakov, *The Heart of A Dog,* 25.

137. His personal papers do not seem to have survived their confiscation by the Nazis in 1938, but in his correspondence with the New York doctor Harry Benjamin there is ample evidence to suggest that contrary to the impression given by his publications, Steinach was experimenting with various, apparently promising methods of female rejuvenation in the 1920s. Born in Berlin in 1885, Benjamin had trained in medicine at Tübingen. He moved to the United States in 1913 and was an energetic advocate of "Steinachism" in the 1920s. Benjamin first visited Steinach in 1921, "became fascinated with his sex-changing experiments in guinea pigs . . . and especially his so-called rejuvenation attempts through vasoligation. I studied with him in Vienna nearly every summer until the late 1930s." H. Benjamin, "Reminiscences," *Journal of Sex Research* 6 no. 1 (1970): 3–9, at 6; and the contributions to the "Memorial for Harry Benjamin," *Archives of Sexual Behavior* 17, no. 1 (1988): 3–31, esp. 13. Benjamin's correspondence with Steinach is available at the Historical Collections of the New York Academy of Medicine. All letters in this collection are filed by date and henceforth will be cited either as "ES to HB" or "HB to ES," followed by the date. My thanks to the Academy for making this collection available to me. On the collection's history, see Ernest Harms, "Forty-Four Years of Correspondence between Eugen Steinach and Harry Benjamin," *Bulletin of the New York Academy of Medicine* 45 (1969): 761–66.

138. Holzknecht, according to Erna Lesky, was one of the founders of "radiology as an independent discipline." See Erna Lesky, *The Vienna Medical School of the 19th century,* trans. L. Williams and I. S. Levij (Baltimore, 1976), 303–4; and Daniela Angetter, *Guido Holzknecht: Leben und Werk des österreichischen Pioniers der Röntgenologie* (Vienna: Werner Eichbauer, 1998).

139. E. Steinach and G. Holzknecht, "Erhöhte Wirkungen der inneren Sekretion bei Hypertrophie der Pubertätsdrüsen," *AEM* 42 (1917): 490–507.

140. Ibid., 500, where the exact dose of radiation and other technical details can also be found.

141. Ibid., 501. Steinach felt far more confident in identifying the hormone-secreting cells of the ovary than other scientists of the time. On the complexities of ovarian histology, as perceived by Steinach's contemporaries, see Alexander Lipschütz, *The Internal Secretions of the Sex Glands: The Problem of the "Puberty Gland"* (Cambridge: Heffer, 1924), 211–83.

142. E. Steinach, *Verjüngung,* 47, 59–60.

143. HB to ES, 12 February 1922; and HB to ES, 27 February 1922. The strong financial incentives attracted Steinach without completely overpowering him: in 1924, Benjamin sent him a woman for the rejuvenative treatment. Noticing that she had severe anaemia, Steinach sent her home with a prescription for iron and asked her to come back after a year for consideration of further treatment. See ES to HB, 17 May 1924.

144. See, for instance, Manfred Fraenkel, "Die Wirkung der Röntgenstrahlen im Hinblick auf Vererbung und Verjüngung," *Archiv für Frauenkunde und Eugenetik* 7 (1921): 254–63; idem, *Die Verjüngung der Frau zugleich ein Beitrag zum Problem der Krebsheilung* (Bern: Bircher, 1924), 38, 40–41; and Hans Thaler, "Über Röntgenbehandlung der Amenorrhöe und anderer auf Unterfunktion der Ovarien beruhender Störungen," *CfG* 46 (1922): 2034–43. For Holzknecht's rejection, see G. Holzknecht, "Gibt es eine Reizwirkung der Röntgenstrahlen?," *MmW* 70 (1923): 761–62. On the contexts of radiotherapeutic interventions in women in early-twentieth-century Central Europe, see Arne Hessenbruch, "Geschlechterverhältnis und rationalisierte Röntgenologie," in Christoph Meinel and Monnika Renneberg, eds., *Geschlechterverhältnisse in Medizin, Naturwissenschaft und Technik* (Stuttgart: Verlag für Geschichte der Naturwissenschaften und der Technik, 1996), 148–58. My thanks to Dr. Hessenbruch for his assistance with this material.

145. H. Benjamin, "The Influence of Röntgen rays on the Endocrine Glands with a Contribution to the Problem of Rejuvenation in Women," *Medical Journal and Record* 120 (1924): 585–89, at 586.

146. Ibid. Benjamin also used the radiation technique in men who would not consent to the Steinach operation or could not, for some medical reason, be operated upon. The results were unimpressive. See HB to ES, 20 February 1924.

147. H. Benjamin, "The Steinach Method as Applied to Women," *New York Medical Journal and Medical Record* 108 (1923): 750–53, at 751. Benjamin wrote in identical terms to Steinach, adding the crucial thought that psychological factors may have aided in her improvement: "Obgleich Frau A eine sehr nuechterne und durchaus nicht histerische [*sic*] Dame ist, kann ich eine psychiatrische Beeinflussung nicht ausschliessen" (HB to ES, 31 May 1922).

148. Benjamin, "Steinach Method as Applied to Women," 752. With another patient, a professional dancer in her late forties, the treatment worked so well that "her husband said she looked as she did twenty years ago. He was so impressed by the change in her appearance that he himself, a man of sixty-one, decided to have the Steinach operation performed" (752–53.)

149. See Emily Wortis Leider, *California's Daughter: Gertrude Atherton and Her Times* (Stanford: Stanford University Press, 1991), 1; Carolyn Forrey, "Gertrude Atherton and the New Woman," *California Historical Society Quarterly* 55 (1976): 194–209; and Margaret Morganroth Gullette, "Creativity, Aging, Gender: A Study of their Intersections, 1910–1935," in Anne M. Wyatt-Brown and Janice Rosen, eds., *Aging and Gender in Literature: Studies in Creativity* (Charlottesville: University Press of Virginia, 1993), 19–48, at 21–22.

150. G. Atherton, *Adventures of a Novelist* (London: Cape, 1932), 538. A virtually identical sentence appears in Atherton's novel *Black Oxen* (New York: Boni and Liveright, 1923), 135.

151. Atherton, *Adventures of a Novelist*, 539.

152. Ibid., 540.

153. H. L. Mencken, "The Gland School," *American Mercury,* quoted by Leider, *California's Daughter*, 299.

154. Leider, *California's Daughter,* 294.

155. Atherton, *Adventures of a Novelist,* 542.

156. Benjamin, "The Influence of Röntgen Rays," 587; HB to ES, 26 November 1923. Some investigators had reported that intensive electrical heating of the testicles of dogs damaged the germinal tissue in ways similar to that seen after the Steinach operation. In humans, the technique supposedly produced significant metabolic changes and the cure of impotence in two men in their forties. See W. Kolmer and P. Liebesny, "Experimentelle Untersuchungen über Diathermie," *WkW* 33 (1920): 945–46; and P. Liebesny, "Beziehungen zwischen Keimdrüsen und Hypophysen und Nachweis der zentralen Regulierung der Keimdrüsen," *Klinische Wochenschrift* 6 (1927): 52–56. Steinach himself was extraordinarily secretive about his diathermy technique, and all he ever published on it was a brief report on its experimental use in castrated male guinea pigs. See E. Steinach, "Zur Geschichte des männlichen Sexualhormons und seiner Wirkungen am Säugetier und beim Menschen," *WkW* 49 (1936): 161–72, 198–205, at 196–98. Steinach's letters to Benjamin do not contain much detail on diathermy (with the partial exception of ES to HB, 23 October 1923) but are full of warnings about the need to maintain utter secrecy: see, for instance, ES to HB, 11 September 1923. Steinach's Berlin disciple Peter Schmidt published an enthusiastic account of the diathermy technique in his 1928 book on rejuvenation, which was translated in 1931 as *The Conquest of Old Age* (see 92–98). Diathermy was often combined with the administration of very small doses of yohimbine and radium, which, Steinach believed, heightened the response of the glandular tissue. See ES to HB, 11 September 1923.

157. See HB to ES, 24 April 1924; HB to ES, 12 November 1924. Steinach warned Benjamin that it was only too easy to bring about total castration and hormonal deficiencies if the dosage was even slightly beyond that required. He himself had not yet had too many impressive cases (ES to HB, 29 March 1922).

158. See ES to HB, 27 October 1923, offering free diathermy treatment to Atherton but warning Benjamin not to reveal the name or the nature of the new treatment to her; see ES to HB, 20 August 1934, for Steinach's appreciative comments on Atherton's referrals. When Steinach began to use his hormonal preparation Progynon for rejuvenating women, Benjamin, on his advice, treated her with high doses of it (see HB to ES, 14 September 1934 and ES to HB, 25 August 1937). There are records of other women patients (at least one of whom was quite well known) in the Benjamin-Steinach correspondence but because of ethical reasons, I have avoided discussing their cases here. Atherton's own open avowals of her treatment(s), of course, free me from such considerations.

159. "Dan Jefferis lay on a table in the clinic of Dr King Heskamp, taking his bi-weekly diathermic treatment of the pituitary gland. His temples, growing hotter every minute, were clasped by electrodes, the high-frequency machine on his left humming cheerfully." Gertrude Atherton, *The Sophisticates* (New York: Horace Liveright, 1931), 129. Diathermy as well as x-ray treatment of the pituitary were considered to be helpful in alleviating menopausal symptoms in women. See "Treatment of Climacteric Conditions by the Action of X Rays on the Hypophysis," *Lancet,* 26 April 1924, 808; and J. B. Porchownik, "Zur Behandlung der

klimakterischen Ausfallserscheinungen mittels Schilddrüsen- u.
Hypophysenbestrahlung nack Borak," *Strahlentherapie* 24 (1927): 701–9.

160. Atherton, *The Sophisticates,* 131.

161. Atherton spoke freely about her own rejuvenation, and in 1935 a reporter
noted that "she moves, acts, and speaks with decision. Her voice has none of the
falsetto quality generally found in 78-year-olds; rather, it tends to huskiness. She talks
rapidly and her swift conversation reflects a brilliant, supple brain." See "Rejuvenation:
78-Year-Old Novelist Feels 30 Years Less," *Newsweek,* 14 December 1935, 40.
Atherton also tried other means of rejuvenation. Shortly after the publication of *Black
Oxen,* she received transplants of sheep ovaries and simultaneous treatment with
pituitary extracts from the New York surgeon H. Lyons Hunt, who was also active in
rejuvenating men with transplants of sheep testicles. The results of this procedure,
however, were disappointing—no bestseller came out of it.

162. Leider, *California's Daughter,* 348.

163. Schmidt, *Conquest of Old Age,* 97–98.

164. Limley, "Science Can Make You Grow Younger," 21. Some renowned
experimental researchers were more hopeful. The gynecologist and endocrinologist
Bernhard Zondek reported that the administration of the follicular hormone to
senile female mice reestablished the sexual cycle. Citing Steinach himself, Zondek
pronounced that the effect of follicular hormone in senile female mice could only be
described as a genuine reactivation of ovarian function. See Bernhard Zondek,
*Die Hormone des Ovariums und des Hypophysenvorderlappens: Untersuchungen zur
Biologie und Klinik der weiblichen Genitalfunktion* (Berlin: Springer, 1931), 101–2.

165. See George F. Corners [pseudonym of George Sylvester Viereck],
Rejuvenation: How Steinach Makes People Young (New York: Seltzer, 1923), 82.

166. See William Wolf, *Endocrinology in Modern Practice,* 2d ed. (Philadelphia:
Saunders, 1939), 230.

167. Schmidt, *Conquest of Old Age,* 31–32, 291.

168. "In the future, aging enchantresses, desirous of retaining their charms, will
combine the [Steinach irradiation] treatment with plastic surgery," predicted the
Steinach enthusiast George Viereck. See G. F. Corners, *Rejuvenation,* 83. On the
history of plastic surgery in general, see Sander L. Gilman, *Making the Body Beautiful:
A Cultural History of Aesthetic Surgery* (Princeton: Princeton University Press, 1999),
esp. 295–328.

169. Harry Benjamin, "The Story of Rejuvenation," *American Mercury,* December
1935, [2]. Cited from independently paginated reprint in the Benjamin-Steinach
correspondence (Box: Eugen Steinach, Biography, Photographs, Articles, Letters
1920–1927; Folder: Biographies of Steinach).

170. Atherton, although she had children, was convinced that she herself did not
possess any "instinct for maternity." See Leider, *California's Daughter,* 44.

171. Atherton, *Black Oxen,* 176.

172. E. E. Cummings, "?," in Cummings, *Complete Poems,* ed. George J. Firmage
(New York: Liveright, 1991), 243. The poem was published in Cummings's 1926
collection *is 5.* See also Thierry Gillyboeuf, "The Famous Doctor Who Inserts

Monkeyglands in Millionaires," *Spring,* 9 (2000), www.gvsu.edu/english/cummings/issue9/Gillybo9.htm, accessed on 21 May 2002. On the problem of capitalizing Cummings's name, see Norman Friedman, "Not 'e. e. cummings'," *Spring* 1 (1992), www.gvsu.edu/english/Cummings/caps.htm, accessed on 21 May 2002.

173. R. G. Hoskins, *The Tides of Life: The Endocrine Glands in Bodily Adjustment* (London: Kegan Paul, Trench, Trubner & Co., 1933), 186–87.

174. H. Benjamin, "Preliminary Communication Regarding Steinach's Method," 687. The confusion continues to this day. See, for a recent example, Lesley A. Hall, *Sex, Gender and Social Change in Britain since 1880* (Basingstoke: Macmillan, 2000), 108.

175. Hamilton, *The Monkey Gland Affair.*

176. *Human* testicular transplantation was attempted by Lydston, as we have seen, and also by L. L. Stanley, the resident physician of the California State Prison at San Quentin. Inspired by the work of Lydston and by the easy availability of human testicles from executed prisoners (about three every year), Stanley and his assistant began to implant testicular tissue in the scrotum of prisoners suffering from conditions ranging from "mental dullness" to "traumatic testicular atrophy" from 1918. Although the grafts did not always survive for long, there were prolonged beneficial effects (subjective improvement, erections, deepening of voice and so forth) which Stanley and his colleague attributed to unknown stimulating substances emanated from the necrotising graft. See L. L. Stanley and G. David Kelker, "Testicle Transplantation," *JAMA* 74 (1920): 1501–3. Stanley later devised a new way of implantation: instead of transplanting the tissue surgically in the scrotum, he injected lightly crushed testicular tissue with a large-bore syringe (to avoid "undue maceration") into subcutaneous space anywhere in the body. Over three hundred prisoners (suffering from a veritable supermarket of ailments from acne to neurasthenia, from tuberculosis to dementia praecox) volunteered for the injections. Many improvements were reported, such as improvement of eyesight, "a feeling of buoyancy," and "increased sexual activity." L. L. Stanley, "Testicular Substance Implantation," *Endocrinology* 5 (1921): 708–14; and idem, "An Analysis of One Thousand Testicular Substance Implantations," *Endocrinology* 6 (1922): 787–94.

177. Hamilton, *Monkey Gland Affair,* 6, 9, 11, 31, 48. Evelyn Bostwick's daughter Joe Carstairs alleged (without any known evidence) that her mother had been murdered by Voronoff for her money. See Kate Summerscale, *The Queen of Whale Cay* (London: Fourth Estate, 1997), 42. My thanks to Lesley Hall for this reference.

178. Hamilton, *Monkey Gland Affair,* 40–42.

179. S. Voronoff, *Life: A Study of the Means of Restoring Vital Energy and Prolonging Life,* trans. E Bostwick Voronoff (New York: E. Dutton, 1920), 58. American physician Herman Rubin would claim in 1935 that "in all the history of the world, there is not one outstanding example of any great work of art, architecture, painting or sculpture accomplished by a eunuch." Herman H. Rubin, *The Glands of Life: The Story of the Mysterious Ductless Glands* (New York: Bellaire, 1935), 60.

180. The grafts were pronounced to be histologically functional by pathologist Edouard Retterer of the École de Médecine. For histological reports on a number of Voronoff's grafts, see Serge Voronoff and George Alexandrescu, *Testicular Grafting from Ape to Man,* trans. Theodore C. Merrill (London: Brentano's, n.d. [1930?]), 33–96.

181. See Voronoff's letter to Lydston, quoted by Max Thorek, *A Surgeon's World: An Autobiography* (Philadelphia: Lippincott, 1943), 183–84.

182. Quoted in Hamilton, *Monkey Gland Affair*, 57.

183. Voronoff and Alexandrescu, *Testicular Grafting*, 2.

184. Quoted by Hamilton, *Monkey Gland Affair*, 76. The monkey testicles were sliced and about three slices grafted directly on to the patient's own testicles. Voronoff insisted on grafting testicles in their natural location. See Voronoff and Alexandrescu, *Testicular Grafting*, 10. See also Kenneth M. Walker, "Hunterian Lecture on Testicular Grafts," *Lancet*, 10 February 1924, 319–26, at 318.

185. Hamilton, *Monkey Gland Affair*, 66, 76, 91.

186. Voronoff and Alexandrescu, *Testicular Grafting*, 27–32.

187. Later, however, he would charge between 500 and 1,000 pounds, comparable to the one recorded instance of the fee for a Steinach operation. See Hamilton, *Monkey Gland Affair*, 51, 62.

188. Thorek, *A Surgeon's World*, 168, 187–88.

189. Ibid., 179. Voronoff offended G. Frank Lydston by failing to acknowledge the Chicago surgeon's priority. See ibid., 180–81.

190. Serge Voronoff, *Life*, 116–17.

191. On the theme of human-ape miscegenation in early science fiction, see Marc Angenot and Nadia Khouri, "An International Bibliography of Prehistoric Fiction," *Science Fiction Studies* 8, no 1 (March 1981), available at http://www.trussel.com/prehist/angenot.htm, accessed on 6 May 2002; and specifically on Champsaur's novel, Brett A. Berliner, "Mephistopheles and Monkeys: Rejuvenation, Race, and Sexuality in Popular Culture in Interwar France," *Journal of the History of Sexuality* 13 (2004): 306–25.

192. See "Voronoff's New Tests Gain Press Support—Paris Newspapers Defend Gland Scientist—Now Seeking to Rejuvenate Women," *New York Times,* 9 October 1922, 4, col. 3. See also "Gland Grafting on Women: Operation was Performed Two Years Ago with Negative Results," *New York Times,* 28 November 1922, 12, col. 1.

193. W. S. Halsted, "Auto- and Isotransplantation in Dogs, of the Parathyroid Glandules," *Journal of Experimental Medicine* 11 (1909): 175–99. For an authoritative endorsement of the principle, see Harvey Cushing, *The Pituitary Body and its Disorders: Clinical States produced by Disorders of the Hypophysis Cerebri* (Philadelphia: Lippincott, 1912), 11.

194. Steinach, *Sex and Life*, 81–82.

195. ES to HB, 28 February 1923. Voronoff, on the other hand, was apparently always tactful about Steinach, while not concealing his differences. "Steinach improves the old horse. I yoke a young horse with the old," he is reported to have observed. See Viereck, *Glimpses of the Great,* 274–75; and Thorek, *A Surgeon's World,* 185–86.

196. R. G. Hoskins, "Studies on Vigor, IV. The Effect of Testicle Grafts on Spontaneous Activity," *Endocrinology* 9 (1925): 277–96, at 288, 290, 294.

197. Voronoff and Alexandrescu, *Testicular Grafting*, 114–19. See also Kenneth M. Walker, "Hunterian Lecture on Testicular Grafts," *Lancet,* 10 February 1924, 319–26, at 325, for an early claim on the longevity of monkey gland grafts.

198. Benno Slotopolsky, "Über Sexualoperationen," 677.

199. Kenneth M. Walker, "Hunterian Lecture on Testicular Grafts," 326.

200. Robert T. Morris, "A Case of Testicle Grafting with Unexpected Results," *JAMA* 67 (1916): 741–42, at 741. Morris now speculated whether in his earlier case of ovarian transplantation, the patient had been able to bear children because the graft had stimulated the production of ova by "latent cell rests in the broad ligaments" (742). For a similar report of a patient's atrophic testes being stimulated by a graft, see G. F. Lydston, "Two Remarkable Cases of Testicle Implantation," *New York Medical Journal* 113 (1921): 232–33, at 232.

201. Robert T. Morris, "A Case of Testicle Grafting," 742.

202. Kenneth M. Walker, "Hunterian Lecture on Testicular Grafts," 326; and idem, *Male Disorders of Sex* (London: Cape, 1930), 71. By the 1940s, Walker had lost his earlier faith in rejuvenative surgery. "Steinach and Voronoff's methods of rejuvenation," he declared, "do not rest on any secure scientific foundation." Walker, *The Circle of Life: A Search for an Attitude to Pain, Disease, Old Age and Death* (London: Cape, 1942), 69.

203. Monkey gland transplants as well as other procedures of glandular rejuvenation were also overshadowed by the gradual availability of potent hormone extracts and then, synthetic sex hormones. See Hamilton, *Monkey Gland Affair*, 128–29, 140.

204. Thorek, *A Surgeon's World*, 186, 190.

205. See the faint allusion to the issue in Medvei, *History of Clinical Endocrinology*, 223.

206. See Patrick M. McGrady Jr, *The Youth Doctors* (New York: Coward-McCann, 1968), 43–44.

207. Schmidt, *Conquest of Old Age*, 238–39; Viereck, *Glimpses of the Great,*249; and G. F. Corners [pseud. of G. S. Viereck], *Rejuvenation*, vii, 1, 95.

208. E. H. Starling, "Hormones," *Nature*, 1 December 1923, 795–98, at 797.

209. Thomas R. Cole, *The Journey of Life: A Cultural History of Aging in America* (Cambridge: Cambridge University Press, 1992), 196.

210. Gerhard van Swieten, quoted in Stefan Schmorrte, "Alter und Medizin: Die Anfänge der Geriatrie in Deutschland," *Archiv für Sozialgeschichte* 30 (1990): 15–41, at 18.

211. On Mechnikov's theories, see Élie Metchnikoff, *The Nature of Man: Studies in Optimistic Philosophy,* trans. C. Mitchell (New York: Putnam's, 1903); idem, *The Prolongation of Life: Optimistic Studies,* trans. C. Mitchell (New York: Putnam's, 1910); Olga Metchnikoff, *Life of Élie Metchnikoff: 1845–1916* (Boston: Houghton Mifflin, 1921); and Chandak Sengoopta, "Rejuvenation and the Prolongation of Life," *Perspectives in Biology and Medicine* 37 (1993): 55–66. Mechnikov's theory of aging could be combined seamlessly with endocrine hypotheses in the early twentieth century. See Leonard Williams, "The Interstitial Gland," *BMJ*, 27 May 1922, 833–35, at 834–35.

212. See T. Brailsford Robinson, *The Chemical Basis of Growth and Senescence*, 169–70.

213. Charles R. Stockard, *The Physical Basis of Personality* (New York: Norton, 1931), 295–96.

214. "The ox," argued Charles Stockard, "does not have a decidedly shorter life than the bull, nor certainly the castrated horse does not die earlier than the mare or the stallion." C. R. Stockard, "Present Status of the Problem of So-Called Rejuvenation," *Bulletin of the New York Academy of Medicine,* 2d ser., 4 (1928): 1241–49, at 1247–48.

215. Artur Biedl, "Organotherapy," *Harvey Lectures* 19 (1923–24): 27–38, 35–36. See also Max Marcuse, "I. Internationale Tagung für Sexualreform, 15.–20. IX. 1921 in Berlin," *DmW* 47 (1921): 1247–48, at 1247.

216. Schmidt, *Conquest of Old Age,* 32. The deficiency-disease concept of aging was not necessarily shared universally. Alfred Scott Warthin asserted, for instance, that "*involution* is a biologic entity equally important with *evolution* in the broad scheme of the immortal procession of life. Its processes are *physiologic* as those of growth . . . no slur or stigma of *pathologic* should be cast upon this process." A. S. Warthin, *Old Age: The Major Involution* (New York: Hoeber, 1929), 163–64, emphases in the original.

217. William J. A. Bailey, quoted in Charles Evans Morris, *Modern Rejuvenation Methods* (New York: Scientific Medical Publishing Co., 1926), 16.

218. Hans Much, "Grundlagen der Organtherapie: Eine kritisch-skeptische Studie," in Max Hirsch, ed., *Handbuch der inneren Sekretion,* 3 vols. in 5 (Leipzig: Kabitzsch, 1928–33), 3, pt. 2 (1933): 1901–78, at 1969–70. Other critics of Steinach had argued earlier that the restitutive effects of the Steinach operation (such as return of appetite, weight gain, and general revitalization) were due not to enhanced hormonal secretions from a supposedly hypertrophied "puberty gland" but to the absorption of the disintegrating germinal cells and their secretions. See B. Romeis, "Untersuchungen zur Verjüngungshypothese Steinachs," *MmW* 68 (1921): 600–603, at 602–3; and Alfred Kohn, "'Verjüngung' und 'Pubertätsdrüse,'" *Medizinische Klinik* 17 (1921): 804–6. For a later argument on the same lines, see Benno Slotopolsky, "Über Sexualoperationen," 679. Even Steinach's supporters emphasized the need for detailed physiological studies to assess the objective biological consequences of the operation. See A. Loewy and Hermann Zondek, "Der Einfluß der Samenstrangunterbindung (Steinach) auf den Stoffwechsel," *DmW* 47 (1921): 349–50.

219. Benno Romeis, "Altern und Verjüngung," 1966.

220. See Harry Benjamin, "The Effects of Vasectomy (Steinach Operation)," *American Medicine,* 28 (1922): 435–43, at 437; and the excellent discussion in Jessica Jahiel, "Rejuvenation Research and the American Medical Association in the Early Twentieth Century: Paradigms in Conflict" (Ph.D. diss., Boston University, 1992), 125–31. For an informative overview of rejuvenation in 1920s America, see Laura Davidow Hirshbein, "The Glandular Solution: Sex, Masculinity, and Aging in the 1920s," *Journal of the History of Sexuality* 9 (2000): 277–304. On the rejuvenation craze in Germany, see Heiko Stoff, *Ewige Jugend: Konzepte der Verjüngung vom späten 19. Jahrhundert bis ins Dritte Reich* (Cologne: Böhlau, 2004).

221. Knud Sand observed that the term was to be deprecated because of "its sensational sound." Sand, "Vasoligature (Epididymectomy) Employed ad mod. Steinach," 387–88. Kenneth Walker pointed out that the word was an unfortunate choice because "it encouraged the raising of extravagant hopes and resulted in the premature exploitation of a method of 'rejuvenation' that had not been sufficiently tested." Kenneth M. Walker and J. A. Lumsden Cook, "Steinach's Rejuvenation

Operation," *Lancet,* 2 February 1924, 223. It was because of its appeal to the more dubious sections of the medical profession that the Steinach operation, the authors asserted, had not been very popular in Britain (ibid., 223).

222. See, for instance, Raymond Pearl's assessment that "however interesting the results [of the Steinach operation] may be from the standpoint of functional rejuvenation in the sexual sphere, the case is not proven that any really significant lengthening of the life span has occurred." R. Pearl, *The Biology of Death* (Philadelphia: Lippincott, 1922), 219.

223. Steinach, *Verjüngung*, 53.

224. Schmidt, *Conquest of Old Age*, 241.

225. Limley, "Science Can Make You Grow Younger," 22.

226. Harry Benjamin, "Preliminary Communication Regarding Steinach's Method of Rejuvenation," 691.

227. See H. Stieve, "Verjüngung durch experimentelle Neubelebung der alternden Pubertätsdrüse von E. Steinach: Ein Referat," *Die Naturwissenschaften* 8 (1920): 643–45, 643; Alfred Kohn, "Einige kritische Bemerkungen zur Verjüngungsfrage," *Medizinische Klinik* 17 (1921): 7–9, at 9; and Hermann Zondek, *Die Krankheiten der endokrinen Drüsen: Ein Lehrbuch für Studierende und Ärzte* (Berlin: Springer, 1926), 340–41.

228. See Kohn, "Einige kritische Bemerkungen zur Verjüngungsfrage."

229. Schmidt, *Conquest of Old Age*, 218–19.

230. See A. S. Warthin, *Old Age: The Major Involution*, 173–74. For another example, see T. Brailsford Robertson, *The Chemical Basis of Growth and Senescence* (Philadelphia: Lippincott, 1923), 172. Under the heading "Dangers of Rejuvenation," a well-known American textbook of geriatrics asserted, "excessive sexual activity after the age of 60 may bring about cerebral changes, coronary artery disease, cerebral apoplexy, and anginal syndrome, due to increased activity." Malford W. Thewlis, *The Care of the Aged (Geriatrics)*, 4th ed. (St. Louis: Mosby, 1942), 400.

231. Heinrich Poll, "Die biologischen Grundlagen der Verjüngungsversuche von Steinach," 919.

232. See Samson Wright, *Applied Physiology*, 6th ed. (London: Oxford University Press, 1936), 238.

233. Leonard Williams, "Truth about the Monkey Gland," *Empire Review* 38 (July–December 1923): 952–60, at 959. Five years later, the Rev. Herbert Davis, writing from a Catholic perspective, insisted that if "the restored powers . . . are used for legitimate purposes in marriage," there could be no objection to the restitution of sexual potency. The public outside the Catholic fold, he pointed out, was "in favour of birth control, which is a practice much more indecent and is positively immoral." H. Davis, "Rejuvenation by Gland-grafting," *The Catholic Medical Guardian* 6 (1928): 102–7, at 103, 105, 107. Much more research is required on religious responses to rejuvenation.

234. Leonard Williams, "Truth about the Monkey Gland," 959.

235. Morris Fishbein, *The New Medical Follies: An Encyclopedia of Cultism and Quackery in These United States* (New York: Boni and Liveright, 1927), 106.

236. James R. Church, "The Modern Ponce de Leon," *Military Surgeon* 48 (1921): 108–11, at 109, quoted by Hirshbein, "The Glandular Solution," 299–300.

237. Hans Much, "Grundlagen der Organtherapie: Eine kritisch-skeptische Studie," 1970; and Alfred Kohn, "Einige kritische Bemerkungen zur Verjüngungsfrage," 9.

238. Kohn, "Einige kritische Bemerkungen zur Verjüngungsfrage," 8.

239. See, for instance, Ernst Payr, "Über die Steinach'sche Verjüngungsoperation," 1138.

240. Lashley replied that Steinach's idea of tying back one of the vas deferens, of which Kammerer was an enthusiastic proponent, had been discredited, and any effect was probably due to "local inflammation." Nadine M. Weidman, *Constructing Scientific Psychology: Karl Lashley's Mind-Brain Debates* (Cambridge: Cambridge University Press, 1999), 167.

241. These remarks were made in 1936, when the Steinach operation was hardly as newsworthy as it had been a decade earlier. See E. Steinach, "Zur Geschichte des männlichen Sexualhormons," 204.

242. G. Stanley Hall, *Senescence*, 317–18.

243. Paul Kammerer, *Rejuvenation and the Prolongation of Human Efficiency*, 184–85.

244. It was, of course, conceded that it was impossible to draw an absolute distinction between senility and its premature form. "The actual age of the patient alone is no criterion; it has to be considered together with the patient's appearance, his symptoms and the results of a general medical examination." Harry Benjamin, "The Effects of Vasectomy (Steinach Operation)," 437.

245. Steinach, *Verjüngung*, 42.

246. Anson Rabinbach, *The Human Motor: Energy, Fatigue, and the Origins of Modernity* (Berkeley and Los Angeles: University of California Press, 1992), 142–44.

247. Peter Schmidt, *Don't be Tired: The Campaign against Fatigue*, trans. Mary Chadwick (London: Putnam's, 1930), 10, 139.

248. See Detlev J. K. Peukert, *The Weimar Republic: The Crisis of Classical Modernity* (London: Penguin, 1993), 7–8.

249. Steinach, *Sex and Life*, 210.

250. Schmidt, *Don't be Tired*, 24.

251. Schmidt, *Theory and Practice of the Steinach Operation*, 136.

252. Van Buren Thorne, "Dr. Steinach and Rejuvenation," *New York Times Book Review and Magazine*, 26 June 1921, 10.

253. Manfred Fraenkel, *Die Verjüngung der Frau*, 42–43.

254. "Mrs. Atherton Causes Amusement in Berlin: Newspapers Ridicule Her Suggestion for Rejuvenation of all Germany's Supermen," *New York Times*, 6 April 1924, sec. 2, 7. Atherton, in spite of her right-wing politics, was unambiguously opposed to Nazism. See Leider, *California's Daughter*.

255. See Paul Weindling, *Health, Race and German Politics between National Unification and Nazism 1870–1945* (Cambridge: Cambridge University Press, 1989), 325.

256. Voronoff was not himself part of any eugenics campaign but, as David Hamilton notes, his "work and attitudes carry the stamp of some of the assumptions of the times, including those of the eugenicists." Hamilton, *The Monkey Gland Affair,* 29. A Catholic moral theologian wrote in 1928 that "rejuvenation by grafting may be called an application of post-natal Eugenics." H. Davis, "Rejuvenation by Gland-grafting," 102.

257. See Weindling, *Health, Race, and German Politics,* 331.

258. See Kammerer, *Rejuvenation,* 189. Viereck, who would later go to jail for disseminating Nazi propaganda in America, was anxious to emphasize: "Rejuvenation must not become the private prerogative of the rich in shekels or in spirit. It must be a jewel attainable by all." Corners [Viereck], *Rejuvenation,* 93. Without mentioning glands, Karel Čapek explored the implications of rejuvenation in his celebrated immortality fantasy, *The Makropulos Secret,* in which an opera singer has lived for more than three hundred years as a young woman, thanks to a potion invented by her doctor father. One character declares that the potion should be used only on the great: "For the mediocre human mob, even the life of an ephemera is too long... there are in this world about ten or twenty thousand men who are irreplaceable. We can preserve them. In these men we can develop superhuman brains and supernormal powers. We can breed ten or twenty thousand supermen, leaders and creators." Karel Čapek, *The Makropulos Secret* (orig. pub. 1922), trans. Yveta Synek Graff and Robert T. Jones, in Peter Kussi, ed., *Toward the Radical Center: A Karel Čapek Reader* (Highland Park, N.J.: Catbird Press, 1990), 110–77, at 169–71. My thanks to Rhodri Hayward for alerting me to Čapek.

259. See Michael Mitterauer, *A History of Youth,* trans. Graeme Dunphy (Oxford: Blackwell, 1992), 212–16.

260. F. Scott Fitzgerald, "Echoes of the Jazz Age," quoted in Michael North, *Reading 1922: A Return to the Scene of the Modern* (New York: Oxford University Press, 1999), 4.

261. Stefan Schmorrte, "Alter und Medizin," 19–33. There is a growing literature on these and related themes. For a stimulating introduction to the issues, see Michael Hau, *The Cult of Health and Beauty in Germany: A Social History* (Chicago: University of Chicago Press, 2003).

262. L. D. Hirshbein, "The Glandular Solution," 298.

263. Limley, "Science Can Make You Grow Younger," 22.

264. Corners [Viereck], *Rejuvenation,* 94. This American concern with energy was far older than rejuvenation. The late-nineteenth-century New York neurologist George Miller Beard's concept of neurasthenia was founded in the conviction that the demands of modern civilization depleted the nervous energy of the individual. See G. M. Beard, *American Nervousness: Its Causes and Consequences* (New York: Putnam's, 1881) and for the contexts, Barbara Sicherman, "The Uses of a Diagnosis: Doctors, Patients, and Neurasthenia," *Journal of the History of Medicine and Allied Sciences* 32 (1977): 33–54; and the essays in Marijke Gijswijt-Hofstra and Roy Porter, eds., *Cultures of Neurasthenia from Beard to the First World War* (Amsterdam: Rodopi, 2001). Although neurasthenia was an almost obsolete concept by the 1920s, there was an occasional attempt (especially in the popular domain) to explain it in endocrine terms. Claiming the condition was simply "an expression of adrenal insufficiency," physician Charles

Evans Morris struck a Beardian note: "America of all countries is the land in which this adrenal reserve is used up with lavish disregard of consequence. The hurry, the bustle, the tumultuous rush, the keen competition, all predispose to the wasting of energy which results in the draining of the adrenal reserve." C. E. Morris, *Modern Rejuvenation Methods,* 28.

265. See Heiko Stoff, "Die hormonelle und die utopische Geschlechterordnung," 246–47.

266. Barbara Spackman, *Decadent Genealogies: The Rhetoric of Sickness from Baudelaire to D'Annunzio* (Ithaca, N.Y.: Cornell University Press, 1989), 30.

267. On these issues, see Chandak Sengoopta, *Otto Weininger: Sex, Science, and Self in Imperial Vienna* (Chicago: University of Chicago Press, 2000).

268. On the German discourse, see Iwan Bloch, *Das Sexualleben unserer Zeit im Lichte der modernen Kultur* (Berlin: Marcus, 1908), 41–56; Annemarie Wettley and Werner Leibbrand, *Von der "Psychopathia sexualis" zur Sexualwissenschaft* (Stuttgart: Enke, 1959), 70–76; and Frank J. Sulloway, *Freud,* 158–60, 292–96. On the British discourse, see Ornella Moscucci, "Hermaphroditism and Sex Difference: The Construction of Gender in Victorian England," in Marina Benjamin, ed., *Science and Sensibility: Gender and Scientific Enquiry* (Oxford: Blackwell, 1991), 174–99. For a suggestive analysis of the theme of universal androgyny, see Lawrence Birken, *Consuming Desire: Sexual Science and the Emergence of a Culture of Abundance, 1871–1914* (Ithaca, N.Y.: Cornell University Press, 1988); and for a comprehensive exploration of one conceptually and clinically crucial theme, see Alice Dreger, *Hermaphrodites and the Medical Invention of Sex.*

269. K. Kraus, "Ich laß es so," *Die Fackel* 400–403 (1914): 68–69. On Mrs. Steinach's opinion, see George F. Corners, *Rejuvenation,* 27–28, and Viereck, *Glimpses of the Great,* 256.

270. Kammerer, *Rejuvenation,* 226.

271. The story was first published in 1923 in the *Strand Magazine* in Britain and later included in *The Case Book of Sherlock Holmes* (1927). All citations here will be to Sir Arthur Conan Doyle, "The Adventure of the Creeping Man," in William S. Baring-Gould, ed., *The Annotated Sherlock Holmes,* 2 vols. (New York: Potter, 1967), 2:751–65.

272. "The Adventure of the Creeping Man," 755, 756, 761.

273. Ibid., 763–64.

274. Ibid., 761, 763–65. In the annals of the Sherlock Holmes fan-literature, there has been some interest in identifying Lowenstein. In the mock-believe tone characteristic of the genre, the Sherlockian scholars J. C. Prager and Albert Silverstein argued that Lowenstein was none other than Eugen Steinach! The detailed—and highly skilled—argument supporting the identification is available in J. C. Prager and A. Silverstein, "Lowenstein of Prague: The Most Maligned Man in the Canon," *Baker Street Journal* 23, no. 4 (December 1973): 220–27. I am indebted to Nicholas Utechin, editor of *The Sherlock Holmes Journal,* for providing me with a photocopy of this article. For a rather less imaginative identification of Lowenstein with some unknown Prague supporter of Brown-Séquard, see Philippe L. Selvais, "The Case of Professor Presbury: A Literary Digression on the Controversial Birth of Endocrinology," *Journal of Medical Biography* 6 (1998): 149–51.

275. Aldous Huxley, *After Many A Summer* (1939; Harmondsworth: Penguin, 1955), 55. In his earlier novel *Antic Hay* (1923), Huxley had referred to a fifteen-year-old monkey, "rejuvenated by the Steinach process . . . shaking the bars that separated him from the green-furred, bald-rumped, bearded young beauty in the next cage. He was gnashing his teeth with thwarted passion." Aldous Huxley, *Antic Hay* (1923; Harmondsworth: Penguin, 1948), 253.

276. See Stephen Jay Gould, *Ontogeny and Phylogeny*, 352–61.

277. Huxley, *After Many A Summer*, 85. Bolk had put it in rather more austere terms: "I would say that man, in his bodily development, is a primate fetus that has become sexually mature." Quoted in Gould, *Ontogeny and Phylogeny*, 361.

278. L. Berman, *The Glands Regulating Personality*, 84.

279. See Knud Sand, "Die Physiologie des Hodens," in Max Hirsch, ed., *Handbuch der innere Sekretion*, 3 vols. in 5 (Leipzig: Kabitzsch, 1928–33), 2, pt. 2 (1933): 2017–2268, at 2073.

280. Of the innumerable studies of concepts of gender in fin-de-siècle Central Europe, I found the following especially useful in analyzing the issues emphasized here: Bloch, *Sexualleben;* Gail Finney, *Women in Modern Drama: Freud, Feminism, and European Theater at the Turn of the Century* (Ithaca, N.Y.: Cornell University Press, 1989); Jacques Le Rider, *Modernity and Crises of Identity: Culture and Society in Fin-de-Siècle Vienna* (New York: Continuum, 1993); and Michael Worbs, *Nervenkunst: Literatur und Psychoanalyse im Wien der Jahrhundertwende* (Frankfurt am Main: Europäische Verlagsanstalt, 1983). The phrase "new womanly man" is from James Joyce, *Ulysses* (New York: Modern Library, 1961), 493.

281. See *Die Fackel*, 389–90 (1913): 38, and for an analysis, Nike Wagner, *Geist und Geschlecht: Karl Kraus und die Erotik der Wiener Moderne* (Frankfurt a.M.: Suhrkamp, 1982), 152–57.

282. See Maurizia Boscagli, *Eye on the Flesh: Fashions of Masculinity in the Early Twentieth Century* (Oxford: Westview Press, 1996), esp. 1–6; and George L. Mosse, *Nationalism and Sexuality: Respectability and Abnormal Sexuality in Modern Europe* (New York: Howard Fertig, 1985).

283. See Kevin J. Mumford, "'Lost Manhood' Found: Male Sexual Impotence and Victorian Culture in the United States," *Journal of the History of Sexuality* 3 (1992): 33–57.

Chapter Four

1. Leonard Williams, "The Interstitial Gland," *BMJ*, 27 May 1922, 833–35, at 834.

2. Arthur Keith, *The Engines of the Human Body, being the Substance of Christmas Lectures given at the Royal Institution of Great Britain, Christmas 1916–1917*, 3d ed. (London: Williams and Norgate, 1926), 229.

3. Julian Huxley, "Sex Biology and Sex Psychology" [1922], in Huxley, *Essays of a Biologist* (Harmondsworth: Penguin, 1939), 111–42, at 122.

4. Ibid.

5. Kenneth M. Walker, "The Internal Secretion of the Testis," *Lancet*, 5 January 1924, 16–21, at 17.

6. E. Steinach, "Pubertätsdrüsen und Zwitterbildung," *AEM* 42 (1916): 307–32, at 309–10.

7. Anne Fausto-Sterling, *Sexing the Body: Gender Politics and the Construction of Sexuality* (New York: Basic Books, 2000), 163.

8. Ibid.

9. Since the female twin did not develop any sexual abnormalities if there was no vascular communication with the male twin, "Nature," Lillie commented, "has thus performed here a perfectly controlled experiment." F. R. Lillie, "Sex-Determination and Sex-Differentiation in Mammals," *Proceedings of the National Academy of Sciences* 3 (1917): 464–70, at 466–67. Human freemartins were prominent in Aldous Huxley's dystopian vision of the future. Since excessive fertility is considered to be a nuisance in the brave new world, only about 30 percent of female embryos are permitted to develop naturally. "The others get a dose of male sex-hormone every twenty-four metres for the rest of the course. Result: they're decanted as freemartins—structurally quite normal (except . . . that they *do* have just the slightest tendency to grow beards), but sterile. Guaranteed sterile." Aldous Huxley, *Brave New World* (1932; London: Flamingo, 1994), 10.

10. F. R. Lillie, "The Theory of the Free-Martin," *Science,* new ser., 43 (1916): 611–13, at 612. See idem, "Sex-Determination and Sex-Differentiation in Mammals," 465, 469. Generally on Lillie, see Ray L. Watterson, "Lillie, Frank Rattray," *Dictionary of Scientific Biography,* 8 (1973): 354–60; and Adele E. Clarke, "Embryology and the Rise of the American Reproductive Sciences, circa 1910–1940," in K. R. Benson, J. Maienschein, and R. Rainger, eds., *The Expansion of American Biology* (New Brunswick, N.J.: Rutgers University Press, 1991), 107–32, esp. 112–14.

11. For biographical information, see Dorothy Price, "Carl Richard Moore, December 5, 1892–October 16, 1955," *Biographical Memoirs of the National Academy of Sciences* 45 (1974): 384–412.

12. For a comprehensive analysis of these experiments and their contexts, see Fausto-Sterling, *Sexing the Body,* 164–66.

13. Dorothy Price, "Feedback Control of Gonadal and Hypophyseal Hormones: Evolution of the Concept," in J. Meites, B. T. Donovan, and S. M. McCann, eds., *Pioneers in Neuroendocrinology,* 2 vols. (New York: Plenum Press, 1975), 1:217–38, at 223. Actually, Moore did investigate the Steinach operation by performing vasoligation in the laboratory on five different animal species, concluding that the procedure did not extirpate the germinal tissue or lead consistently to a proliferation of the interstitial cells. "Up to the present time there has been no indication," Moore insisted in 1931, "that such an operation in any way modifies the rate of hormone secretion or is advantageous in any other respect aside from a sterilizing operation." C. R. Moore, "The Regulation of Production and the Function of the Male Sex Hormone," *JAMA* 97 (1931): 518–22, at 519.

14. Carl R. Moore, "On the Physiological Properties of the Gonads as Controllers of Somatic and Psychical Characteristics, III. Artificial Hermaphroditism in Rats," *Journal of Experimental Zoology* 33 (1921): 129–71, at 130–31.

15. Anne Fausto-Sterling has questioned whether Moore's testis grafts were even functional—unless they were histologically shown to be rich in interstitial cells (and

276 • Notes to Pages 121–124

therefore capable of secreting hormones), they could not be considered active by Steinach's criteria. See Fausto-Sterling, *Sexing the Body,* 168–69.

16. On Halsted's Law, see W. S. Halsted, "Auto- and Isotransplantation in Dogs, of the Parathyroid Glandules," *Journal of Experimental Medicine* 11 (1909): 175–98.

17. Moore, "On the Physiological Properties of the Gonads as Controllers of Somatic and Psychical Characteristics, III," 134, 165, 167.

18. Ibid., 168.

19. F. R. Lillie, "The Free-Martin: A Study of the Action of Sex Hormones in the Foetal Life of Cattle," *Journal of Experimental Zoology* 23 (1917): 371–452, at 402–4.

20. The riddle of the freemartin, according to Anne Fausto-Sterling, remains unsolved to this day, although most of Lillie's hypotheses still "offer a 'best fit.'" See Fausto-Sterling, *Sexing the Body,* 333, n. 77.

21. Alexander Lipschütz, "Is There an Antagonism Between the Male and the Female Sex-Endocrine Gland?," *Endocrinology* 9 (1925): 109–16, at 115–16.

22. The problem of grafting, on the other hand, was probably only partly hormonal and partly due to unspecific, trophic factors—it was, in any case, an experimental phenomenon of relatively low biological interest. See E. Steinach and H. Kun, "Antagonistische Wirkungen der Keimdrüsen-Hormone," *Biologia Generalis* 2 (1926): 815–34, esp. 820. Although putatively co-authored, this paper constantly refers to "my" work.

23. Ibid., 815–17, 819–20.

24. Ibid., 816.

25. Ibid., 821–22, 823–26.

26. Welcoming the availability of extracts, Moore observed: "The success of the biochemists in obtaining effective preparations of the male hormone, and the development of effective and quickly applied procedures for determining its presence or absence in the organism . . . unite in providing instruments with which real advances may be made in the study of the production and the function of the internal secretion of the testis." Carl R. Moore, "The Regulation of Production and the Function of the Male Sex Hormone," 519.

27. Carl R. Moore and Dorothy Price, "Gonad Hormone Functions and the Reciprocal Influence between Gonads and Hypophysis with its Bearing on the Problem of Sex Hormone Antagonism," *American Journal of Anatomy* 50 (1932): 13–67, at 17.

28. Carl R. Moore, "A Critique of Sex Hormone Antagonism," in A. W. Greenwood, ed., *Proceedings of the Second International Congress for Sex Research, London, 1930* (Edinburgh: Oliver and Boyd, 1931), 293–303, at 294–95.

29. Moore and Price, "Gonad Hormone Functions and the Reciprocal Influence between Gonads and Hypophysis," 29, 31–33.

30. Ibid., 21–22.

31. Ibid., 34–36, 41.

32. Dorothy Price, "Feedback Control of Gonadal and Hypophyseal Hormones," 228.

33. Carl R. Moore, "A Critique of Sex Hormone Antagonism," 296.

34. Harvey Cushing, "Neurohypophysial Mechanisms from a Clinical Standpoint," in H. Cushing, *Papers relating to the Pituitary Body, Hypothalamus and Parasympathetic Nervous System* (London: Baillière, Tindall and Cox, 1932), 1–57, at 5.

35. The *Oxford English Dictionary* defined the "pituitary" in 1909 as "a small bilobed body of unknown function attached . . . to the brain." Quoted by Harvey Cushing, "Neurohypophysial Mechanisms from a Clinical Standpoint," 10. Humphry Davy Rolleston observed in 1936 that "until nearly the beginning of this [the twentieth] century it was regarded as little more than a vestigial relic; now it has become the head stone of the corner." Humphry Davy Rolleston, *The Endocrine Organs in Health and Disease with an Historical Review* (London: Oxford University Press, 1936), 55.

36. See H. Cushing, "The Hypophysis Cerebri: Clinical Aspects of Hyperpituitarism and of Hypopituitarism," *JAMA* 53 (1909): 249–55, at 249; and idem, "Disorders of the Pituitary Gland: Retrospective and Prophetic," *JAMA* 76 (1921): 1721–26, at 1723.

37. This, of course, is a rough and ready distinction. For a technical description of the differences between and nomenclature of different regions of the two lobes, see Solly Zuckerman, "The Secretions of the Brain: Relation of Hypothalamus to Pituitary Gland," *Lancet*, 17 April 1954, 789–95, at 789.

38. Walter Langdon Brown, "Recent Observations on the Pituitary Body," *The Practitioner* 127 (1931): 614–25, at 614. Three decades later, another expert commented that Brown had referred to "'the leader in the endocrine orchestra,' having in mind the pituitary as the first fiddle and not as the conductor. If we wish we may still reserve the rostrum and the baton for the hypothalamus." Douglas Hubble, "The Endocrine Orchestra," *BMJ*, 25 February 1961, 523–28, at 523. On early research on hypothalamic influence on the pituitary, see Humphry Davy Rolleston, *The Endocrine Organs in Health and Disease*, 56.

39. Kenneth Walker, *Diagnosis of Man*, rev. ed. (Harmondsworth: Penguin, 1962), 37.

40. The posterior lobe was important enough, but it was the anterior lobe that seemed to control growth, metabolism, sex "and to be bound up in ways which as yet are symptomatically obscure with the function probably of all the other ductless glands of the body." Harvey Cushing, "The Hypophysis Cerebri: Clinical Aspects of Hyperpituitarism and of Hypopituitarism," 252.

41. In what follows, I shall use the simpler expression "pituitary" to refer to the anterior lobe of the gland.

42. Quoted by Van Buren Thorne, "The Craze for Rejuvenation," review of Benjamin Harrow, *Glands and Health and Disease, New York Times*, 4 June 1922, section 3, 18. American researcher Roy Hoskins commented in 1933 that so little was known about the pituitary and so apparently complex its nature and functions that "to those who, in the present state of our knowledge, would glibly re-write physiology and psychology in terms of pituitary functions the timorousness of the proverbial angel is commended." R. G. Hoskins, *The Tides of Life: The Endocrine Glands in Bodily Adjustment* (London: Kegan Paul, Trench, Trubner and Co., 1933), 169.

43. Louis Berman, *The Glands Regulating Personality: A Study of the Glands of Internal Secretion in relation to the Types of Human Nature* (New York: Macmillan, 1922), 67.

44. Berman, *Glands Regulating Personality*, 177–78. Even in 1962, surgeon, former monkey-gland enthusiast, and aspiring philosopher Kenneth Walker repeated Berman's notions of gland-centered human types with only half-hearted disclaimers. See K. Walker, *Diagnosis of Man*, 44.

45. Moreover, Napoleon's sex life was "abnormal," characterized by "explosive periodicity"—"another mark of some pituitary-centered personalities." Urinary difficulties, too, suggested the same diagnosis. "Besides," Berman added, "his insatiable energy indicated an excellent thyroid, his pugnacity, animality and genius for practical affairs a superb adrenal." But all was not well with the emperor's pituitary. He became obese and indecisive in later life, losing much of his military brilliance. At postmortem, his body revealed a feminine pattern of obesity around the hips and upper thighs, a lack of body hair, and minuscule genitals—all signs of pituitary insufficiency. Clearly, therefore, an unbalanced pituitary had led him to glory and then collapsed, leaving him and his empire floundering. See Berman, *Glands Regulating Personality*, 232–34. The British physician and popularizer of endocrinology Ivo Geikie Cobb concurred: "The gland which ruled the man who ruled Europe was the pituitary." See Ivo Geikie Cobb, *The Glands of Destiny: A Study of the Personality*, 3d ed. (London: William Heinemann Medical Books, 1947), 202, 238. Decades later, Kenneth Walker recapitulated Berman's analysis of Napoleon with great respect, concluding that "Waterloo was lost to France . . . when a tiny gland, the size of a pea, began to show ominous signs of disrepair." K. Walker, *Diagnosis of Man*, 48.

46. Harvey Cushing, *The Pituitary Body and its Disorders: Clinical States produced by Disorders of the Hypophysis Cerebri* (Philadelphia: Lippincott, 1912), 271.

47. Ibid., 190.

48. L. R. Broster, *Endocrine Man: A Study in the Surgery of Sex* (London: William Heinemann Medical Books, 1944), 61.

49. Cushing, *The Pituitary Body and its Disorders*, ix–x. There is a large literature on Cushing and his versatile career. The Latest biography is Michael Bliss, *Harvey Cushing: A Life in Surgery* (New York: Oxford University Press, 2005). On his work on the pituitary, see Sam L. Teichman and Peter Aldea, "Pioneers in Pituitary Physiology: Harvey Cushing and Nicolas Paulescu," *Journal of the History of Medicine and Allied Sciences* 40 (1985): 68–72; and D. C. Aron, "The Path to the Soul: Harvey Cushing and Surgery on the Pituitary and its Environs in 1916," *Perspectives in Biology and Medicine* 37 (1993–94): 551–65.

50. Harvey Cushing, *The Pituitary Body and its Disorders*, 1. Cushing did not, of course, deny the importance of the work already done on the pituitary. Oliver and Schäfer, for instance, had shown that the injection of posterior pituitary extract raised blood pressure, while Henry Dale, William Blair-Bell, and Pantland Hick had demonstrated the extract's ability to produce contractions of the intestines, urinary bladder, and the uterus. Even more recently, it had been shown that extracts of the posterior lobe stimulated the secretion of milk. The anterior lobe's influence on growth, too, had long been suspected, especially from clinical studies of acromegaly (ibid., 10). In 1921, Herbert Maclean Evans and J. A. Long would report the experimental

production of gigantism in laboratory rats by injections of pituitary extract. The isolation of pituitary growth hormone would follow in a fascinating story that we cannot, unfortunately, go into here, but see Herbert M. Evans, "The Function of the Anterior Hypophysis," *Harvey Lectures* 19 (1923–24): 212–35; idem, "Present Position of Our Knowledge of Anterior Pituitary Function," *JAMA* 101 (1933): 425–32; and idem, "Clinical Manifestations of the Dysfunction of the Anterior Pituitary," *JAMA* 104 (9 February 1935): 464–72. On Evans and the contexts of his work, see his long interview with Alan Parkes in *Journal of Reproduction and Fertility* 19 (1969): 1–49.

51. Surgery, Cushing justly emphasized, was crucial in elucidating ductless gland functions, whether in the laboratory or in the clinic. "Had it not been for the neighborhood pressure symptoms, especially those of tracheal distortion, the early operations for goitre (1883–1886) by Kocher and Reverdin would not, in all probability, have been undertaken. . . . Had it not been for the obvious tumor of the pituitary body," argued Cushing, "which likewise causes its own characteristic neighborhood symptoms," acromegaly or the adiposogenital syndrome would probably not have been ascribed to "this supposedly unimportant and vestigial structure." Cushing, *The Pituitary Body and its Disorders,* 23.

52. Bernhard Aschner (1883–1960), gynecologist at Schauta's clinic in Vienna, denied that the loss of the pituitary was lethal—it was the damage to surrounding brain areas during the removal of the gland, he argued, that was responsible for the fatalities. See B. Aschner, "Über die Beziehungen zwischen Hypophysis und Genitale," *AfG* 97 (1912): 200–228, at 202. Aschner conducted numerous experiments on dogs, some involving total removal of the pituitary and others in which only certain parts were extirpated. Using innovative techniques, Aschner removed the gland without damaging the brain tissue around it, and although some of his animals succumbed to intercurrent infection, most survived the operation. The youngest animals did, however, show characteristic growth abnormalities—their skeletal systems and many other organs remained infantile.

53. Harvey Cushing, *The Pituitary Body and its Disorders,* 13. Again, Bernhard Aschner dissented, claiming that the partial extirpation of the pituitary in *adults* did not lead to gross genital atrophy in his own careful experiments, although pregnancy seemed to be unattainable in the absence of the pituitary. Cushing's findings, he suggested, were probably due again to the damage he had caused to the brain tissue while removing the gland. It was only in infant animals that partial removals of the pituitary caused profound gonadal and genital abnormalities. See Aschner, "Über die Beziehungen zwischen Hypophysis und Genitale," 205–7, 210–19. Aschner's observation was generally ignored until the 1920s, and then, many of the symptoms he had attributed to general brain damage were explained as being due to damage to specific areas of the hypothalamus. See Humphry Davy Rolleston, *The Endocrine Organs in Health and Disease,* 57.

54. See Harvey Cushing, *The Pituitary Body and its Disorders,* 44. The adiposogenital syndrome was first reported by Viennese clinician Alfred Fröhlich (1871–1953). See his landmark paper, "Ein Fall von Tumor der Hypophysis cerebri ohne Akromegalie," *Wiener klinische Rundschau* 15 (1901): 883–906.

55. Harvey Cushing, *The Pituitary Body and its Disorders,* x, 275. The pituitary-gonad link, he also speculated, was "a subject which deserves the close

attention of the psycho-analysts of the Freudian school. . . . It is quite probable that the psychopathology of everyday life hinges largely upon the effect of ductless gland discharge upon the nervous system" H. Cushing, "Psychic Disturbances associated with Disorders of the Ductless Glands," *American Journal of Insanity* 69 (1912–13): 965–90, at 988–89.

56. Philip E. Smith, "Hastening Development of Female Genital System by Daily Homoplastic Pituitary Transplants," *Proceedings of the Society for Experimental Biology and Medicine* 24 (1926–27): 131–32. In immature male rats, the same procedure resulted in comparable outcomes, although the process was slower. Although there was no increase in the dimensions of the testicles, the rest of the genital system increased considerably in size. Prior castration prevented the pituitary effect. See P. E. Smith, "Genital System Responses to Daily Pituitary Transplants," *Proceedings of the Society for Experimental Biology and Medicine* 24 (1926–27): 337–38.

57. Philip E. Smith, "Induction of Precocious Sexual Maturity in the Mouse by Daily Pituitary Homoeo- and Heterotransplants," *Proceedings of the Society for Experimental Biology and Medicine* 24 (1926–27): 561–62.

58. Joseph Schleidt, "Über die Hypophyse bei feminierten Männchen und maskulierten Weibchen," *Zentralblatt für Physiologie* 27 (1914): 1170–72, at 1171–72.

59. See Steinach, "Zur Geschichte des männlichen Sexualhormons und seiner Wirkungen am Säugetier und beim Menschen," *WkW* 49 (1936): 161–72; 198–205, at 198. Steinach would claim in this article that Schleidt's work was the first demonstration that pituitary function was not sex-specific. Schleidt, he argued, had established the centripetal connection between gonads and pituitary, while the centrifugal (gonadotropic) links were later ascertained by Aschner, Smith, and Steinach himself in collaboration with Kun.

60. In the popular realm, Louis Berman disseminated similar gonadocentric views in the early 1920s, albeit in more colorful language. Gonadal stimulation of the pituitary was essential to the development of judgment—in the prepubertal years, memory might be excellent but judgment was always poor and eunuchs had poor intellect and judgment. "Thus Abelard, who was punished with castration by his uncle for his love affair with Hélöise, never composed a verse of poetry thereafter," he concluded. The anterior pituitary, then, could not function adequately without appropriate stimulation from the gonads. See L. Berman, *Glands Regulating Personality,* 177–78, 183–84.

61. "Foreword: Bernhard Zondek," *Journal of Reproduction and Fertility* 12 (1966): 1.

62. Bernhard Zondek, *Die Hormone des Ovariums und des Hypophysenvorderlappens: Untersuchungen zur Biologie und Klinik der weiblichen Genitalfunktion* (Berlin: Springer, 1931), 107.

63. B. Zondek and S. Aschheim, "Das Hormon des Hypophysenvorderlappens, I. Testobjekt zum Nachweis des Hormons," *KW* 6 (1927): 248–52, 249.

64. B. Zondek, *Die Hormone des Ovariums und des Hypophysenvorderlappens,* 17.

65. Zondek and Aschheim termed this phase Hypophysenvorderlappenreaktion I, or HVR-I.

66. These stages were called Hypophysenvorderlappenreaktion (HVR) II and HVR III. See B. Zondek and S. Aschheim, "Das Hormon des Hypophysenvorderlappens:

Darstellung, chemische Eigenschaften, biologische Wirkungen," *KW* 7 (1928): 831–35, at 831.

67. Ibid., 834.

68. B. Zondek and S. Aschheim, "Das Hormon des Hypophysenvorderlappens, I. Testobjekt zum Nachweis des Hormons," 251.

69. Later, Zondek assigned the name Prolan to the pituitary hormone marketed under his guidance by the IG Farbenindustrie-Leverkusen and used the abbreviation HVH (for Hypophysevorderlappenhormon) in his research publications, with the follicle stimulating hormone being named HVH-A and the luteinizing hormone as HVH-B. See B. Zondek, *Die Hormone des Ovariums und des Hypophysenvorderlappens*, 107, n. 1; and idem, "Über die Hormone des Hypophysenvorderlappens, I. Wachstumshormon, Follikelreifungshormon (Prolan A), Luteinisierungshormon (Prolan B), Stoffwechselhormon?," *KW* 9 (1930): 245–48, at 245.

70. B. Zondek and S. Aschheim, "Das Hormon des Hypophysenvorderlappens, I. Testobjekt zum Nachweis des Hormons," 250.

71. Bernhard Zondek, "The Relation of the Anterior Lobe of the Hypophysis to Genital Function," *American Journal of Obstetrics and Gynecology* 24 (1932): 836–43, at 837.

72. B. Zondek and S. Aschheim, "Das Hormon des Hypophysenvorderlappens, I. Testobjekt zum Nachweis des Hormons," 251.

73. Bernhard Zondek, "Über die Hormone des Hypophysenvorderlappens, V: Die Ausscheidung des Follikelreifungshormons (HVH-A) im mensuellen Cyclus," *KW* 10 (1931): 2121–23, at 2122.

74. See H. Cushing, *The Pituitary Body and its Disorders*, 234–36.

75. The same hormone, claimed Aschheim and Zondek, was present in enormous quantities in the placenta: the implantation of even a tiny piece of human placenta, especially from the earlier months, into an infant mouse evoked estrus and the other signs of a positive reaction. See B. Zondek and S. Aschheim, "Das Hormon des Hypophysenvorderlappens: Darstellung, chemische Eigenschaften, biologische Wirkungen," 832.

76. See chapter 5 for the Allen-Doisy reaction and its significance.

77. S. Aschheim and B. Zondek, "Schwangerschaftsdiagnose aus dem Harn (durch Hormonnachweis)," *KW* 7 (1928): 8–9, at 8.

78. The morning urine of the woman being tested was injected subcutaneously in six doses into each of five infant mice—three injections on the first day and three more on the second day. Vaginal smears were taken on the fourth and fifth days from the mice and they were killed ninety-six hours after commencing the test, their ovaries being examined for corpora lutea and blood points. The latter were usually visible to the naked eye, and microscopy was rarely needed. A similar quantity of urine was also injected into an adult castrated mouse to ascertain whether any of the hormonal effects on the first test animal were due to ovarian hormone in the urine. The test did not actually detect the presence of a live fetus—what it indicated was a connection between maternal and chorionic circulation. Hence, it was strongly positive in conditions like hydatidiform mole, where there was no viable fetus but plenty of chorionic epithelium. Another major problem with the test was that it took about four days to obtain results,

although a later American modification (involving the intravenous injection of the urine into rabbits) reduced that to one to two days. See Aschheim, "Pregnancy Tests," *JAMA* 104 (1935): 1324–29, at 1326–27.

79. Aschheim and Zondek, "Schwangerschaftsdiagnose aus dem Harn (durch Hormonnachweis)," 9. See also S. Aschheim, "Pregnancy Tests."

80. Pregnancy urine was also suitable for the chemical extraction of the anterior pituitary hormone. See B. Zondek and S. Aschheim, "Das Hormon des Hypophysenvorderlappens: Darstellung, chemische Eigenschaften, biologische Wirkungen," 832. It was also ineffective when administered orally to experimental animals. See Zondek, "Über die Hormone des Hypophysenvorderlappens, I. Wachstumshormon, Follikelreifungshormon (Prolan A), Luteinisierungshormon (Prolan B), Stoffwechselhormon?," 246. The ovarian hormone was very different: it was resistant to high temperatures, insensitive to acids and alkalis, and soluble in lipid solvents. See Zondek and Aschheim, "Das Hormon des Hypophysenvorderlappens: Darstellung, chemische Eigenschaften, biologische Wirkungen," 832–33.

81. For an overview with references, see C. F. Fluhmann, "The Interrelationship of the Anterior Hypophysis and the Ovaries," *American Journal of Obstetrics and Gynecology* 26 (1933): 764–75, esp. 765, 767–78. It was clear, however, that prolan was quite distinct from the substance in anterior pituitary extract that had caused skeletal growth in rats in the experiments of Herbert Evans. Indeed, Evans was convinced that the growth-promoting substance was antagonistic to the gonad-stimulating hormone. Puberty occurred *after* the growth hormone had done its job and the latter was free to act. See Bernhard Zondek, "Über die Hormone des Hypophysenvorderlappens, I. Wachstumshormon, Follikelreifungshormon (Prolan A), Luteinisierungshormon (Prolan B), Stoffwechselhormon?," 245; and H. M. Evans and M. E. Simpson, "Antagonism of Growth and Sex Hormones of the Anterior Hypophysis," *JAMA* 91 (1928): 1337–38. The growth hormone of the anterior pituitary stimulated the development of all the cells of the soma, while the sex-specific growth hormone—the follicular hormone in females and the testicular hormone in males—induced the development of those tissues that were concerned with the reproductive processes. For influential endorsements of this hypothesis, see Ludwig Seitz, "Das Follikelhormon als geschlechtsspezifisches Wachstumshormon und seine Beziehungen zum allgemeinen Körperwachstumshormon," *Monatsschrift für Geburtshülfe und Gynäkologie* 103 (1936): 185–93, at 188–90; and Robert T. Frank, *The Female Sex Hormone* (London: Baillière, Tindall and Cox, 1929), 17.

82. Walter Langdon-Brown, *The Integration of the Endocrine System*, 20–22.

83. See Zondek, "Über die Hormone des Hypophysenvorderlappens, I. Wachstumshormon, Follikelreifungshormon (Prolan A), Luteinisierungshormon (Prolan B), Stoffwechselhormon?," 245.

84. Ibid., 245, 246. Just as prolan had at least two components, the ovarian hormone, too, Zondek had to acknowledge by 1930, was not a single agent. First of all, it was perplexing why the pituitary would secrete two distinct hormones to stimulate the secretion of the single ovarian hormone. Nor was it obvious why there should be two histologically different kinds of secretory cells in the ovary—the granulosa cells and the theca cells—if they produced the same secretion. On using the follicular hormone as replacement treatment in oophorectomized women, moreover, the uterine mucosa

proliferated but never reached the secretory phase. George W. Corner's recent work on the luteal hormone had shown that it could induce the secretory phase. Consequently, it was clear that one should refer not only to anterior pituitary hormones (the two Prolans) but also to the ovarian hormones : a follicular hormone (which Zondek called "folliculin"), the secretion of which was stimulated by Prolan A and which induced the uterine mucosa to proliferate, and Corner's luteal hormone ("lutin," according to Zondek), which was secreted by the lutein cells produced from granulosa cells under the influence of Prolan B. The rhythm of ovarian function—which, of course, determined the female reproductive cycle from menstruation to pregnancy—was determined proximately by the two ovarian hormones but ultimately by the secretory succession of the two prolans. See ibid., 247–48. For details of Corner's experiments, see George W. Corner, "Physiology of the Corpus Luteum, I. The Effect of the Very Early Ablation of the Corpus Luteum upon Embryos and Uterus," *American Journal of Physiology* 86 (1928): 74–81; and George W. Corner and Willard M. Allen, "Physiology of the Corpus Luteum, II. Production of a Special Uterine Reaction (Progestational Proliferation) by Extracts of the Corpus Luteum," *American Journal of Physiology* 88 (1929): 326–39. For contemporary German research on corpus luteum, see Carl Clauberg, "Über das Hormon des Corpus luteum," in A. W. Greenwood, ed., *Proceedings of the Second International Congress for Sex Research, London 1930* (Edinburgh: Oliver and Boyd, 1931), 345–51.

85. When he succeeded in obtaining extracts of (predominantly) Prolan B, its prolonged administration led to a monstrous enlargement of the ovaries in adult mice and remarkable proliferation of corpora lutea. Normally, a mouse ovary had 2–8 of the latter—when treated for long with Prolan B, however, the number rose to as high as 80. Ultimately, the ovary became almost one huge corpus luteum and all follicles disappeared. The animal was now hormonally sterilized—ovulation and pregnancy were ruled out. See B. Zondek, *Die Hormone des Ovariums und des Hypophysenvorderlappens,* 174–75, 180.

86. Bernhard Zondek, "Über die Hormone des Hypophysenvorderlappens, II. Follikelreifungshormon (Prolan A)—Klimakterium—Kastration," *KW* 9 (1930): 393–96, at 396.

87. Ibid., 393–94; and Bernhard Zondek, "Über die Hormone des Hypophysenvorderlappens, V: Die Ausscheidung des Follikelreifungshormons (HVH-A) im mensuellen Cyclus."

88. Oophorectomy also resulted in the elevation of Prolan A in the urine about ten days after the operation, the levels remaining high for as long as a year. This finding dovetailed neatly with the observation that the pituitary gland enlarged after oophorectomy in women and also in eunuchs. Zondek also speculated that similar elevations of Prolan A might also occur in men suffering from functional deficiencies of the testicular secretions. See Bernhard Zondek, "Über die Hormone des Hypophysenvorderlappens, II. Follikelreifungshormon (Prolan A)—Klimakterium—Kastration," 394–96. Zondek and Aschheim had found that the urine exhibited a positive Prolan A type reaction (i.e., mostly HVR-I and very rarely II and III) in the majority of patients with pelvic malignancies. So frequent was this result that a positive reaction virtually pinpointed a pelvic malignancy. In non-pelvic malignancies, the reaction was positive in about 36 percent of women and in only about 13 percent of

men. Although Zondek had identified Prolan A in tumor cells, he was unsure whether the hormone was actually produced by the tumor or simply overproduced by the pituitary and concentrated in the tumor. See Bernhard Zondek, "Über die Hormone des Hypophysenvorderlappens, III. Follikelreifungshormon (Prolan A) und Tumoren," *KW* 9 (1930): 679–82, at 680, 682; and idem, "Maligne Hodentumoren und Hypophysenvorderlappenhormone: Hormonale Diagnostik aus Harn, Hydrocelenflüssigkeit und Tumorgewebe," *KW* 11 (1932): 274–79, at 274.

89. Philip E. Smith, "Genital System Responses to Daily Pituitary Transplants," 337–38.

90. B. Zondek and S. Aschheim, "Das Hormon des Hypophysenvorderlappens: Darstellung, chemische Eigenschaften, biologische Wirkungen," 834.

91. B. Zondek, "Über die Hormone des Hypophysenvorderlappens, I. Wachstumshormon, Follikelreifungshormon (Prolan A), Luteinisierungshormon (Prolan B), Stoffwechselhormon?," 248, note.

92. B. Zondek, "Maligne Hodentumoren und Hypophysenvorderlappenhormone: Hormonale Diagnostik aus Harn, Hydrocelenflüssigkeit und Tumorgewebe," 275, 277. The production of prolan reactions in mice ovaries after actual implantation of tumor tissue could, Zondek suggested, be the first step toward a whole new method of hormonal tissue diagnosis (*Hormonale Gewebsdiagnostik*). See Bernhard Zondek, "The Relation of the Anterior Lobe of the Hypophysis to Genital Function," *American Journal of Obstetrics and Gynecology* 24 (1932): 836–43, at 841. Later, it was argued that different types of testicular tumor elevated urinary gonadotropin levels to different extents and that these differences were so consistent that urine analysis could indicate the type of tumor. The removal of the tumor, however, did not seem to reduce the levels of gonadotropic hormone in the urine. See E. C. Dodds, "Hormones in Cancer," *Vitamins and Hormones* 2 (1944): 353–59, at 357.

93. E. Steinach and H. Kun, "Die entwicklungsmechanische Bedeutung der Hypophysis als Aktivator der Keimdrüseninkretion: Versuche an infantilen, eunuchoiden und senilen Männchen," *Medizinische Klinik* 24 (1928): 524–29, at 524.

94. He used an aqueous extract of sheep anterior pituitary, about the preparation of which he never seems to have published any details. Daily injections of this extract in infant male rats brought about the descent of the testicles; the enlargement of the scrotum, penis, seminal vesicles and prostate; and the arousal of libido and sexual potency in about ten or twelve days. Microscopic examination of the testicles revealed a striking proliferation of the interstitial cells and the degeneration of the germinal tissue in long-continued injections of pituitary extract. Cases of sexual precocity in humans too, Steinach speculated, were probably due to premature activation of the gonads by the pituitary. See E. Steinach and H. Kun, "Die entwicklungsmechanische Bedeutung der Hypophysis," 524, 526. The same extract, produced at the Schering laboratories, was used in later experiments on luteinization, discussed below. See E. Steinach and H. Kun, "Luteingewebe und männliche Geschlechtscharaktere," *PAgP* 227 (1931): 266–78, at 270.

95. Steinach and Kun, "Die entwicklungsmechanische Bedeutung der Hypophysis," 528. In Mikhail Bulgakov's 1925 novel, a dog is transplanted with a human pituitary in the hope of rejuvenation. Unfortunately, the graft, instead of rejuvenating the dog, leads to its complete humanization. See M. Bulgakov, *The Heart*

of A Dog, translated by Michael Glenny (London: Harvill, 1968), esp. 63–64, 66, 109. In 1936, American physician Henry Smith Williams speculated that "the next 'lost virility' treatment to be popularized will involve the use of the pituitary hormone. The name 'hypophesis' [*sic*] . . . will perhaps become a household word, with something of the charm of mystery that attached to 'interstitial gland' a few years ago." H. S. Williams, *Your Glands and You* (New York: McBride, 1936), 139.

96. Steinach and Kun, "Die entwicklungsmechanische Bedeutung der Hypophysis," 528–29. Harry Benjamin surmised that the beneficial effects of the Steinach operation were not to be attributed to the simple enhancement of sex hormone secretion but also to the effects of those secretions on the pituitary. See Harry Benjamin, "The Pituitary as Activator of the Gonadal Function: Result of the Latest Researches of Steinach and Kun, Vienna," *Medical Journal and Record* 128 (1928): 227–29, at 228–29.

97. B. Zondek, "The Relation of the Anterior Lobe of the Hypophysis to Genital Function," 836; and idem, *Die Hormone des Ovariums und des Hypophysenvorderlappens,* 230.

98. See Zondek, *Die Hormone des Ovariums und des Hypophysenvorderlappens,* 232–33.

99. Dorothy Price, "Feedback Control of Gonadal and Hypophyseal Hormones: Evolution of the Concept," 228; and C. R. Moore and D. Price, "Gonad Hormone Functions and the Reciprocal Influence between Gonads and Hypophysis," 56.

100. "I did not," she emphasized, "call this brainchild of mine a theory or a hypophysis [*sic*] and I certainly did not anticipate that it would come to be known as a negative feedback system. I thought it a beautiful and logical scheme, and, furthermore, Moore's results fitted in like a dream." Dorothy Price, "Feedback Control of Gonadal and Hypophyseal Hormones," 228.

101. See, for instance, E. T. Engle, "Pituitary-Gonadal Mechanism and Heterosexual Ovarian Grafts," *American Journal of Anatomy* 44 (1929): 121–39; idem, "The Effect of Daily Transplants of the Anterior Lobe from Gonadectomized Rats on Immature Test Animals," *American Journal of Physiology* 88 (1929): 101–6; H. M. Evans and Miriam E. Simpson, "A Comparison of Anterior Hypophysial Implants from Normal and Gonadectomized Animals with Reference to their Capacity to Stimulate the Immature Ovary," *American Journal of Physiology* 89 (1929): 371–74.

102. Dorothy Price, "Feedback Control of Gonadal and Hypophyseal Hormones," 229–30.

103. Moore and Price, "Gonad Hormone Functions and the Reciprocal Influence between Gonads and Hypophysis," 29, 31.

104. Ibid., 48–49; P. E. Smith and E. T. Engle, "Experimental Evidences regarding the Rôle of the Anterior Pituitary in the Development and Regulation of the Genital System," *American Journal of Anatomy* 40 (1927): 159–217; and P. E. Smith, "The Induction of Precocious Sexual Maturity by Pituitary Homeotransplants," *American Journal of Physiology* 80 (1927): 114–25. See also C. F. Fluhmann, "Anterior Pituitary Hormone in the Blood of Women with Ovarian Deficiency," *JAMA* 93 (1929): 672–74.

105. Moore and Price, "Gonad Hormone Functions and the Reciprocal Influence between Gonads and Hypophysis," 53; and R. K. Meyer, S. L. Leonard, F. L. Hisaw,

and S. J. Martin, "Effects of Oestrin on Gonad Stimulating Power of the Hypophysis," *Proceedings of the Society for Experimental Biology and Medicine* 27 (1930): 702–4.

106. Moore and Price, "Gonad Hormone Functions and the Reciprocal Influence between Gonads and Hypophysis," 31.

107. Just in case anybody didn't notice, Moore and Price spelt out the implications of this for the Steinach operation or for monkey-gland transplants. "One may transplant additional testes to normal males, and ovaries to normal females, but even if . . . the grafts 'take'—one is not permitted the assumption that more hormone is being secreted." Moore and Price, "Gonad Hormone Functions and the Reciprocal Influence between Gonads and Hypophysis," 59.

108. Ibid., 51.

109. Ibid., 61. "We have advanced the working hypothesis that gonadal hormones suppress the hypophysis, making available to the organism a reduced amount of gonad stimulating secretion; that hypophysis hormone is necessary for gametogenesis or hormone production by the gonads; that sex hormones act directly upon homologous accessories with or without the presence of hypophysis secretion and have no effect upon heterologous accessories; and that sex hormones have no direct action upon the gonads." Carl R. Moore and Dorothy Price, "The Question of Sex Hormone Antagonism," *Proceedings of the Society for Experimental Biology and Medicine* 28 (1930–31): 38–40, at 40.

110. The same finding had been reported in 1931 by gynecologist Harald Siebke of the University Women's Clinic at Kiel, although Moore and Price do not seem to have been familiar with that report. In Siebke's experiments, when the female hormone was administered to capons along with male hormone, the growth of the comb was not inhibited. Conversely, the male hormone did not interfere with the development of estrus in oophorectomized mice when both male and female hormones were given together. The male and female hormones had their respective functions, and those were fulfilled regardless of the sex of the experimental subject: the female hormone stimulated breast development in males as well as females, while the male hormone induced growth of the comb in cocks, hens, or capons. The only inhibitory influence of the female hormone in males was exerted on the testes—because it interfered with testicular function, it could indirectly affect the secondary sex characters of the male. That, however, was not direct hormonal antagonism. See H. Siebke, "Thelykinin und Androkinin, das weibliche und männliche Sexualhormon, im Körper der Frau," *AfG* 146 (1931): 417–62, at 453–54. For a similar study, see J. M. Kabak, "Männliches Geschlechtshormon aus Frauenurin," *Endokrinologie* 10 (1932): 12–15.

111. There was, however, a potential problem with this line of argument. During pregnancy, the urine was full of estrin, which meant that the pituitary should be inactive. "But it is also pregnancy urine that yields such great amounts of the gonadal-stimulating hormone, and the source of this substance has been held by some to be the hypophysis." This, however, was far from proven. It seemed "very certain that oestrin of pregnancy urine does not come from the ovary" and nor did the gonad-stimulating substance. All the available evidence suggested, in fact, that "the hypophysis of pregnancy is less active in producing this gonadal-stimulating substance than the hypophysis of non-pregnant forms, hence it is probable that the urinary

gonadal-stimulating substance is not of hypophyseal origin." Moore and Price, "Gonad Hormone Functions and the Reciprocal Influence between Gonads and Hypophysis," 54. See also B. Zondek and S. Aschheim, "Hypophysenvorderlappen und Ovarium," *AfG* 130 (1927): 1–45; Alfons Bacon, "A Comparative Study of the Anterior Hypophysis in the Pregnant and Non-Pregnant States," *American Journal of Obstetrics and Gynecology* 19 (1931): 352; and E. Philipp, "Hypophysenvorderlappen und Placenta," *CfG* 54 (1930): 450–53.

112. Dorothy Price, "Feedback Control of Gonadal and Hypophyseal Hormones," 231.

113. Moore and Price, "Gonad Hormone Functions and the Reciprocal Influence between Gonads and Hypophysis," 52.

114. The concept of a high or a low dose was partly impressionistic at the time, especially if the reference was to a clinical dose. As Hans Simmer reminds us, the doses used by clinicians in the early 1930s were so small that they could not possibly have had any therapeutic effect on humans. For details, see Hans H. Simmer, "Der 'Hohlweg-Effekt': Anspruch und Wirklichkeit bei der Entstehung eines Eponyms (Teil I)," *Medizinhistorisches Journal* 30 (1995): 167–83, esp. 171–73, 180–83.

115. Walter Hohlweg, "The Regulatory Centers of Endocrine Glands in the Hypothalamus," in Joseph Meites, Bernard T. Donovan, and Samuel M. McCann, *Pioneers in Neuroendocrinology* (New York: Plenum, 1975), 161–72, at 167. Not all of his findings, however, were as novel as Hohlweg claimed. In 1930, long before "the Hohlweg effect" had been demonstrated, two American scientists and, independently of them and of each other, two Italian researchers had postulated a positive feedback loop between the ovary and the pituitary. For details, see Hans H. Simmer, "Der 'Hohlweg-Effekt': Anspruch und Wirklichkeit," 167–83.

116. See H. Kun, "Psychische Feminierung und Hermaphrodisierung von Männchen durch weibliches Sexualhormon," *Endokrinologie* 13 (1934): 311–23, at 321.

117. Steinach and Kun, "Antagonistische Wirkungen der Keimdrüsen-Hormone," 833.

118. Ibid., 833–34; Ernst Laqueur, P. C. Hart and S. E. de Jongh, "Über weibliches Sexualhormon (Menformon), das Hormon des östrischen Zyklus. III: Bemerkungen zur Eichung, reaktivierender Einfluß auf senile Mäuse; antimaskuline Wirkung," *DmW* 52 (1926): 1247–50.

119. Anne Fausto-Sterling, *Sexing the Body,* 168–69.

120. Paul de Kruif, *The Male Hormone* (New York: Harcourt, Brace, 1945), 116.

121. Fausto-Sterling, *Sexing the Body,* 169.

122. Ibid., 168.

123. Weiert Velle, "Urinary Oestrogens in the Male," *Journal of Reproduction and Fertility* 12 (1966): 65–73, at 65–66.

124. O. O. Fellner, "Über die Wirkung des Placentar- und Hodenlipoids auf die männlichen und weiblichen Sexualorgane," *PAgP* 189 (1921): 199–214, esp. 205–12, at 211–12; and Fellner's comments in E. Herrmann and M. Stein, "Über den Einfluß eines Hormones des Corpus luteum auf die Entwicklung männlicher Geschlechtsmerkmale," *WkW* 29 (1916): 177.

125. N. Oudshoorn, *Beyond the Natural Body: An Archeology of Sex Hormones* (London; Routledge, 1994), 28.

126. E. Laqueur, E. Dingemanse, P. C. Hart and S. E. De Jongh, "Über das Vorkommen weiblichen Sexualhormons (Menformon) im Harn von Männern," *KW* 6 (1927): 1859.

127. Max Dohrn of the Schering Laboratory in Berlin warned against assuming that any substance producing such a result was necessarily an ovarian hormone. Testicular extracts, he claimed, could produce the same result in an oophorectomized mouse. Only further research could reveal whether the ovarian and testicular secretions were chemically identical or simply similar, but it was already quite clear that one should not regard a positive Allen-Doisy reaction as being specific to the female sex hormone. See M. Dohrn, "Ist der Allen-Doisy-Test spezifisch für das weibliche Sexualhormon?," *KW* 6 (1927): 359–60.

128. S. Loewe, H. E. Voss, F. Lange and A. Wähner, "Sexualhormonbefunde im männlichen Harn," *KW* 7 (1928): 1376–77. For references to other work in the area, see Egon Diczfalusy and Christoph Lauritzen, *Oestrogene beim Menschen* (Berlin: Springer, 1961). For recent scientific opinion on the functions of estrogen in males, see R. M. Sharpe, "The Roles of Estrogen in the Male," *Trends in Endocrinology and Metabolism* 9 (1998): 371–77.

129. See H. Hirsch, "Brunsthormon im Blute des Mannes," *KW* 7 (1928): 313–14.

130. B Zondek, "Mass Excretion of Oestrogenic Hormone in the Urine of the Stallion," *Nature* 133 (1934): 209–10, at 209, emphasis in original. The mass excretion of female hormones in male urine seemed, Zondek reported, to be a peculiarity of equine species. Although the stallion was at the top, other equines such as the zebra or the Asiatic wild ass came close. The bull, by contrast, excreted very modest amounts of female hormones.

131. "After five daily injections each of 0.5 c.c. of stallion's urine, the weight of the uterus of a young rabbit weighing 1200 gm rose from 0.47 gm to 1.8 gm. On introducing larger quantities of urine . . . the weight of the uterus rose from 0.47 gm to 5.48 gm and the weight of the vagina from 0.15 gm to 2.48 gm." B Zondek, "Mass Excretion of Oestrogenic Hormone," 209–10.

132. B. Zondek, "Oestrogenic Hormone in the Urine of the Stallion," *Nature* 133 (1934): 494. Zondek's finding led to a flurry of research. Nelly Oudshoorn states that from nine in 1928, the number of publications on female hormones in males rose to forty-two in the 1936 *Quarterly Cumulative Index Medicus,* but declined to fifteen in 1937. See Oudshoorn, *Beyond the Natural Body,* 157, n. 16.

133. B. Zondek, "Oestrogenic Hormone in the Urine of the Stallion." In spite of other potential explanations, Zondek, in purely Whiggish terms, was correct. "It is remarkable that Zondek already in 1934, before anything was known about the biochemical reactions involved, stated that the female hormone present in the male organism quite possibly is a conversion product of the male hormone. . . . This has indeed turned out to be the case," wrote Weiert Velle ("Urinary Oestrogens in the Male," 68).

134. One study in the early 1960s claimed to have found the final frontier between the sexes by showing that a specific form of female hormone (ring B-unsaturated

estrogens) was missing from the stallion's urine whereas it was characteristic of the urine of the pregnant mare. "A marked sex difference seemed to exist in the horse," the researchers concluded, but one of them later admitted ruefully that even the ring B-unsaturated estrogens did occur in stallion's urine in trace amounts. See Weiert Velle, "Urinary Oestrogens in the Male," 67. Astoundingly, Zondek himself, even in late life, continued to embrace the very concept of antagonism that he had supposedly demolished in the 1930s. "To this day," he confessed in 1966, "I don't understand how it is that the high concentration of oestrogen in stallion testes and blood does not exert an emasculating effect." See Michael Finkelstein, "Professor Bernhard Zondek: An Interview," *Journal of Reproduction and Fertility* 12 (1966): 3–19, at 11. Earlier, Zondek had cited Steinach's experiments, emphasizing that the antimasculine effects of female sex hormones was a fact beyond dispute (*außer Zweifel steht*). See B. Zondek, *Die Hormone des Ovariums und des Hypophysenvorderlappens*, 103.

135. On food-related explanations of female sex hormone in males and critiques of these hypotheses, see Nelly Oudshoorn, *Beyond the Natural Body*, 27, and Weiert Velle, "Urinary Oestrogens in the Male," 70.

136. See Oudshoorn, *Beyond the Natural Body*, 28, and C. D. Kochakian, "Excretion and Fate of Androgens: Conversion of Androgens to Estrogens," *Endocrinology* 23 (1938): 463–67.

137. E. Steinach, H. Kun, and O. Peczenik, "Beiträge zur Analyse der Sexualhormonwirkungen: Tierexperimentelle und klinische Untersuchungen," *WkW* 49 (1936): 899–903, at 902.

138. Ibid.

139. E. Steinach and H. Kun, "Umwandlung des männlichen Sexualhormons in einen Stoff mit weiblicher Hormonwirkung: Untersuchungen an jungen und alten Männern," *Anzeiger der Akademie der Wissenschaften in Wien: Mathematisch-naturwissenschaftliche Klasse* 74 (1937): 121–23. A translation was published as: E. Steinach and H. Kun, "Transformation of Male Sex Hormones into a Substance with the Action of a Female Hormone," *Lancet*, 9 October 1937, 845.

140. The physiological male hormone, therefore, was far superior in its nonsexual effects than the chemically pure injectable preparations such as testosterone propionate or androsterone benzoate. For details of the experiments, see E. Steinach, H. Kun, and O. Peczenik, "Beiträge zur Analyse der Sexualhormonwirkungen."

141. Ibid., 900.

142. The rejuvenative effects of the operation, Steinach emphasized, might be replicated more reliably with hormone injections, if a small quantity of the female hormone were used to supplement the larger doses of the male hormone. In castrated male rats, a combination of male and female hormones restored potency and libido at a smaller dose than with male hormones alone. See Steinach, Kun, and Peczenik, "Beiträge zur Analyse der Sexualhormonwirkungen," 901–3. In the few cases treated experimentally with such a combination, there were dramatic improvements, generalized as well as psychic. In one case of impotence in a twenty-seven-year-old man, almost entirely refractory to treatment with androsterone, the combination of androsterone with female hormone led to the sustained restoration of erectile function. See ibid., and Steinach and Kun, "Umwandlung des männlichen Sexualhormons in einen Stoff mit weiblicher Hormonwirkung," 123.

143. Nellie Oudshoorn points out, however, that the phenomenon was less prominently reported in the medical literature. Between 1927 and 1939, fifty-three articles were indexed by the *Quarterly Cumulative Index Medicus* on female hormones in males but only fourteen on male hormones in females. See Oudshoorn, *Beyond the Natural Body,* 157, n. 18.

144. Robert Meyer, "Beitrag zur Frage der Funktion von Tumoren der Ovarien, insbesondere solcher, die zur Entweiblichung und zur Vermännlichung führen. Arrhenoblastome," *CfG* 54 (1930): 2374–89, at 2378.

145. See Harald Siebke, "Thelykinin und Androkinin, das weibliche und männliche Sexualhormon, im Körper der Frau," *AfG* 146 (1931): 417–62, esp. 438, 454–55.

146. E. Steinach and H. Kun, "Luteingewebe und männliche Geschlechtscharaktere," *PAgP* 227 (1931): 266–78.

147. The same results could also be produced by injections of anterior pituitary extracts over extended periods. Using a pituitary extract produced at the Schering-Kahlbaum laboratories, Steinach and Kun succeeded in masculinizing uncastrated, normal female guinea pigs. In controls treated with extracts of follicular extract, these changes were not seen. See ibid., 269–70.

148. Ibid., 271–73.

149. Ibid., 274–76.

150. H. Kun, "Psychische Feminierung und Hermaphrodisierung von Männchen," 314–15.

151. Some human conditions, such as hermaphroditism or masculinization in ovarian tumors, could well be due to an elevation in the levels of the masculinizing hormone of the corpora lutea, Steinach speculated. See Steinach and Kun, "Luteingewebe und männliche Geschlechtscharaktere," 278.

152. See R. T. Hill, "Ovaries Secrete Male Hormone, I: Restoration of the Castrate Type of Seminal Vesicle and Prostate Glands to Normal by Grafts of Ovaries in Mice," *Endocrinology* 21 (1937): 495–502.

153. In 1932, Alexander Lipschütz had reported a case in which a castrated male guinea pig with an ovarian transplant showed proliferation of breast tissue (and lactation) but no loss of masculine sexual characters: the penis, the seminal vesicles, the prostate, and the vas deferens were seen to have retained their normal dimensions on autopsy thirty-four months after castration. The ovarian graft had proliferated into a tumor-like mass and the adrenal glands had also undergone enlargement. Although Lipschütz suggested that the likeliest reason for the retention of male characters was the secretion of male hormone(s) from the ovarian graft, he nevertheless took it as a singular, pathological instance and did not suggest that male hormones were *normally* secreted by ovaries. See A. Lipschütz, "Wiedervermännlichung eines kastrierten männlichen Meerschweinchens nach Eierstocksverpflanzung," *Virchows Archiv für pathologische Anatomie und Physiologie und für klinische Medizin* 285 (1932): 35–45.

154. R. T. Hill, "Ovaries Secrete Male Hormone, I."

155. A. S. Parkes, "Ambisexual Activity of the Gonads," in L. Brouha, ed., *Les hormones sexuelles, I: Les propriétés des hormones sexuelles* (Paris: Hermann, 1938),

67–87, at 71. Parkes's own studies suggested that although ovarian extracts possessed androgenic properties, not all the androgens in the female were secreted by the ovaries themselves. See A. S. Parkes, "Androgenic Activity of Ovarian Extracts," *Nature* 139 (1937): 965.

156. For this history, which began effectively in 1856 with controversial extirpation experiments by none other than Charles-Edouard Brown-Séquard, see Humphry Davy Rolleston, *The Endocrine Organs in Health and Disease with an Historical Review* (London: Oxford University Press, 1936), 319–21.

157. It had been known since the eighteenth century that the bodies of women with adrenal tumors could become hairy and masculine. See Rolleston, *The Endocrine Organs in Health and Disease*, 361–71.

158. Berman, *Glands Regulating Personality*, 71.

159. Arthur Keith, *The Engines of the Human Body*, 3d ed. (London: Williams and Norgate, 1926), 229. Many cases of adrenal hypersecretion, however, led only to mild symptoms that patients did not even complain about since, as William Blair-Bell would have it, "it is a question whether the shame of a moustache or pride in masculinity has the greater weight with some women." See W. Blair-Bell, "The Relation of the Internal Secretions to the Female Characteristics and Functions in Health and Disease," *Proceedings of the Royal Society of Medicine* 7 (1913), pt. 2, Obstetrical and Gynaecological Section, 47–100, at 75.

160. Quoted in Max A. Goldzieher, *The Adrenal Glands in Health and Disease* (Philadelphia: F. A. Davis, 1944), 478. Such patients, Humphry Davy Rolleston thought, had long existed but had been labeled merely as monstrosities. See Rolleston, *Endocrine Organs in Health and Disease*, 368.

161. The condition was not usually fatal, unless it was due to a malignant tumor of the adrenal. See Rolleston, *Endocrine Organs in Health and Disease*, 362.

162. Ernest E. Glynn, "The Adrenal Cortex, its Rests and Tumors; Its Relation to Other Ductless Glands, and especially to Sex," *Quarterly Journal of Medicine* 5 (1912): 157–92, at 186–88.

163. See S. Levy Simpson, P. de Fremery and Alison Macbeth, "The Presence of an Excess of 'Male' (Comb-Growth and Prostate-Stimulating) Hormone in Virilism and Pseudo-hermaphroditism," *Endocrinology* 20 (1936): 363–72. This study was conducted on eleven women diagnosed with adrenogenital syndrome, Cushing's syndrome, and Achard-Thiers syndrome (a not very clearly defined diagnosis usually reserved for postmenopausal women with Cushing's syndrome: "the fat, bearded, diabetic woman") and on three "young girls exhibiting pseudohermaphroditism."

164. Ibid., 367, 372.

165. Glynn, "The Adrenal Cortex, its Rests and Tumours," 181. Glynn's article contains an excellent overview of the details I leave out here.

166. Jean Paul Pratt, "Endocrine Disorders in Sex Function in Man," in Edgar Allen, ed., *Sex and Internal Secretions: A Survey of Recent Research* (Baltimore: Williams and Wilkins, 1932), 880–911, at 905. One dissenter from this consensus was Alexander Lipschütz, who thought that the adrenal was unlikely to have a direct masculinizing action, but, otherwise, the adrenal maintained its masculine profile. See Alexander Lipschütz, *The Internal Secretions of the Sex Glands: The Problem of the*

"Puberty Gland" (Cambridge: Heffer, 1924), 378. My thanks to Helen Blackman for alerting me to this argument.

167. See L. R. Broster, *Endocrine Man,* 85.

168. See Max A. Goldzieher, *The Adrenal Glands in Health and Disease,* 464–67.

169. The latter was hardly satisfactory since, as Broster pointed out, it was "capable of removing only part of the cause," but the adrenal cortices being essential to life, the removal of both adrenals was quite out of the question. See Broster, *Endocrine Man,* 88.

170. Ibid., 88–89. See also Rolleston, *Endocrine Organs in Health and Disease,* 366.

171. L. Berman, *Glands Regulating Personality,* 80, 139, 210. These—or at least similar—notions were shared by many. Some considered only the adrenal cortex to be male, and the adrenal medulla as female. American gynecologist Samuel Bandler, for example, argued that women were more emotional than men because the adrenal medulla was overdeveloped in the female. The adrenal medulla, the posterior pituitary, and the ovary were closely related, generating "characteristics of coyness and self-display" and ensuring that woman was "not the aggressor." In men, the anterior pituitary, the testicles, and the adrenal cortex were similarly related, making man more intellectual but also "more brutal, more criminal and more coarse." S. W. Bandler, *The Endocrines* (Philadelphia: Bandler, 1921), 248. For another portrayal of the adrenals as male glands, see Leonard Williams, "The Interstitial Gland," *BMJ,* 27 May 1922, 833–35, at 834.

172. "The degree of masculine trend in a woman is a crude measure of adrenal domination, the degree of feminine deviation in a man is roughly proportional to the amount of pituitary influences in his make-up," observed Berman, who insisted that all human beings were biologically male as well as female—"we are all, more or less, partial hermaphrodites." Berman, *Glands Regulating Personality,* 139, 143, 210.

173. Glynn, "The Adrenal Cortex, its Rests and Tumours," 181.

174. Julian Huxley, "Sex Biology and Sex Psychology" [1922], 122–23.

175. Rolleston, *The Endocrine Organs in Health and Disease,* 369.

176. A. S. Parkes, "Ambisexual Activity of the Gonads," 72.

177. Julius Baur, "Homosexuality as an Endocrinological, Psychological, and Genetic Problem," *Journal of Criminal Psychopathology* 2 (1940–41): 188–97, at 194.

178. A. S. Parkes, "Ambisexual Activity of the Gonads," 73.

179. L. R. Broster, *Endocrine Man,* 90. See also Broster, "The Adrenogenital Syndrome," *Lancet,* 21 April 1934, 830–34, at 834.

180. Broster, *Endocrine Man,* 90. See also L. R. Broster and H. W. C. Vines, *The Adrenal Cortex: A Surgical and Pathological Study* (London: Lewis, 1933). Although some patients of the syndrome showed "alterations in the psychological outlook towards the male type," there was "a return to the finer points of feminine interest" after the removal of the pathological adrenal. Broster, "The Adrenogenital Syndrome," 830, 831, 834.

181. J. Botella-Llusiá, "Nebennierenrinde und Keimdrüsen," in Tassilo Antoine, ed., *Klinische Fortschritte: "Gynäkologie"* (Vienna: Urban & Schwarzenberg, 1954), 281–95.

182. Berman, *Glands Regulating Personality*, 80, 205.

183. L. R. Broster, "The Adrenogenital Syndrome," 834.

184. Ibid.

185. There is an enormous literature on the history of eugenics in England. See, for instance, Daniel J. Kevles, *In the Name of Eugenics: Genetics and the Uses of Human Diversity* (New York: Knopf, 1985); Pauline M. H. Mazumdar, *Eugenics, Human Genetics, and Human Failings: The Eugenics Society, its Sources and its Critics in Britain* (London: Routledge, 1992); and for the broader political context, G. R. Searle, *Eugenics and Politics in Britain, 1900–1914* (Leyden: Noordhoff, 1976); Richard A. Soloway, *Demography and Degeneration: Eugenics and the Declining Birthrate in Twentieth-Century Britain* (Chapel Hill: University of North Carolina Press, 1990); and Michael Freeden, *The New Liberalism: An Ideology of Social Reform* (Oxford: Clarendon Press, 1978).

186. Lothar von Frankl-Hochwart, "Über Diagnose der Zirbeldrüsentumoren," *Deutsche Zeitschrift für Nervenheilkunde* 37 (1909): 455–65. The syndrome was very rare—only twenty-two cases were on record in the mid-1930s. See H. D. Rolleston, *Endocrine Organs in Health and Disease*, 467.

187. H. D. Rolleston, *Endocrine Organs in Health and Disease*, 467.

188. Ibid., 451–54.

189. L von Frankl-Hochwart, "Über Diagnose der Zirbeldrüsentumoren."

190. H. Cushing, *The Pituitary Body and its Disorders*, 228–29.

191. He pointed out, however, that the sexualizing consequences of pineal tumors could conceivably be due to secondary pressure effects on the pituitary gland. See ibid., 229–30, 283.

192. Artur Biedl, *The Internal Secretory Organs: Their Physiology and Pathology*, trans. Linda Forster (London: Bale, Sons and Danielsson, 1913), 334–35.

193. C. Posner, in *Therapie der Gegenwart* 8 (1916), quoted by Rolleston, *Endocrine Organs in Health and Disease*, 461.

194. H. D. Rolleston, *Endocrine Organs in Health and Disease*, 460–61.

195. L. G. Rowntree, J. H. Clark, A. Steinberg, and A. M. Hanson, "Biologic Effects of Pineal Extract (Hanson)," *JAMA* 106 (1936): 370–73.

196. See the overview in August A. Werner, *Endocrinology: Clinical Application and Treatment*, 2d ed. (London: Henry Kimpton, 1942), 785–86.

197. H. D. Rolleston, *Endocrine Organs in Health and Disease*, 467.

198. The influence of the pineal on human sexual functions has never been clarified to the satisfaction of scientists. It is known to secrete melatonin (which has often been implicated in the maintenance of circadian rhythms, menstruation, and ovulation) but also other active substances. Pineal extract, when freed of melatonin, has been shown to inhibit the gonadotropic hormones of the pituitary, and the association of *some* pineal tumors with sexual precocity is unquestioned. Confusingly, however, some other kinds of pineal tumor retard sexual development. See V. C. Medvei, *The History of Clinical Endocrinology* (Carnforth, Lancs: Parthenon, 1993), 264–65.

199. Rolleston, *Endocrine Organs in Health and Disease*, 435–40.

200. G. Keynes, "The Physiology of the Thymus Gland," *BMJ*, 18 September 1954, 659–63.

201. Rolleston, *Endocrine Organs in Health and Disease,* 439.

202. See Ann Dally, "Status Lymphaticus: Sudden Death in Children from 'Visitation of God' to Cot Death," *Medical History* 41 (1997): 70–85, esp. 76; Warren G. Guntheroth, "The Thymus, Suffocation, and Sudden Death Syndrome: Social Agenda or Hubris?," *Perspectives in Biology and Medicine* 37 (1993): 2–13; and Rolleston, *Endocrine Organs in Health and Disease,* 447–48.

203. Ann Dally, "Status Lymphaticus," 78, 84. Only in the 1960s did this practice finally die out. See ibid., 84.

204. Rolleston, *Endocrine Organs in Health and Disease,* 441, 443; and James Henderson, "On the Relationship of the Thymus to the Sexual Organs," *Journal of Physiology* 31 (1904): 222–29.

205. D. Noel Paton, "The Relationship of the Thymus to the Sexual Organs, II: The Influence of Removal of the Thymus on the Growth of the Sexual Organs," *Journal of Physiology* 32 (1905): 28–32.

206. See E. T. Halnan and F. H. A. Marshall, "On the Relation between the Thymus and the Generative Organs and the Influence of these Organs upon Growth," *Proceedings of the Royal Society of London B* 88 (1914): 68–89, esp. 85–89.

207. W. Blair-Bell, "The Relation of the Internal Secretions to the Female Characteristics and Functions," 50.

208. Berman, *Glands Regulating Personality,* 137.

209. Ibid., 217–18.

210. "If the pituitary and the thyroid can enlarge to compensate for their defects, they may become the queer brilliants, the eccentric geniuses of the arts and sciences" he added. "Should they not, mental deficiency and delinquency are their portion.... Should they survive all other hazards, suicide may still be their most frequent fate" (ibid., 223). In 1933, the endocrine researcher Roy Hoskins referred to "respectable statistics which seem to indicate unusual prevalence of status thymicolymphaticus [*sic*] in suicides, drug addicts and victims of caisson disease. Whether we are dealing here with mere coincidence or with an actual biological relationship is a question that would seem to deserve further research." R. G. Hoskins, *The Tides of Life: The Endocrine Glands in Bodily Adjustment,* 254–55.

211. Swale Vincent, *Internal Secretion and the Ductless Glands* (London: Arnold, 1912), 365.

212. See contribution of R. Hofstätter to F. v. Mikulicz-Radecki et al., "Die praktische Bedeutung der Hormonbehandlung in der Gynäkologie," *Medizinische Klinik* 30 (1934): 959–65; 1022–26, 1486–90, at 959–60.

213. See L. G. Rowntree, J. H. Clark, and A. M. Hanson, "The Biologic Effects of Thymus Extract (Hanson): Accruing Acceleration in Growth and Development in Successive Generation of Rats under Continuous Treatment with Thymus Extract," *JAMA* 103 (1934): 1425–30; and H. D. Rolleston, *The Endocrine Organs in Health and Disease,* 442.

214. Nelly Oudshoorn, *Beyond the Natural Body,* 28–29.

Chapter Five

1. H. Cushing, "Disorders of the Pituitary Gland: Retrospective and Prophetic," *JAMA* 76 (1921): 1721–26, at 1726.

2. Of course, the *average* gynecological practitioner had always been a consumer rather than a scientific innovator. In 1921, American gynecologist Samuel Bandler had declared that in his practice, "endocrine therapy has displaced and replaced the old time drugs, so that . . . practically 90 per cent. of all my prescriptions for internal use consist almost entirely, if not wholly, of endocrine extracts." Samuel Wyllis Bandler, *The Endocrines* (Philadelphia: Saunders, 1921), 314–15. By 1940, however, *most* gynecologists had become consumers.

3. Corner, "The Early History of the Oestrogenic Hormones," *Journal of Endocrinology* 31 (1964–65): iii–xvii, at ix–x. See also H. H. Simmer, "On the History of Hormonal Contraception, II: Otfried Otto Fellner (1873–19??) and Estrogens as Antifertility Hormones," *Contraception* 3, no. 1 (1971): 1–20. Iscovesco as well as Fellner are remarkably elusive figures; neither Corner nor Simmer succeeded in discovering much about their lives or their work, beyond that reported in their published papers.

4. E. Steinach, M. Dohrn, W. Schoeller, W. Hohlweg and W. Faure, "Über die biologischen Wirkungen des weiblichen Sexualhormons," *PAgP* 219 (1928): 306–36. For the first time, a Steinach paper was announced not merely as being from the "Biologischen Versuchsanstalt" of the Academy of Sciences but also from the Laboratory of Schering-Kahlbaum, Berlin. The Schering website seems to be the only prominent locus where the name of Steinach is still remembered with appreciation. See http://www.schering.de/unternehmen/Historie/ChronikderScheringForschung2.htm (my thanks to Angela Dahrmann of the Schering Museum for her assistance). Steinach's connection with Schering was long, intimate, and apparently lucrative—when in exile in Switzerland after 1938, he was supported financially by the company. Much remains to be elucidated about this link, which, while it is obviously similar to the relationship of Ernst Laqueur with the Dutch firm of Organon, analyzed by Nelly Oudshoorn (*Beyond the Natural Body: An Archeology of the Sex Hormones* [London: Routledge, 1994], 82–111), seems, prima facie, to be distinctive in its contexts. See E. Steinach, H. Kun, and W. Hohlweg, "Reaktivierung des senilen Ovars und des weiblichen Gesamtorganismus auf hormonalem Wege," *PAgP* 219 (1928): 325–36, at 336 for the claim that the hormone injections were being used in humans for preclimacteric and climacteric symptoms and associated signs of old age with encouraging results.

5. Later, according to a gynecologist who used it extensively, Progynon was enriched with sex hormones extracted from the urine of pregnant women. See Erwin Last, "Weitere Erfahrungen mit Progynon-Steinach," *Medizinische Klinik* 27 (1931): 766–67, at 766.

6. E. Steinach, H. Heinlein, and B. P. Wiesner, "Auslösung des Sexualzyklus, Entwicklung der Geschlechtsmerkmale, reaktivierende Wirkung auf den senilen weiblichen Organismus durch Ovar- und Placentaextrakt: Versuche an Ratten und Meerschweinchen," *PAgP* 210 (1925): 598–611.

7. Ibid., 603.

8. Ibid., 605–6.

9. Ibid., 604.

10. Steinach, Kun, and Hohlweg, "Reaktivierung des senilen Ovars," 331–35. As more became known of the role of the pituitary in stimulating the gonads, Steinach claimed that the injections of female hormones worked their magic on the ovary not by direct stimulation but by stimulating the pituitary. This hypothesis would be demolished by Moore and Price's demonstration in the 1930s of the negative feedback between the pituitary and the ovary. It had, however, been widely believed in the early days of organotherapy that extracts of any particular organ stimulated that organ. This was supposed to be a virtually invariable phenomenon and referred to as Hallion's Law (after a French physician who first enunciated it). See Henry R. Harrower, *Practical Hormone Therapy: A Manual of Organotherapy for General Practitioners* (London: Baillière, Tindall and Cox, 1914), 24. For Moore's categorical dismissal of the concept that an undersecreting gland could be stimulated by the administration of its own secretions, see Carl R. Moore, "The Regulation of Production and the Function of the Male Sex Hormone," *JAMA* 97 (1931): 518–22, at 521.

11. Steinach, Heinlein, and Wiesner, "Auslösung des Sexualzyklus," 604. See also Steinach, Kun, and Hohlweg, "Reaktivierung des senilen Ovars," 335.

12. Erwin Last, "Weitere Erfahrungen mit Progynon-Steinach." For a brief earlier report, see J. Novak and E. Last, "Klinische Erfahrungen mit Progynon, einem hochwertigen weiblichen Sexualhormon," *Medizinische Klinik* 24 (1928): 1715–18.

13. On the use of pituitary irradiation in Holzknecht's radiology clinic in Vienna "with a view to alleviating the worrying conditions of the climacterium," see the anonymous report, "Treatment of Climacteric Condition [*sic*] by the Action of X Rays on the Hypophysis," *Lancet,* 26 April 1924, 868. The rationale of the treatment "was based on the resemblance of climacteric conditions to vasomotor hypersensitiveness due to endocrine disturbance. . . . The effect seems to persist for some months and is regarded by Holzknecht rather as the stimulation of a hypo-functionating gland than as the depression of a hyper-function." If accurate, this report contradicts Holzknecht's well-known conviction that x-rays always destroyed tissues and never stimulated them. See my discussion of this issue in chapter 3. Well into the 1930s, many leading clinicians continued to teach that low-dose irradiation of the ovaries or the pituitary was valuable in some cases of menstrual disturbance. See, for an example, Manfred Fraenkel, *Die Verjüngung der Frau zugleich ein Beitrag zum Problem der Krebsheilung* (Bern: Bircher, 1924), 29. For a powerful American endorsement of this thesis with many supporting references, see Charles Mazer and Leopold Goldstein, *Clinical Endocrinology of the Female* (Philadelphia: Saunders, 1932), 278–91, 298–99, 306–10.

14. Last, "Weitere Erfahrungen mit Progynon-Steinach," 767.

15. Ibid., 766.

16. Corner, "Early History of the Oestrogenic Hormones," xi.

17. Ibid.

18. See Arthur T. Hertig, "Allen and Doisy's 'An Ovarian Hormone,'" *JAMA* 250 (1983): 2684–88, at 2685.

19. Edgar Allen and Edward A. Doisy, "An Ovarian Hormone: Preliminary Report on its Localization, Extraction and Partial Purification, and Action in Test Animals," *JAMA* 81 (1923): 819–21.

20. J. Süß, "Die Ein-Hormon-Hypothese: Eine aufschlußreiche Episode in der Geschichte der Ovarialendokrinologie," *Geburtshilfe und Frauenheilkunde* 47 (1987), 134–37. See also C. R. Stockard and G. N. Papanicolaou, "The Existence of a Typical Oestrous Cycle in the Guinea Pig with a Study of Its Histological and Physiological Changes," *American Journal of Anatomy* 22 (1917): 225.

21. Süß, "Die Ein-Hormon-Hypothese," 135.

22. Corner, "Early History of the Oestrogenic Hormones," xi.

23. H. Siebke, "Thelykinin und Androkinin, das weibliche und männliche Sexualhormon, im Körper der Frau," *AfG* 146 (1931): 417–62, at 439.

24. Robert T. Frank, *The Female Sex Hormone* (London: Baillière, Tindall and Cox, 1929), Foreword [unpaginated].

25. HB to ES, 3 November 1939. See also Diana Long Hall, "Biology, Sex Hormones and Sexism in the 1920s," *Philosophical Forum* 5, nos. 1–2 (1973/74): 81–96, on the "chemicalization" of endocrine physiology from the late 1920s. When Steinach died at the age of 84 in Territet near Montreux in 1944, he was frail (he had never had a Steinach Operation himself, as he lamented to Benjamin while recovering from a severe illness), lonely (his wife had died in 1938 and they had never had any children) and almost forgotten (see the later letters in the Benjamin-Steinach correspondence, especially HB to ES, 25 March 1940; and ES to HB, 19 August 1941). In spite of Benjamin's sincere efforts, the call to America had never come and only the *New York Times* remained loyal enough to mark Steinach's passing with a tribute. Gone, however, were the heroic tones of the 1920s. "The trouble with Steinach," observed the *Times,* "was that his operation was based on a hypothesis which proved to be wrong and which he declined to change. The procession had passed him by." See "Dr. Eugen Steinach," *New York Times,* 16 May 1944, 20, col. 3. I have been unable to locate any obituary in the major British, American, and Central European medical journals of the time.

26. George Sylvester Viereck, *Glimpses of the Great* (New York: Macaulay, 1930), 264.

27. Carl R. Moore, "The Regulation of Production and the Function of the Male Sex Hormone," *JAMA* 97 (1931): 518–22, at 519.

28. Regretting this separation, two German authors emphasized in the early 1960s that the era of pioneer investigations was not necessarily over, but that it was imperative that co-ordination and communication between the laboratory scientist and the bedside doctor be improved. See Egon Diczfalusy and Christian Lauritzen, *Oestrogene beim Menschen* (Berlin: Springer-Verlag, 1961), 473.

29. A. S. Parkes, "The Rise of Reproductive Endocrinology, 1926–1940," Proceedings of the Society for Endocrinology, *Journal of Endocrinology* 34 (1966): ix–xxxii.

30. Emil Novak, "The Therapeutic Use of Estrogenic Substances," *JAMA* 104 (1935): 1815–21, at 1816.

31. F. Siegert, "Erfahrungen und Ergebnisse mit der Hormonbehandlung weiblicher Genitalstörungen," *Zeitschrift für Geburtshilfe und Gynäkologie* 107 (1934): 117–57, at 117–18.

32. Ibid., 119.

33. Emil Novak, "The Therapeutic Use of Estrogenic Substances," 1817.

34. Obviously, the absence of menstruation could be due to numerous causes, with endocrine anomalies being only one of them, and ovarian deficiency simply one of the latter. At the very least, one had to exclude diabetes and thyroid deficiency. Ideally, recommended the Edinburgh gynecologist Thomas MacGregor in 1938, treatment with sex hormones should not be attempted before estimation of urinary levels of estrogen metabolites *and* the clinical establishment of an undeveloped uterus. If the uterus was fully developed, then the amenorrhea was unlikely to be caused by a simple deficiency of estrogenic hormone. See T. N MacGregor, "Amenorrhoea: Its Aetiology and Treatment," *BMJ*, 2 April 1938, 717–22, at 717–18.

35. Emil Novak, "The Therapeutic Use of Estrogenic Substances," 1818.

36. Ibid., 1816. This attentiveness to the findings of experimental physiologists, however, was not particularly common among contemporary gynecologists. In 1934, Friedrich Geller of Ludwig Fraenkel's Gynaecology Clinic had appealed to clinical experience to argue that externally administered hormones could indeed stimulate the glands producing them. This, he argued, was because all endocrine glands secreted multiple, functionally interrelated hormones. When the production of one was hampered by some pathological process, the entire gland suffered and the therapeutic administration of that hormone restored the hormonal balance within the gland, thereby restoring all its secretions to normal levels. See F. C. Geller, "Die Hormontherapie in der Gynäkologie," *Therapie der Gegenwart* 75 (1934): 121–26, at 121.

37. Emil Novak, "The Therapeutic Use of Estrogenic Substances," 1817. He added that regular mensturation could now, in principle, be simulated by administering luteal hormone in sequence, although a commercially available luteal hormone remained a distant prospect. From German reports of such treatment, it seemed, however, that the dosage required might be astronomical in comparison to the doses used in everyday practice in America. Even those small American doses were found expensive by "all except a few patients' (ibid., 1818). Although a luteal hormone was already available in Central Europe, it was so expensive that even an enthusiastic advocate of hormone treatment like Otfried Otto Fellner used it only rarely. See Fellner's contribution to F. v. Mikulicz-Radecki et al., "Die praktische Bedeutung der Hormonbehandlung in der Gynäkologie," *Medizinische Klinik* 30 (1934): 959–65, 1022–26, 1486–90, at 960–61. Those who did try it reported mixed results; see, for examples, the contributions to ibid., 1023, 1490.

38. Emil Novak, "The Therapeutic Use of Estrogenic Substances," 1818. Other clinicians were happy enough to use the urine of pregnant women directly, as long as the urine did not come from women with tuberculosis or syphilis and had been filtered before use. The urine was superior to the commercially available—and very expensive—pituitary products like Antuitrin-S because the latter deteriorated quickly in storage while urine could be obtained "almost daily at no cost to the patient," asserted a well-known American guide to gynecological endocrinology. Even repeated

injections caused fewer local reactions than the commercial products. See Charles Mazer and Leopold Goldstein, *Clinical Endocrinology of the Female,* 272.

39. Emil Novak, "The Therapeutic Use of Estrogenic Substances," 1817.

40. Contribution of R. Hofstätter to F. v. Mikulicz-Radecki et al., "Die praktische Bedeutung der Hormonbehandlung," 959–60.

41. See, for example, Franz Siegert's report on two cases of primary amenorrhea, in which treatment with gonadotropins, ovarian hormone, and thyroid hormone in various combinations had failed to reestablish menstruation: F. Siegert, "Erfahrungen und Ergebnisse mit der Hormonbehandlung," 122–23. Earlier, Ludwig Fraenkel, too, had emphasized the complete ineffectiveness of hormonal treatment in primary amenorrhea. See Ludwig Fraenkel and E. Fels, "Die Beeinflussung der Geschlechtsfunktion durch Hormontherapie," in A. W. Greenwood, ed., *Proceedings of the Second International Congress for Sex Research, London, 1930* (Edinburgh: Oliver and Boyd, 1931), 467–82, at 482.

42. In Siegert's cases, menstruation was reestablished, sometimes temporarily, only in about 30 percent of cases of secondary amenorrhea. See F. Siegert, "Erfahrungen und Ergebnisse mit der Hormonbehandlung," 124–27. The difference between success and failure might, however, depend merely on dosage. Otfried Otto Fellner recalled that in a series of 296 patients with secondary amenorrhea treated with low-dose hormones, ninety-one had failed to show any response. In a later series of a hundred patients receiving hormones in high doses, there had been only six failures. See Fellner's contribution to F. v. Mikulicz-Radecki et al., "Die praktische Bedeutung der Hormonbehandlung," 960–61.

43. Emil Novak, "The Therapeutic Use of Estrogenic Substances," 1818. The situation was not much clearer in the hormonal treatment of dysmenorrhea (pain and acute discomfort during menstruation). Ovarian hormones were used extensively in this condition, but since the estrogens were supposed to stimulate the contractility of the uterus, their use to relieve pain during menstruation, Novak pointed out, seemed to be irrational. See Novak, ibid., 1820. But once again, assessments varied between clinicians. Otfried Otto Fellner, for instance, was particularly pleased with results of estrogenic treatment in dysmenorrhea. See Fellner's contribution to F v. Mikulicz-Radecki et al., "Die praktische Bedeutung der Hormonbehandlung," 961. For a later, British perspective on the question, see Kenneth Bowes, "The Oestrogens: Synthetic and Natural," *The Practitioner* 169 (1952): 243–52, at 251.

44. Robert T. Frank, *The Female Sex Hormone,* 281.

45. Frank had added, however: "Whether this will in any way help to re-establish the normal menstrual function appears doubtful to me, as all evidence favors the view that the ovaries themselves are unaffected by the female sex hormone. It is much more likely that when we obtain a potent anterior lobe pituitary preparation, real and valuable stimulation of the ovaries may be brought about. . . . Perhaps by this means the ovarian function may be permanently stimulated." Frank, *The Female Sex Hormone,* 281.

46. See the contribution of Richard Hofstätter in F. v. Mikulicz-Radecki et al., "Die praktische Bedeutung der Hormonbehandlung," 959–60.

47. Ibid.

48. F. Siegert, "Erfahrungen und Ergebnisse mit der Hormonbehandlung," 155.

49. Ibid., 155–56. Friedrich von Mikulicz-Radecki claimed that the older substitution therapies—whether with ovarian substance or ovarian implants—were often no less efficacious than most modern hormone preparations. See F v. Mikulicz-Radecki et al., "Die praktische Bedeutung der Hormonbehandlung," 959.

50. See Hans Schmid's contribution to Mikulicz-Radecki et al., "Die praktische Bedeutung der Hormonbehandlung," 1025.

51. As one German gynecologist observed, the coming of hormonal treatment had ended the age of purely uterine gynecology ("die Zeit der reinen 'Uterusgynäkologie'"). See the contribution of Egon Weinzierl to Mikulicz-Radecki et al., "Die praktische Bedeutung der Hormonbehandlung," 1490.

52. See Mikulicz-Radecki et al., "Die praktische Bedeutung der Hormonbehandlung," 965, 1023, 1024. Bernhard Zondek warned users of his estrogenic product, Folliculin, that one must not expect hormones to do the impossible. B. Zondek, "Folliculin," *KW* 8 (1929): 2229–32, at 2230.

53. Otfried Otto Fellner argued strongly for the therapeutic superiority of the newer hormone preparations. Most recent failures were due, said Fellner, to inadequate dosage or the use of tablets, which were quickly inactivated by the digestive hormones. See Fellner's contribution to Mikulicz-Radecki et al., "Die praktische Bedeutung der Hormonbehandlung," 960.

54. See, for instance, Geller, "Die Hormontherapie in der Gynäkologie," 125.

55. See Emil Novak, "The Therapeutic Use of Estrogenic Substances," 1817. See also "Estrogenic Substances: Theelin. Report of the Council on Pharmacy and Chemistry, *JAMA* 100 (1933): 1331–38. On the uncertainties of dosage as viewed by Central European gynecologists, see the different contributions to Mikulicz-Radecki et al., "Die praktische Bedeutung der Hormonbehandlung," especially that by Herbert Buschbeck (1022–23).

56. Hans-Ulrich Hirsch-Hoffmann, "Über hormonale Therapie in der Gynäkologie mit besonderer Berücksichtigung der Frage: Hormontherapie—Organtherapie," *AfG* 168 (1939): 295–305, at 295.

57. Ibid. In fact, it was not even clear that therapeutically administered hormones acted directly on the cells of the target organs. It was possible that the hormones affected the sympathetic and parasympathetic nervous systems and the actions were produced by consequent changes in nervous stimuli. See ibid., 296.

58. Ibid., 297; and F. C. Geller, "Die Hormontherapie in der Gynäkologie," 125. A decade earlier, Ludwig Fraenkel had called the ovary "a luxury organ" ("ein Luxusorgan") because its disorders did not endanger life. See Ludwig Fraenkel and E. Fels, "Die Beeinflussung der Geschlechtsfunktion durch Hormontherapie," 478. Prices of ovarian extracts were high everywhere. In America, Robert Frank had reported in the late 1920s that ten tablets of Progynon (each with 250 mouse units) had cost him $9.50. Three ampules of Folliculin (each containing 40 mouse units of estrogen), prepared according to the procedure of Bernhard Zondek, cost $5.00. The American product Amniotin (made by Squibb) cost $3.00 for 50 units and $6.00 for 100 units. Frank described these prices as "almost prohibitive." See Robert T. Frank, *The Female Sex Hormone*, 276.

59. Hirsch-Hoffmann, "Über hormonale Therapie in der Gynäkologie," 299. Earlier, Ludwig Fraenkel had pointed out that defining *the* female hormone as the

substance that produced estrus in experimental animals was most unsatisfactory: it was almost as if scientists had forgotten that the female hormone's ultimate biological function was not simply to produce estrus in castrated mice but to develop and maintain the female sexual characters and sexual functions in the widest sense. See Fraenkel and Fels, "Die Beeinflussung der Geschlechtsfunktion durch Hormontherapie," 481.

60. Hirsch-Hoffmann, "Über hormonale Therapie in der Gynäkologie," 300.

61. The vasomotor symptoms of the climacteric were treated by a bewildering variety of agents before a professional consensus was reached on the beneficial effects of the pure hormones. For a glimpse into the problems and the options before that consensus was reached, see L. Schoenholz and C. Werner, "Zur Behandlung der vasomotorischen Störungen in Klimakterium," *KW* 8 (1929): 2232–35.

62. Hirsch-Hoffmann, "Über hormonale Therapie in der Gynäkologie," 300.

63. Ibid., 302.

64. For a political history of these themes in modern Britain, see Naomi Pfeffer, *The Stork and the Syringe: A Political History of Reproductive Medicine* (Oxford: Polity, 1993).

65. Ibid., 69. Treatments for infertility, of course, were not the only organotherapeutic products on sale. They comprised just one section of an endless array of "hormone beauty baths, bust development glands, slimming glands, gland tablets for male and female [*sic*] impotence." See ibid., 69–71.

66. Emil Novak, "The Therapeutic Use of Estrogenic Substances," 1820.

67. F. C. Geller, "Die Hormontherapie in der Gynäkologie," 123.

68. Hans Otto Neumann, "Hormonale Behandlung der weiblichen Sterilität," *Medizinische Klinik* 33 (1937): 861–64, at 861, 862.

69. Ibid., 862. He admitted, however, that in nine other cases, he had had no success at all.

70. Ibid.

71. Naomi Pfeffer provides some interesting details on how semen was collected. Since masturbation was frowned upon, men were asked to have sexual intercourse with their wives using a condom, but when it was found that the talcum powder used in the condoms was spermicidal, the couple was asked to have coitus and the semen was collected from the region around the woman's cervix. See Pfeffer, *The Stork and the Syringe,* 55–56.

72. Neumann, "Hormonale Behandlung der weiblichen Sterilität," 862.

73. The uterine mucosa, of course, proliferated under hormonal stimulus and recent work indicated that the transport of the ova through the uterine tubes might be facilitated by hormones, but this was scarcely sufficient explanation. Neumann wondered, for instance, whether the hormones had stimulated the ovary itself, but this was not because he was ignorant of Moore and Price's work. As Hohlweg had demonstrated recently, a massive dose of follicular hormone could produce corpora lutea in the ovary and it was possible that large doses of follicular hormones suppressed the secretion of the follicle-stimulating gonadotropin but enhanced the secretion of the luteinizing gonadotropin. See Neumann, "Hormonale Behandlung der weiblichen Sterilität," 863.

302 · Notes to Pages 164–165

74. Pfeffer, *The Stork and the Syringe,* 70–71.

75. Kenneth Bowes, "The Oestrogens: Synthetic and Natural," 252.

76. See H. H. Simmer, "On the History of Hormonal Contraception, I. Ludwig Haberlandt (1885–1932) and His Concept of 'Hormonal Contraception,'" *Contraception* 1, no. 1 (January 1970): 3–27.

77. Ibid., 5.

78. There was nothing inevitable in the association of the corpus luteum with artificial sterilization. In 1922, Otfried Otto Fellner reported successful sterilization by injections of *estrogenic* extracts of the ovary. According to Fellner, the follicular hormone—which he named "Feminin"—caused infertility by damaging the ova but did so only in very large doses. It was eventually clarified that high-dose estrogen acted, for the most part, by inhibiting the pituitary secretion of Follicle Stimulating Hormone, thereby suppressing ovulation. Needless to say, even the most carefully prepared "estrogenic" extracts of the 1920s—Fellner's Feminin, Zondek and Aschheim's "Folliculin"—were never purely estrogenic. With hindsight, we can see that they almost invariably contained progesterone or substances other than estrogen. Even toward the end of the decade, it was not yet established whether the ovary secreted just one hormone or more than one. See H. H. Simmer, "On the History of Hormonal Contraception, II: Otfried Otto Fellner (1873–19??)," 7, 8.

79. H. H. Simmer, "On the History of Hormonal Contraception, I.," 8–9.

80. Josef Halban, "Innersekretorische Fragen in der Gynäkologie," *MmW* 68 (1921): 1314–17.

81. Simmer, "On the History of Hormonal Contraception, I," 9.

82. Ibid., 10–11.

83. Ibid., 1, 11.

84. Ibid., 11.

85. See Ludwig Haberlandt, "Über hormonale Sterilisierung des weiblichen Tierkörpers," *MmW* 68 (1921): 1577–78; idem, "Über hormonale Sterilisierung weiblicher Tiere durch subcutane Transplantation von Ovarien trächtiger Weibchen," *PAgP* 194 (1922): 235–70; and idem, "Über hormonale Sterilisierung weiblicher Tiere, II. Injectionsversuche mit Corpus luteum-, Ovarial- und Placentra-Opton," *PAgP* 202 (1924): 1–13. The grafting, Haberlandt argued, would induce the interstitial cells of the host's ovaries to proliferate, resulting in the secretion of the ovulation suppressing substance, the nature of which was, of course, unknown in chemical terms but which, physiologically, was supposed to be identical to the luteal secretion. The luteal extracts were prepared for Haberlandt by the Merck Company, but he himself supplied the ovaries procured from "slaughterhouses in Innsbruck, Graz and Vienna." See H. H. Simmer, "On the History of Hormonal Contraception, I.," 11, 13.

86. The broader conceptual background remained a confusing one. The very same placental extracts Haberlandt used for temporary sterilization, Hans Simmer points out, "had been used to treat sterility by stimulation of a hypoplastic uterus. The same objection, of course, could be raised regarding the ovarian extracts." Simmer, "On the History of Hormonal Contraception, I.," 14.

87. Unexpectedly, however, some of the experimental animals became *permanently* infertile. Haberlandt emphasized the phenomenon but could not explain why such an

outcome did not result from injections of those same extracts. See Ludwig Haberlandt, "Über hormonale Sterilisierung weiblicher Tiere (Fütterungsversuche mit Ovarial- und Plazenta-Opton)," *MmW* 74 (1927): 49; idem, "Hormonale Sterilisierung weiblicher Tiere," *CfG,* 51 (1927): 1418–20; and idem, "Über hormonale Sterilisierung weiblicher Tiere, III. Fütterungsversuche mit Ovarial- und Plazenta-Opton," *PAgP* 216 (1927): 525–33. See also H. H. Simmer, "On the History of Hormonal Contraception, I," 14.

88. Ludwig Haberlandt, "Die hormonale Sterilisierung des weiblichen Organismus," *MmW* 77 (1930): 2064–65, at 2065.

89. See Robert Jütte, *Lust ohne Last: Geschichte der Empfängnisverhüttung von der Antike bis zur Gegenwart* (Munich: Beck, 2003), 311–12.

90. Johann Ude, "Hormontabletten und Geburtenrückgang," *Medizinische Welt* 2 (1928): 959–61, at 960.

91. The first contraceptive pill in Hungary, released in 1966, was called Infecudin in honor of Haberlandt. See Jütte, *Lust ohne Last,* 312.

92. See F. L. Hisaw and S. L. Leonard, "Relation of the Follicular and Corpus Luteum Hormones in the Production of Progestational Proliferation of the Rabbit Uterus," *American Journal of Physiology* 92 (1930): 574–82; C. Bachman, "Oestrogenic Hormone and the Mechanism of Corpus Luteum Formation in the Rabbit," *Proceedings of the Society for Experimental Biology and Medicine* 33 (1936): 551–54.

93. See, however, Germaine Greer, *The Change: Women, Ageing and the Menopause* (London: Penguin, 1992); Lois W. Banner, *In Full Flower: Aging Women, Power, and Sexuality* (New York: Vintage, 1993); Frances B. McCrea, "The Politics of Menopause: The 'Discovery' of a Deficiency Disease," *Social Problems* 31 (1983): 111–23; Ornella Moscucci, "Medicine, Age and Gender: The Menopause in History," *Journal of the British Menopause Society* 5 (1999): 149–53; Susan E. Bell, "Changing Ideas: The Medicalization of Menopause," *Social Science and Medicine* 24 (1987): 535–42; Nelly Oudshoorn, "Menopause, Only for Women? The Social Construction of Menopause as an exclusively Female Condition," *Journal of Psychosomatic Obstetrics and Gynaecology* 18 (1997):137–44; M. N. Dukes, "The Menopause and the Pharmaceutical Industry," *Journal of Psychosomatic Obstetrics and Gynaecology* 18 (1997): 181–88; and Ruth Formanek, *The Meanings of Menopause: Historical, Medical and Clinical Perspectives* (Hillsdale, NJ: Analytic Press, 1989).

94. Emil Novak, "The Therapeutic Use of Estrogenic Substances," 1819. Franz Siegert acknowledged that one could oppose the idea of "treating" what was, after all, a perfectly natural process. See F. Siegert, "Erfahrungen und Ergebnisse mit der Hormonbehandlung weiblicher Genitalstörungen," 148. Friedrich Geller recommended that uterine bleeding in the menopausal years should be controlled by irradiating the ovaries into quiescence rather than by stimulating the reproductive system with hormones at an age when that was contrary to nature. See F. C. Geller, "Die Hormontherapie in der Gynäkologie," 124. Nelly Oudshoorn argues from an examination of Dutch sources that it was only in the 1930s that the female menopause became a medical preoccupation. See N. Oudshoorn, "Menopause, Only for Women?" This argument fits very well with Susan Bell's earlier claim (in Bell, "Changing Ideas: The Medicalization of Menopause") that it was only with the

introduction of the synthetic estrogen, diethylstilboestrol (DES) that menopause came to be seen by American physicians as a treatable disease. My sources, however, do not entirely support this interpretation. The menopausal female had long been regarded as pathological and, in principle, treatable by hormonal substances. Such treatment did, of course, become routine only in the 1930s, but classic organotherapy with ovarian extracts had long been used extensively in menopausal symptoms. See my discussion above; H. Kopera, "Zur Geschichte der hormonellen Behandlung klimakterischer Krankheitserscheinungen," *WmedW* 141 (1991): 346–48; and, above all, H. H. Simmer, "Organotherapie mit Ovarialpräparaten in der Mitte der neunziger Jahre des 19. Jahrhunderts: Medizinische und pharmazeutische Probleme," in Erika Hickel and Gerald Schröder, eds., *Neue Beiträge zur Arzneimittelgeschichte: Festschrift für Wolfgang Schneider zum 70. Geburtstag* (Stuttgart: Wissenschaftliche Verlagsgesellschaft, 1982), 229–64.

95. Emil Novak, "The Therapeutic Use of Estrogenic Substances," 1819.

96. Ibid. See also B. Zondek, "Folliculin," *KW* 8 (1929): 2229–32, at 2231–32. Some menopausal symptoms, Zondek emphasized, could be due to an *excess* of follicular hormone or of pituitary gonadotropins. See Bernhard Zondek, "Polyhormonale Krankheitsbilder: Funktionelle Betrachtung gynäkologischer Erkrankungen," *CfG* 54 (1930): 1–7.

97. In 1929, Robert Frank had stated that the vasomotor symptoms of menopause were best treated by sedatives. "Attempts to combat these symptoms by means of female sex hormone have failed in my hands almost in every instance, although others report successes with every imaginable organotherapeutic preparation." Frank, *The Female Sex Hormone*, 282. Siegert pointed out that menopausal women were extraordinarily suggestible, traveling from doctor to doctor, always hoping for more efficacious remedies. This meant that no clinical researcher could hope to have enough long-term patients to study. See F. Siegert, "Erfahrungen und Ergebnisse mit der Hormonbehandlung," 149.

98. Emil Novak, "The Therapeutic Use of Estrogenic Substances," 1819.

99. See Fraenkel and Fels, "Die Beeinflussung der Geschlechtsfunktion durch Hormontherapie," 468.

100. P. M. F. Bishop, "The Menopause," *BMJ*, 17 April 1937, 819–21, at 819.

101. In order to determine which symptoms were genuinely induced by hormone deficiency, Bishop recommended beginning the treatment with a course of medication apparently identical to the hormones—whether in the color of the tablets, the route of administration, or the dosage—but which was completely inert. "This procedure serves to differentiate the more fanciful complaints, which tend to disappear, from the genuine symptoms of ovarian deficiency or gonadotropic excess which persist at the conclusion of such a course" (ibid).

102. Ibid., 821.

103. Ibid., 820.

104. William Hunter, "The Management of the Climacteric," *The Practitioner* 170 (1953): 386–90, at 389.

105. Bowes, "The Oestrogens: Synthetic and Natural," 247. There must have been some demand from women patients to justify such medical assertions, but that issue must be left for future exploration.

106. See Watkins, "Dispensing with Aging," 24.

107. B. M. Caldwell, R. I. Watson, and W. B. Kountz, "Psychologic Effects of Estrogen Therapy in Postmenopausal Women," abstract in *Journal of Gerontology* 5 (1950): 384.

108. John Esben Kirk, "Steroid Hormones and Aging: A Review," *Journal of Gerontology* 6 (1951): 253–63, at 258–59. In the late 1940s, several papers were published on the skin-rejuvenating actions of the sex hormones. The estrogens were found to be more potent in improving the resiliency of the skin than testosterone. See, for instance, J. W. Goldzieher, "The Direct Effect of Steroids on the Senile Human Skin," *Journal of Gerontology* 4 (1949): 104–12.

109. Kirk, "Steroid Hormones and Aging," 259–60. See also Fuller Albright, "Osteoporosis," *Annals of Internal Medicine* 27 (1947): 861–82.

110. See L. J. G. Gooren, "The Age-Related Decline of Androgen Levels in Men: Clinically Significant?," *British Journal of Urology* 78 (1996): 763–68, at 765; and J. S. Finkelstein, A. Klibanski and R. M. Neer, "Increase in Bone Density During Treatment of Men with Idiopathic Hypogonadotropic Hypogonadism," *Journal of Clinical Endocrinology and Metabolism* 69 (1989): 776–83.

111. See Egon Diczfalusy and Christian Lauritzen, *Oestrogene beim Menschen* (Berlin: Springer-Verlag, 1961), 162.

112. See contribution of R. Hofstätter to F. v. Mikulicz-Radecki et al., "Die praktische Bedeutung der Hormonbehandlung in der Gynäkologie' 959–60.

113. William Hunter, "The Management of the Climacteric," 390.

114. Ibid.

115. See P. A. van Keep, "The History and Rationale of Hormone Replacement Therapy," *Maturitas,* 12 (1990): 163–70.

116. Robert A. Wilson and Thelma A. Wilson, "The Fate of the Nontreated Postmenopausal Woman: A Plea for the Maintenance of Adequate Estrogen from Puberty to Grave," *Journal of the American Geriatrics Society* 11 (1963): 347–62; Elizabeth Siegel Watkins, "Dispensing with Aging: Changing Rationales for Long-term Hormone Replacement Therapy, 1960–2000," *Pharmacy in History* 43 (2001): 23–37, at 25; Frances B. McCrea and Gerald E. Markle, "The Estrogen Replacement Controversy in the USA and UK: Different Answers to the Same Question?," *Social Studies in Science* 14 (1984): 1–26; and Judith A. Houck, "'What Do These Women Want?': Feminist Responses to *Feminine Forever,* 1963–1980," *Bulletin of the History of Medicine* 77 (2003): 103–32.

117. Watkins, "Dispensing with Aging," 25.

118. Robert A. Wilson and Thelma A. Wilson, "The Fate of the Nontreated Postmenopausal Woman," 355.

119. For an argument that women are being compelled by male doctors to aspire to invulnerability through HRT, see Jennifer Harding, "Bodies at Risk: Sex, Surveillance and Hormone Replacement Therapy," in Alan Petersen and Robin Bunton, eds., *Foucault, Health and Medicine* (London: Routledge, 1997), 134–50, at 142.

120. Watkins, "Dispensing with Aging," 25.

121. Greer, *The Change,* 5–6.

122. See M. N. Dukes, "The Menopause and the Pharmaceutical Industry."

123. Saffron Whitehead, "Menopause," in Colin Blakemore and Sheila Jennett, eds., *The Oxford Companion to the Body* (Oxford: Oxford University Press, 2001), 457.

124. Francis Skae, "Climacteric Insanity," *Edinburgh Medical Journal* 10 (1864–65): 703–16, at 703.

125. W. J. Conklin, "Some Neuroses of the Menopause," *Transactions of the American Association of Obstetrics and Gynecology* 2 (1889): 301–11, quoted in Peter J. Schmidt and David R. Rubinow, "Menopause-Related Affective Disorders: A Justification for Further Study," *American Journal of Psychiatry* 148 (1991): 844–52, at 844.

126. Skae, "Climacteric Insanity," 704, 706, 707, 708.

127. Ibid., 716.

128. Charles B. Molony, "Endocrine Therapy and the Psychoses," *Journal of Mental Science* 73 (1927): 64–80, at 70, 78.

129. Ibid., 67. He was convinced that this organotherapy was not simply substitutive but also stimulated the "functional activity of the same organ in the patient to whom it is administered." See the discussion of Hallion's Law, above.

130. August A. Werner, Louis H. Kohler, C. C. Ault, and Emmett F. Hoctor, "Involutional Melancholia: Probable Etiology and Treatment," *Archives of Neurology and Psychiatry* 35 (1936): 1076–80, at 1077.

131. Werner and his team emphasized that involutional melancholia could mask other psychiatric conditions and that the symptoms of the latter might become prominent once the melancholia has been successfully treated. See ibid., 1079–80. At the Royal Edinburgh Hospital for Mental and Nervous Diseases, seventeen involutional melancholics were treated with estradiol injections, again with good results, although not as impressive as those of Werner's team. See M. S. Jones, T. N. MacGregor and H. Tod, "Oestradiol Benzoate Therapy in Depressions at the Menopause," *Lancet,* 6 February 1937, 320–22.

132. C. C. Ault, Emmett F. Hoctor, and August A. Werner, "Theelin Therapy in the Psychoses," *JAMA* 109 (1937): 1786–88, at 1788.

133. Ivo Geikie Cobb, *The Glands of Destiny: A Study of the Personality,* 3d ed. (London: William Heinemann Medical Books, 1947), 197.

134. R. Gibson, "Involutional Melancholia: A Study of Twenty Cases Treated with Theelin," *Journal of Mental Science* 89 (1943): 278–83, esp. 279–80. See also W. B. Titley, "Prepsychotic Personality of Patients with Involutional Melancholia," *Archives of Neurology and Psychiatry* 36 (1936): 19–33.

135. Gibson, "Involutional Melancholia: A Study of Twenty Cases," 282–83.

136. Purcell G. Schube, M. C. McManamy, C. E. Trapp, and G. F. Houser, "Involutional Melancholia: Treatment with Theelin," *Archives of Neurology and Psychiatry* 38 (1937): 505–12, at 507.

137. Ibid., 510.

138. Ibid., 512. There was not a scrap of discussion in the paper as to why estrogens were administered to male patients suffering from involutional melancholia.

139. See H. S. Ripley, E. Shorr, and G. N. Papanicolaou, "The Effect of Treatment of Depression in the Menopause with Estrogenic Hormone," *American Journal of Psychiatry* 96 (1940): 905–11.

140. W. Malamud, S. L. Sands, and I. Malamud, "The Involutional Psychoses: A Socio-Psychiatric Study," *Psychosomatic Medicine* 3 (1941): 410–26. Later, it was argued by some psychiatrists that menopause merely unmasked preexisting personality problems and psychological inadequacies. See Germaine Greer, *The Change*, 109.

141. Saul Rosenthal, "The Involutional Depressive Syndrome," *American Journal of Psychiatry* 124, Supplement (May 1968): 21–35, at 26–27.

142. A. E. Bennett and C. B. Wilbur, "Convulsive Shock Therapy in Involutional States after Complete Failure with Previous Estrogenic Treatment," *American Journal of Medical Science* 208 (1944): 170–76.

143. In the 1950s, it was recommended that ordinary menopausal depressions responded best to combined treatment with estrogens and androgens. The latter were used in small-enough doses so as not to cause overt masculinization, and their use enabled the doctor to use estrogens in low doses and thereby avoid side-effects such as uterine bleeding. See Gilbert S. Gordan and Karl M. Bowman, "The Central Nervous System," in Max A. Goldzieher and Joseph W. Goldzieher, eds., *Endocrine Treatment in General Practice* (New York: Springer, 1953), 202–3.

144. Rosenthal, "The Involutional Depressive Syndrome," 32. More recently, two American investigators endorsed that skepticism, pointing out that even with today's refined tools and measurements, "the extent to which gonadal hormones, gonadotropins, catecholamines, and neuropeptides may interact in the development of climacteric/menopause-related mood and behavioral disorders is purely a matter for speculation." Peter J. Schmidt and David R. Rubinow, "Menopause-Related Affective Disorders: A Justification for Further Study," *American Journal of Psychiatry* 148 (1991): 844–52, at 845, 849–50.

145. Carroll LaFleur Birch, "Hemophilia and the Female Sex Hormone," *JAMA* 97 (1931): 244.

146. The urine samples were assayed on oophorectomized female rats. Only five haemophiliacs were tested. For details, see ibid.

147. Birch reported that she had used a range of ovarian preparations to treat her cases: "I have used theelin, corpus luteum, and whole ovarian substance. . . . I have also used fresh pig ovaries, raw and served on toast. I have a long list of preparations left to try. I have obtained my best results with whole ovarian substance." She emphasized that the supposedly pure follicular hormone—theelin—had proved less efficacious than whole ovarian substance. See Carroll Lafleur Birch, "Haemophilia," *JAMA* 99 (1932): 1566–72, at 1572.

148. Interestingly, these authors claimed that female hormone was not found in the urine of all normal males but only in isolated cases. See Jacob Brem and Jerome S. Leopold, "Ovarian Therapy: Relationship of the Female Sex Hormone to Hemophilia," *JAMA* 102 (1934): 200–202.

149. See Anne Fausto-Sterling, *Sexing the Body: Gender Politics and the Construction of Sexuality* (New York: Basic Books, 2000), 193.

150. Raymond Greene, "Current Therapeutics, XVIII: Androgens," *The Practitioner* 162 (1949): 512–19, at 516–17.

151. L. Seitz, "Das Follikelhormon als geschlechtsspezifisches Wachstumshormon und seine Beziehungen zum allgemeinen Körperwachstumshormon," *Monatsschrift für Geburtshülfe und Gynäkologie* 103 (1936): 185–93, at 185.

152. Frank, *The Female Sex Hormone*, 1.

153. See Edward C. Reifenstein and Fuller Albright, "The Metabolic Effects of Steroid Hormones in Osteoporosis," *Journal of Clinical Investigation* 26 (1947): 24–56.

154. Bowes, "The Oestrogens: Synthetic and Natural," 247.

155. Viereck, *Glimpses of the Great*, 264.

156. Paul de Kruif, *The Male Hormone* (Garden City, N.Y.: Garden City Publishing, 1945), 188.

157. Gerald J. Newerla, "The History of the Discovery and Isolation of the Male Hormone," *New England Journal of Medicine* 228, no. 2 (14 January 1943): 39–47.

158. They were prescribed by innumerable clinicians, but few, if any, were endorsed as physiologically active by medical scientists. The active male hormone, it came to be known later, was not stored in the testicle and no matter how genuine the testicular extract, it could not conceivably be rich in hormone. The extraction of the active male sex hormone, however, would depend on the discovery of sources richer in it than testicular tissue itself. See T. F. Gallagher and F. C. Koch, "The Testicular Hormone," *Journal of Biological Chemistry* 84 (1929): 495–500; and Harry Benjamin, "The Latest Endocrine Advance: The Male Hormone. Preliminary Communication with Remarks on Hormone Isolation in General," *Medical Journal and Record* 131 no. 11 (4 June 1930): 545–48, at 545.

159. On the gender dimensions of this assay, see Nelly Oudshoorn, "On Measuring Sex Hormones: The Role of Biological Assays in Sexualizing Chemical Substances," *Bulletin of the History of Medicine* 64 (1990): 243–61.

160. See Gallagher and Koch, "The Testicular Hormone," 496, 497.

161. Paul de Kruif, *The Male Hormone*, 188.

162. The quotation is from C. Funk, B. Harrow and A. Lejwa, "The Male Hormone," *American Journal of Physiology* 92 (1930): 440–49, at 440, who clarified that by the phrase "the male hormone," "all we mean to imply ... is that we have an extract which influences the growth of comb and wattles, and this extract, for purposes of convenience, we designate the 'male hormone'" (441).

163. Funk, Harrow, and Lejwa, "The Male Hormone," 440.

164. One "cock unit" of the male hormone, which Funk and Harrow soon referred to as "testiculin," was that amount which, "when injected daily into each of 6 castrated cocks, will give an average increase in comb of 10mm. in 10 days"—this amount could be obtained from 75 cc of a young man's urine. See Funk, Harrow, and Lejwa, "The Male Hormone," 440; and E. C. Dodds, A. W. Greenwood, and E. J. Gallimore, "Note on a Water-Soluble Active Principle Isolated from the Mammalian Testis and Urine, and its Relation to Oestrin," *Lancet*, 29 March 1930, 683–85. See also C. Funk and B. Harrow, "The Male Hormone," *Proceedings of the Society for Experimental Biology and Medicine* 26 (1928–29): 325–26, at 326.

165. Parkes, "Rise of Reproductive Endocrinology," xxv.

166. K. David, E. Dingemanse, J. Freud, and E. Laqueur, "Über krystallinisches männliches Hormon aus Hoden (Testosteron), wirksamer als aus Harn oder aus Cholestrin bereitetes Androsteron," *Zeitschrift für physiologische Chemie* 233 (1935): 281; Parkes, "Rise of Reproductive Endocrinology," xxv.

167. Erica R. Freeman, David A. Bloom and Edward J. McGuire, "A Brief History of Testosterone," *Journal of Urology* 165 (2001): 371–73. The chemists presented their successes as pioneering feats without paying any attention to the physiological phase of endocrine research. Virtually all we have talked about in the pages preceding this section, the chemists declined even to mention, let alone salute. Only the names of Brown-Séquard and Berthold were mentioned as forebears—otherwise, history, they implied, began with them. These points were made by Eugen Steinach in 1936. See E. Steinach, "'Zur Geschichte des männlichen Sexualhormons und seiner Wirkungen am Säugetier und beim Menschen," *WkW* 49 (1936): 161–72, 198–205, at 199. Again, the pertinence of Steinach's critique is not weakened by its clear and obvious self-interest. See also John Hoberman, *Testosterone Dreams: Rejuvenation, Aphrodisia, Doping* (Berkeley and Los Angeles: University of California Press, 2005), which, unfortunately, was published too late for me to use it at any depth.

168. E. P. McCullagh and H. R. Rossmiller, "Methyl Testosterone, I. Androgenic Effects; II. Calorogenic Activity; III. Body Weight and Growth," *Journal of Clinical Endocrinology* 1 (1941): 496, 503, 507.

169. Initially, the different sex hormones were supposed to be better-suited for different tasks. The male hormone androsterone, for instance, was supposed by one author to be a good tonic for the organism as a whole and also for its beneficial effects on mood, psyche, and self-esteem. The actions of testosterone seemed limited, on the other hand, to the sexual sphere—hence it was considered more appropriate for the treatment of disturbances of sexual development, such as delayed puberty. See Gerhard Venzmer, "Die Bekämpfung vorzeitiger Alterserscheinungen mit synthetischem männlichen Hormon," *DmW* 63 (1937): 1402–4, at 1402.

170. Androgens had long been supposed to improve blood circulation, as we know from the explanations for rejuvenation by the Steinach operation. The circulatory effects of androgens were highlighted after the introduction of synthetic hormone preparations, and some physicians found them useful in relieving angina pectoris and hypertension. See, for example, E. Steinach, O. Peczenik and H. Kun, "Über hormonale Hyperämisierung, insbesonders über den Einfluss der männlichen Sexualhormone und ihrer Kombination mit weiblichem Hormon auf erhöhten Blutdruck und Hypertonus," *WkW* 51 (1938): 65–67, 102–6, 134–39; H. Arndt, "Zur Therapie extragenitaler Störungen mit Sexualhormonen," *WmedW* 89 (1939): 222–27; and Maurice A. Lesser, "Testosterone Propionate Therapy in One Hundred Cases of Angina Pectoris," *Journal of Clinical Endocrinology* 6 (1946): 549–57. For a negative verdict, see S. A. Levine and W. B. Likoff, "The Therapeutic Value of Testosterone Propionate in Angina Pectoris," *New England Journal of Medicine* 229 (1943): 770–72.

171. Greene, "Current Therapeutics, XVIII: Androgens," 515.

172. C. D. Creevy and C. E. Rea, "The Treatment of Impotence by Male Sex Hormone," *Endocrinology* 27 (1940): 392–94, at 394.

173. A. W. Spence, "Testosterone Propionate in Functional Impotence," *BMJ*, 28 September 1940, 411–13, at 411.

174. Ibid., 412; and Carl G. Heller and William O. Maddock, "The Clinical Uses of Testosterone in the Male," *Vitamins and Hormones* 5 (1947): 393–432, at 419.

175. See Creevy and Rea, "The Treatment of Impotence by Male Sex Hormone," 392.

176. It proved useless in the treatment of male infertility—indeed, testosterone treatment was found to reduce the number of viable spermatozoa. See W. O. Thompson, "Uses and Abuses of the Male Sex Hormone," *JAMA* 132 (1946): 185–88, at 187. Recently, however this effect has led to serious proposals for the use of androgens as male contraceptives. See John M. Hoberman and Charles E. Yesalis, "The History of Synthetic Testosterone," *Scientific American* 272, no. 2 (February 1995): 60–65, at 64; and Nelly Oudshoorn, *The Male Pill: A Biography of Technology in the Making* (Durham, N.C.: Duke University Press, 2003).

177. L. R. Broster, *Endocrine Man: A Study in the Surgery of Sex* (London: William Heinemann Medical Books, 1944), 78.

178. Ibid.

179. Heller and Madock, "The Clinical Uses of Testosterone in the Male," 407–8. The only harmful effect noted by Raymond Greene was "the excessive sexual activity of the eunuch to whom a too high dosage is given." But, he added, the effect was transient "and the eunuch does not mind much." See Greene, "Current Therapeutics, XVIII: Androgens," 514.

180. Considerably more research is needed on this topic, but see Mike Featherstone and Mike Hepworth, "The History of the Male Menopause 1848–1936," *Maturitas* 7 (1985): 249–57.

181. The Victorian surgeon William Acton, for instance, had considered all manifestations of sexual desire in old age to be pathological. On Acton and the change of life in men, see ibid., 253.

182. Henry Halford, "On the Climacteric Disease," in Halford, *Essays and Orations read and delivered at the Royal College of Physicians* (London: John Murray, 1931), 1–15.

183. Ibid., 4–7.

184. Ibid., 8–11.

185. Not every twentieth-century physician considered the male climacteric to be an exact analogue of menopause. In Britain, A. P. Cawadias commented: "In contrast to women, however, men do not suffer from acute psychological trauma from the gradual wane of sexual potency and libido. Sex is not the centre of life for them, whereas women see by the menopausis [*sic*] that their sexual life is at an end and feel that living itself is empty. Men, however, may experience fear of the advancing years, dread of chronic invalidism, of economic dependence, of loss of position or dominance in the family or profession. . . . Male climacteric disease [is] a less common and a less clamorous condition than its female counterpart." A. P. Cawadias, *Clinical Endocrinology and Constitutional Medicine* (London: Frederick Muller, 1947), 166.

186. "It is not very improbable," Halford had written, "that this important change in the condition of the constitution is connected with a deficiency in the energy of the

brain itself, and an irregular supply of the nervous influence to the heart." Henry Halford, "On the Climacteric Disease," 13–14.

187. Archibald Church, "Nervous and Mental Disturbances of the Male Climacteric," *JAMA* 55 (23 July 1910): 301–3, at 301. Steinach, as we know, took an identical view of the aging sexual body.

188. Church cited Harry Campbell, *Differences in the Nervous Organization of Man and Woman* (1891) and an anonymous "German observer [who] carefully followed the temperature in a number of soldiers living under uniform conditions, and found that, as in the case of women there was a distinct monthly variation or curve." Church, "Nervous and Mental Disturbances of the Male Climacteric," 301–2. On the interest among contemporary scientists and physicians on periodicity in biological processes and functions, see Frank J. Sulloway, *Freud, Biologist of the Mind—Beyond the Psychoanalytic Legend* (London: Fontana, 1980), 152–58.

189. Church, "Nervous and Mental Disturbances of the Male Climacteric," 303.

190. Ibid.

191. Ibid.

192. See Kurt Mendel, "Die Wechseljahre des Mannes (Climacterium virile)," *Neurologisches Centralblatt* 29 (1910): 1124–36, at 1124.

193. Until the onset of the climacteric symptoms, most of his patients had been in good health and had never had any history of nervous symptoms. See ibid., 1125–26.

194. Ibid., 1126.

195. Although not as constant as these, other symptoms resembling those of menopausal women—headaches, paraesthesias, daytime sleepiness combined with nocturnal insomnia—were also very common. See ibid., 1127.

196. Physically, they continued to be in excellent shape and signs of arteriosclerosis were consistently absent. See ibid., 1128.

197. Ibid., 1129.

198. Gynecologist Mathias Vaerting asserted in 1918 that while more men died in *any* age group than women, it was between forty and sixty that male mortality was at its highest. The ovaries merely stopped releasing ova at menopause but there was no abrupt change in production—all the ova had been produced long before this date. In climacteric men, on the other hand, there was a fall in the production as well as the maturation of spermatozoa, and the numbers involved were astronomical. The decline in male gonadal function, therefore, was much more drastic. Psychologically, too, the climacteric could be more traumatic for men than for women. Women did not lose the capacity to have sexual intercourse after menopause, while the potency of the aging man was weak at best. The male climacteric thus was far more critical and dangerous than the female climacteric. See Mathias Vaerting, "Wechseljahre und Altern bei Mann und Weib," *Neurologisches Centralblatt* 37 (1918): 306–15. See also Kurt Mendel, "Die Wechseljahre des Mannes (Climacterium virile)," *Zentralblatt für die gesmate Neurologie und Psychiatrie* 29 (1922): 385–93, at 386. Hereafter, to differentiate the two articles with the same title, the citations will specify 1910 and 1922, respectively.

199. Mendel, who thought that average educated laymen were more perceptive on sexual matters than doctors, had asked some of his patients whether they had noticed any similarities between their condition and those of women going through menopause.

They had all found it to be a very apt analogy. See Mendel, "Die Wechseljahre des Mannes (Climacterium virile)" (1910), 1133–34. The urologist and sexologist Max Marcuse, writing in response to Mendel's paper, suggested that rather than being a manifestation of the deficiency of testicular secretions alone, the male climacteric reflected a deficiency of the entire endocrine system. See Max Marcuse, "Zur Kenntnis des Climacterium virile, insbesondere über urosexuelle Störungen und Veränderungen der Prostata bei ihm," *Neurologisches Centralblatt* 35 (1916): 577–91, at 579, 581, 590.

200. Even in menopausal women, Mendel suspected, ovarian substance produced most of its reported benefits by suggestion. See Mendel, "Die Wechseljahre des Mannes (Climacterium virile)" (1910), 1134. Sexologist Max Marcuse was a bit more impressed with the actions of testicular extracts in male climacteric, but even he was hardly an enthusiast. See Marcuse, "Zur Kenntnis des Climacterium virile," 591.

201. See Mendel, "Die Wechseljahre des Mannes (Climacterium virile)" (1910), 1135.

202. Bernard Hollander, following Mendel, "Die Wechseljahre des Mannes (Climacterium virile)" (1910), 1282–86.

203. Ibid., 1283.

204. Ibid., 1285.

205. Sedatives by mouth, galvanic currents to the head, and faradic stimulation of the whole body were the best initial treatments, followed by tonics such as damiana, phosphorus, and strychnine. A stimulating diet was especially important in England, where the average middle-class diet was bland and unexciting. This kind of regimen gave splendid results, and Hollander's patients never relapsed. See ibid., 1286.

206. Guthrie Rankin, "The Climacteric of Life," *BMJ*, 18 January 1919, 63–67, at 63.

207. The patient, Rankin counseled, must be helped to face the fact "the years are beginning to tell their tale, and that a new and more subdued plan of life must be found and courageously adopted in order to escape the danger of a too early advent of the time when 'the grasshopper shall be a burden, and desire shall fail'" (ibid., 63–64, 67).

208. Kurt Mendel, "Die Wechseljahre des Mannes (Climacterium virile)" (1922), 358.

209. Kurt Mendel, "Die Wechseljahre des Mannes (Climacterium virile)" (1910), 1131.

210. Kurt Mendel, "Die Wechseljahre des Mannes (Climacterium virile)" (1922), 392. Urologist Max Marcuse agreed that the strain of the war and the abnormal sexual conditions it entailed (abstinence followed by excesses) could well expedite the onset of the male climacteric. See Marcuse, "Zur Kenntnis des Climacterium virile," 584.

211. Kurt Mendel, "Die Wechseljahre des Mannes (Climacterium virile)" (1922), 393.

212. K. Mendel, "Zur Beurteilung der Steinachschen Verjüngungsoperation," *DmW* 47 (1921): 986–89.

213. See ibid.; and Mendel, "Die Wechseljahre des Mannes (Climacterium virile)" (1922), 393. Mendel speculated that the operation might lead to a toxic accumulation

of the testicular secretions in the blood (which he called "Testitoxikose"). Mendel, "Zur Beurteilung der Steinachschen Verjüngungsoperation."

214. Kenneth Walker, "The Accidents of the Male Climacteric," *BMJ*, 9 January 1932, 50–53, at 50.

215. Ibid.

216. The sexual difficulties of the male climacteric were not trivial, however, and could be expressed in virulent forms: "the eminent and elderly gentleman arrested in Hyde Park for indecent behaviour should be treated as an invalid rather than as a criminal. . . . His exhibitionism is as certainly the psychological consequence of an upset of the endocrine balance" (ibid., 52). In 1948, A. P. Cawadias observed that the male climacteric could commence with a pathological intensification of libido: "Disturbances of behaviour, sexual frivolities, are the result and constitute the 'démon du midi,' an inelegant swan-song of sexual life. Libido," he added, "may take a homosexual direction through arteriosclerotic lesions of the brain." Cawadias, *Clinical Endocrinology and Constitutional Medicine*, 167.

217. Walker, "The Accidents of the Male Climacteric," 53. In 1944, the eminent endocrinologist Julius Bauer had declared that the male climacteric was a misnomer: although the "functional activity of the testicles slackens with advance in years," there was never "generally nor individually a definite age at which the testicles stop functioning as do the ovaries." Julius Bauer, "The Male Climacteric—A Misnomer," *JAMA* 126 (1944): 914. An almost "natural' and self-limiting condition such as the male climacteric was, perhaps, less attractive to the new surgery of virility that took the stage in the 1920s.

218. The "premature senility" treated by the Steinach operation, no doubt, was not quite the same as Mendel's concept of the male climacteric, but it is nonetheless tempting to speculate that the paucity of publications on the latter during the 1920s and 1930s was attributable to the medical eagerness to combat "premature senility" by the methods of Voronoff, Steinach, and others. Despite the similarities between the two constructs, the classical descriptions of the climacteric all emphasize emotionality and depressive symptoms to a degree that one does not encounter in the literature on rejuvenation.

219. Cawadias, *Clinical Endocrinology and Constitutional Medicine*, 165.

220. August A. Werner, "The Male Climacteric," *JAMA* 112 (1939): 1441–43.

221. A. R. Schmidt, ed., *Research in Endocrinology by August A. Werner, MD, and Associates* (Belleville, Illinois: Belleville Daily Advocate, 1952), 51.

222. Werner, "The Male Climacteric," 1442. Involutional melancholia had never been earmarked for women. In Britain, Francis Skae had reported on sixty cases of male climacteric insanity in 1865, characterized by profound depression in men aged 48 to 60, often with paroxysmal agitation and a "determined suicidal tendency"—symptoms "identical with those seen in climacteric insanity in the female." The condition, Skae had asserted, was "the result of the effect upon the brain of a constitutional disturbance accompanying a great climacteric change." See Francis Skae, "Climacteric Insanity in the Male," *Edinburgh Medical Journal* 11 (1865–66): 232–44, at 233–34. In 1927, Charles Molony had declared that men between 50 and 60 suffered from a condition virtually identical to female involutional melancholia. See

Charles B. Molony, "Endocrine Therapy and the Psychoses," *Journal of Mental Science* 73 (1927): 64–80, at 78.

223. Werner, "The Male Climacteric," 1441. Werner argued that the fact that men did not have a *meno*pause in the strict sense was not as important a distinction as it might seem. A woman without a uterus would not menstruate and "can and will feel perfectly normal until the climacteric is reached." And then, her symptoms, if sufficiently serious, would resemble the male climacteric syndrome. See August A. Werner, "The Male Climacteric: Report of Two Hundred and Seventy-Three cases," *JAMA* 132 (1946): 188–94, at 194.

224. N. E. Miller, Gilbert Hubert, and J. B. Hamilton, "Mental and Behavior Changes following Male Hormone Treatment of Adult Castration, Hypogonadism and Psychic Impotence," *Proceedings of the Society for Experimental Biology and Medicine* 38 (1938): 538–40.

225. Werner, "The Male Climacteric," 1442.

226. This study did not, however, notice any elevation of potency or libido in the majority of the patients—the hormones, the author argued, were not aphrodisiacs but exerted a general regulating influence on the unbalanced endocrine system. Hence, they should work in women, and testosterone propionate injections were also prescribed to twenty-nine middle-aged women complaining of similar psychic symptoms. Twenty-eight of them recovered fully. See O. L. Weiss, "Behandlung psychischer Alterserscheinungen bei Männern und Frauen mit synthetischem Testeshormon," *DmW* 65 (1939): 261–62, at 261.

227. The efficacy of the treatment was underscored when the patient relapsed three months after the end of the course of injections. See Werner, "The Male Climacteric," 1443.

228. August A. Werner, "The Male Climacteric: Report of Two Hundred and Seventy-Three Cases," *JAMA* 132 (1946): 188–94, at 194.

229. Ibid., 188.

230. H. B. Thomas and R. T. Hill, "Testosterone Propionate and the Male Climacteric," *Endocrinology* 26 (1940): 953–54.

231. "It is a psychosis," August Werner had claimed, "but of definitely endocrinous origin, and should be entitled 'climacteric psychosis.'" Werner, "The Male Climacteric: Report of Two Hundred and Seventy-Three Cases," 190.

232. "Abstract of Discussion" following Werner, "The Male Climacteric," 194.

233. Ibid.

234. Werner, "The Male Climacteric," 188.

235. "Abstract of Discussion" following Werner, "The Male Climacteric," 194.

236. Richard L. Landau, "The Concept of the Male Climacteric," *Medical Clinics of North America* 35 (1951): 279–88, at 282.

237. Ibid., 285. Nevertheless, the assumption that gonadotropin secretion went up as "a compensatory mechanism" when gonad hormone secretion went down was widely made. See, for instance, John Kirk, "Steroid Hormones and Aging," 255.

238. Landau, "The Concept of the Male Climacteric," 285. On the lack of a precise, invariable correlation between androgen secretion and the levels of urinary

gonadotropins, see also E. P. McCullagh and J. F. Hruby, "Testis-Pituitary Interrelationship: The Relative Inability of Testosterone to Reduce Urinary Gonadotrophin in Eunuchoid Men," *Journal of Clinical Endocrinology* 9 (1949): 113–25; and R. P. Howard, R. C. Sniffen, F. A. Simmons, and F. Albright, "Testicular Deficiency: A Clinical and Pathological Study," *Journal of Endocrinology* 10 (1950): 121–86.

239. Landau, "The Concept of the Male Climacteric," 286.

240. Kirk, "Steroid Hormones and Aging: A Review," 255–56.

241. Greene, "Current Therapeutics, XVIII: Androgens," 513.

242. Kirk, "Steroid Hormones and Aging: A Review," 260.

243. Reiter reported on a trial of combined estrogen/androgen pellets in twenty-seven patients between forty-four and sixty-seven years of age, all of whom were "true cases of male climacteric exhibiting most of the significant symptoms and signs." Within weeks of implantation, every single symptom—the irritability, the fatigue, and even the libido—responded positively. Astonishingly, Reiter claimed that "without exception the potency was restored to a level substantially beyond that prevailing for several years before the onset." The combination treatment, therefore, relieved the symptoms of the male climacteric far more effectively than testosterone propionate alone. See Tiberius Reiter, "Treatment of the Male Climacteric by Combined Implantation," *The Practitioner* 170 (1953): 181–84, at 182.

244. In castrated male rats, a combination of male and female hormones had restored potency and libido at much smaller doses than with male hormones alone. See Steinach, Kun, and Peczenik, "Beiträge zur Analyse der Sexualhormonwirkungen," 901–2.

245. Max A. Goldzieher and Joseph W. Goldzieher, "The Male Climacteric," in Max A. Goldzieher and Joseph W. Goldzieher, eds., *Endocrine Treatment in General Practice*, 280–84, at 281. As far as one can see, it was an empirical suggestion the ostensible benefits of which did not induce any significant rethinking of gender boundaries. Nor did the treatment of the male climacteric with androgens and estrogens ever, to my knowledge, become popular, and I have yet to find any detailed analysis of its possible physiological basis. Reiter claimed that the benefits of the combination treatment were estimated in comparison with results "previously achieved with other methods over about fourteen years with efficient gonad hormone preparations." He did not, however, cite any of those earlier results. See Tiberius Reiter, "Treatment of the Male Climacteric by Combined Implantation," 182, 184. Today, advocates of hormone replacement in aging males emphasize that "testosterone replacement therapy should maintain not only physiological levels of serum testosterone, but also the metabolites of testosterone, including dihydrotestosterone and estradiol, to optimize maintenance of libido, virilization and sexual function." Alvaro Morales, Jeremy P. W. Heaton, and Culley C. Carlson III, "Andropause: A Misnomer for a True Clinical Entity," *Journal of Urology* 163 (2000): 705–23, at 712.

246. Kirk, "Steroid Hormones and Aging: A Review," 260. This old theme was constantly being resurrected. See, for another instance, Max A. Goldzieher and Joseph W. Goldzieher, "Senescence," in Max A. Goldzieher and Joseph W. Goldzieher, eds., *Endocrine Treatment in General Practice*, 23–30, at 30.

247. Kirk, "Steroid Hormones and Aging: A Review," 260–61.

248. From the 1990s, the male climacteric (now usually called the andropause) is again being discussed widely in the medical literature. While we cannot go into this in detail, many investigators claim that androgen levels do decline in old age but this decline is rather modest, and there is no consensus on its clinical or pathological significance. See L. J. G. Gooren, "The Age-Related Decline of Androgen Levels in Men," 763; Steven W. J. Lamberts, Annewieke W. van den Beld, and Aart-Jan van der Lely, "The Endocrinology of Aging," *Science* 278 (1997): 419–31; Alvaro Morales, Jeremy P. W. Heaton, and Culley C. Carlson III, "Andropause: A Misnomer for a True Clinical Entity"; and A. Gray, A. Feldman, J. B. McKinlay, and C. Longcope, "Age, Disease, and Changing Sex Hormone Levels in Middle-Aged Men: Results of the Massachusetts Male Aging Study," *Journal of Clinical Endocrinology and Metabolism* 73 (1991): 1016–25. There is, however, considerable variability between "normal" androgen levels as measured by different commercially available kits used in laboratories. See L. R. Boots, S. Potter, D. Potter, and R. Azziz, "Measurement of Total Serum Testosterone Levels Using Commercially Available Kits: High Degree of Between-Kit Variability," *Fertility and Sterility* 69 (1998): 286–92; and Andrew R. Hoffman, "Editorial: Should We Treat the Andropause?," *American Journal of Medicine* 111 (2001): 322–23. Despite these uncertainties, a recent feature in the *New Yorker* magazine reports that American doctors are being asked in advertisements (paid for by the manufacturers of a gel-form of testosterone that simply has to be rubbed in to the skin once a day) to "screen for symptoms of low testosterone" and restore any deficiencies they might detect. See Jerome Groopman, "Hormones for Men: Is Male Menopause a Question of Medicine or of Marketing?," *The New Yorker,* 29 July 2002, 34–38. My thanks to Sam Alberti for this reference.

249. Ivo Geikie Cobb, *The Glands of Destiny,* 3d ed., 83.

250. B. Brahn, "Haben homosexuelle Männer mehr Ovarial-Hormon in ihrem Harn als Normale?," *KW* 10 (1931): 504–5. The author indicated the need for the comparison of hormone levels in blood but I have not found any such study.

251. On American scientific and medical involvement in the study of homosexuality, see Jennifer Terry, *An American Obsession: Science, Medicine, and Homosexuality in Modern Society* (Chicago: University of Chicago Press, 1999).

252. Frank, *The Female Sex Hormone,* 115.

253. On Wright, see Stephanie Kenen, "Who Counts When You're Counting Homosexuals? Hormones and Homosexuality in Mid-Twentieth-Century America," in Vernon A. Rosario, ed., *Science and Homosexualities* (New York: Routledge, 1997), 197–218.

254. See Clifford A. Wright, "Endocrine Aspects of Homosexuality: A Preliminary Report," *Medical Record* 142 (1935): 407–10.

255. Clifford A. Wright, "Further Studies of Endocrine Aspects of Homosexuality," *Medical Record* 147 (1938): 449–52, at 452. Wright, for all his innovative approach to hormonal ratios, did not reject earlier notions of the morphological distinctiveness of homosexuals. "Homosexuals," he observed, "have many common characteristics, particularly do they frequently exhibit the mannerisms, gait and manner of speech of the opposite sex, as well as heterosexual body characteristics." The first case he reported on was a twenty-four-year-old Japanese

man "definitely feminine in type, had mannerisms and the high pitched voice of the female. . . . This boy dressed as a girl, plucked his eyebrows, colored his nails and in other ways imitated women." Wright, "Endocrine Aspects of Homosexuality: A Preliminary Report," 408.

256. Wright, "Further Studies of Endocrine Aspects of Homosexuality," 449, emphasis in the original.

257. Clifford A. Wright, "The Sex Offender's Endocrines," *Medical Record* 149 (1939): 399–402, at 400.

258. S. J. Glass, H. J. Deuel, and C. A. Wright, "Sex Hormone Studies in Male Homosexuality," *Endocrinology* 26 (1940): 590–94, at 593.

259. Ibid., 594.

260. Wright, "Endocrine Aspects of Homosexuality: A Preliminary Report," 407. As the Director of the Child Guidance Home of the Cincinnati Jewish Hospital would point out later, "every homosexual act is a delinquent act and is so labeled by society. The homosexual very often also becomes involved in criminal acts of the most sordid type. Any treatment that offers relief in these apparently hopeless cases is worthy of a trial." Louis A. Lurie, "The Endocrine Factor in Homosexuality: Report of Treatment of 4 Cases with Androgen Hormone," *American Journal of the Medical Sciences* 208 (1944): 176–86, at 178.

261. Wright, "The Sex Offender's Endocrines," 400.

262. Wright, "Endocrine Aspects of Homosexuality: A Preliminary Report," 408.

263. Ibid., 409; Wright, "Further Studies of Endocrine Aspects of Homosexuality," 452. By the end of the 1930s, however, he seems to have forgotten Moore: he was now advocating the use of testosterone in homosexuals because it was a "strongly masculinizing factor." See Clifford A. Wright, "Results of Endocrine Treatment in a Controlled Group of Homosexual Men," *Medical Record* 154 (1941): 60–61, at 60.

264. Among the critics of Wright's hormonal approach was the well-known endocrinologist Julius Baur. Since levels of female hormones could vary widely in different urine samples from the same individual, Baur warned against hasty diagnoses. Baur was convinced that sex—as well as all the characters of sex, including sexual orientation—were ultimately determined by the chromosomes. The hormones acted merely as "stimulants and a kind of lubricant for certain cerebral centers which represent the site of sexual urge. The direction of this urge, however, at least in human beings, does not seem to depend upon the hormones, contrary to the results obtained in animal experiments." Julius Baur, "Homosexuality as an Endocrinological, Psychological, and Genetic Problem," *Journal of Criminal Psychopathology* 2 (1940–41): 188–97, at 192.

265. Jean Paul Pratt, "Sex Functions in Man," in Edgar Allen et al., eds., *Sex and Internal Secretions*, 2d ed. (Baltimore: Williams and Wilkins, 1939), 1263–1334, at 1277. Pratt was a gynecologist based in Detroit. The comparable chapter in the first edition of the compendium had not even mentioned homosexuality. See Jean Paul Pratt, "Endocrine Disorders in Sex Function in Man," in Edgar Allen, ed., *Sex and Internal Secretions* (Baltimore: Williams and Wilkins, 1932), 880–911.

266. "Either the androgens are normal and the estrogens very high (13 cases), or the androgens are low and the estrogens high or very high (13 cases). . . . Other

material, such as the urine of masturbators, transvesticists [*sic*], impotent males, does not show this type of hormone constitution, and neither does the urine of the normal male." During the discussion, Myerson explained: "The masturbator of the inveterate type has a high male and female hormonic content. . . . The transvestite on the whole has a male hormonic content." A. Myerson and R. Neustadt, "Sex Hormones in the Urine of the Child," *Transactions of the American Neurological Association,* 66th annual meeting, 1940, 115–20, at 117–18.

267. A. Myerson and R. Neustadt, "Sex Hormones in the Urine of the Child," 118. This confidence was undermined, however, by the somewhat sheepish admission by the same authors in a slightly later paper that although "males usually excrete more androgens and females more estrogens than the opposite sex," both sexes excreted both kinds of hormones in amounts so large that "it is often difficult to determine whether a given urine specimen is from a male or a female individual." They also found that androgen injections—or even more reliably, ultraviolet irradiation of the testicles—changed the urinary hormone levels without always producing any "essential change in the homosexual feelings or conduct." See Rudolph Neustadt and Abraham Myerson, "Quantitative Sex Hormone Studies in Homosexuality, Childhood, and Various Neuropsychiatric Disturbances," *American Journal of Psychiatry* 97 (1940): 524–51, at 527 and 534.

268. Abraham Myerson and Rudolph Neustadt, "The Bisexuality of Man," *Journal of the Mt Sinai Hospital, New York* 9 (1942): 668–78, at 668.

269. Saul Rosenzweig and R. G. Hoskins, "A Note on the Ineffectualness of Sex-Hormone Medication in a Case of Pronounced Homosexuality," *Psychosomatic Medicine* 3 (1941): 87–89, at 87. African-Americans, as Jennifer Terry has pointed out, were supposed to be particularly prone to inverted sexuality and many American doctors spent much time warning against such inverts' predilection for white lovers. More broadly, there was a distinct tradition in the United States to regard homosexuality as a nonwhite vice that was spreading into white communities. See Jennifer Terry, *An American Obsession,* 87–97.

270. Rosenzweig and Hoskins, "A Note on the Ineffectualness of Sex-Hormone Medication in a Case of Pronounced Homosexuality," 89.

271. S. J. Glass and Roswell H. Johnson, "Limitations and Complications of Organotherapy in Male Homosexuality," *Journal of Clinical Endocrinology* 4 (1944): 540–44.

272. Ibid., 541.

273. Since there were no controls, the authors admitted that it was hard to exclude any effect of suggestion, but those "who have attempted the difficult feat of treating homosexuality by psychotherapy alone," they pointed out, "will perhaps be most impressed with the positive results that were obtained." See Glass and Johnson, "Limitations and Complications of Organotherapy in Male Homosexuality," 542.

274. Ibid., 542.

275. Ibid., 543.

276. Alfred C. Kinsey, "Criteria for a Hormonal Explanation of the Homosexual," *Journal of Clinical Endocrinology* 1 (1941): 424–28, at 425. See also Stephanie Kenen, "Who Counts When You're Counting Homosexuals?," 205–10. On the biographical

contexts (including Kinsey's own bisexual practices), see James H. Jones, *Alfred Kinsey: A Public/Private Life* (New York: Norton, 1997); and Jonathan Gathorne-Hardy, *Alfred C. Kinsey: Sex, the Measure of All Things* (London: Chatto and Windus, 1998).

277. Kinsey, "Criteria for a Hormonal Explanation of the Homosexual," 427.

278. "If intergradation were to be thought of as vitiating classification the medical sciences would require revolution," protested Glass and Johnson. "In actuality, intergradation commonly reveals merely the concurrent operation of several factors and not the lack of validity of any one." Also, they pointed out, Kinsey had underestimated the statistical weight of some studies by using inappropriate methods of comparison. See Glass and Johnson, "Limitations and Complications of Organotherapy in Male Homosexuality," 542.

279. Thomas V. Moore, "The Pathogenesis and Treatment of Homosexual Disorders: A Digest of Some Pertinent Evidence," *Journal of Personality* 14 (1945): 47–83, at 68. The study cited by Moore was H. Fischer, "Über Eunuchoidismus, insbesondere über seine Genese und seine Beziehungen zur Reifung und zum Altern," *Zeitschrift für die gesmate Neurologie und Psychiatrie* 87 (1923): 323–24.

280. Moore, "The Pathogenesis and Treatment of Homosexual Disorders," 68.

281. "It would seem to follow," they added, "that vigorous androgenic treatment would tend to increase the power of the sex drive in both normal and homosexual males without influencing the direction of the sex drive in either case." This, of course, was also the argument of Glass and Johnson (see above). See Carl G. Heller and William O. Maddock, "The Clinical Uses of Testosterone in the Male," *Vitamins and Hormones* 5 (1947): 393–432, at 422.

282. It was revived, however, in the early 1970s, when it was reported that male homosexuals excreted less testosterone in their urine than heterosexual men while lesbians excreted more than "normal" women. Many other studies followed, but their results contradicted each other so greatly that the only possible conclusion was that it was "highly unlikely that deviations in testosterone levels and production in adulthood can be held responsible for the development of male homosexuality in general." See Heino F. L. Meyer-Bahlburg, "Sex Hormones and Male Homosexuality in Comparative Perspective," *Archives of Sexual Behavior* 6 (1977): 297–325, at 301–11. Also in the 1970s, it was claimed that the artificial production of androgen-deficiency during certain critical phases of embryonic development could lead to "homosexual mating behavior" in experimental animals. Whether humans, too, were similarly disposed and whether human homosexuality could be prevented by the administration of androgens to the mother became legitimate research questions, especially in the work of Gunter Dörner, a student of Walter Hohlweg, one of Steinach's own later associates. See G. Dörner, Ingrid Poppe, F. Stahl, J. Kölzsch and R. Uebelhack, "Gene- and Environment-Dependent Neuroendocrine Etiogenesis of Homosexuality and Transsexualism," *Experimental and Clinical Endocrinology* 98 (1991): 141–50, at 142, 143, 146; G. Dörner, W. Rohde, F. Stahl, L. Krell, and W. G. Masius, "A Neuroendocrine Predisposition for Homosexuality in Men," *Archives of Sexual Behavior* 4 (1975): 1–8; G. Dörner and G. Hinz, "Induction and Prevention of Male Homosexuality by Androgen," *Journal of Endocrinology* 40 (1968): 387–88; and G. Dörner, "Hormonal Induction and Prevention of Female Homosexuality," *Journal of*

Endocrinology 42 (1968): 163–64. For a brief historical critique of Dörner's work, see Rainer Herrn, "On the History of Biological Theories of Homosexuality," in John P. De Cecco and David Allen Parker, eds., *Sex, Cells, and Same-Sex Desire: The Biology of Sexual Preference* (New York: Harrington Park Press, 1995), 31–56, at 47.

283. Even more comprehensive was cyproterone acetate, which suppressed gonadotropic activity and also interfered with the action of androgens at their sites of action. See Heino F. L. Meyer-Bahlburg, "Sex Hormones and Male Homosexuality," 299.

284. One well-known "patient" treated with estrogens in the early 1950s was the British mathematician and computer pioneer Alan Turing. "It is supposed to reduce sexual urge whilst it goes on," wrote Turing to a friend, "but one is supposed to return to normal when it is over. I hope they're right." See Andrew Hodges, *Alan Turing: The Enigma of Intelligence* (London: Unwin Hyman, 1985), 456–59, 468–74. On the hormonal treatment of homosexuality in Britain, see F. L. Golla and R. Sessions Hodge, "Hormone Treatment of the Sexual Offender," *Lancet,* 11 June 1949, 1006–7; and Dalton E. Sands, "Further Studies on Endocrine Treatment in Adolescence and Early Adult Life," *Journal of Mental Science* 100 (1954): 211–19.

285. There were many other midcentury treatments of homosexuality that we cannot discuss here, of which psychoanalysis and behavior therapy were probably the most noteworthy. On these, see Jennifer Terry, *An American Obsession;* and M. B. King and A. Bartlett, "British Psychiatry and Homosexuality," *British Journal of Psychiatry* 174 (1999): 106–13.

286. See Meyer-Bahlburg, "Sex Hormones and Female Homosexuality: A Critical Examination," *Archives of Sexual Behavior* 8 (1979): 101–119, at 103.

287. Some female homosexuals and female-to-male transexuals excreted more androgens in their urine than one might expect, but vast numbers of others from those same categories did not. See ibid., 108–9.

288. Ibid., 112–13. The fortunate result of this lack of interest, of course, is that there has never been any endocrinological experimentation on lesbians. Meyer-Bahlburg could not find any "concrete clinical reports on actual attempts at 'curing' lesbianism by compensatory administration of estrogens or by antiandrogen treatment" (112).

289. "Hormone treatment," Dillon added, "whether for disease of the gonads, or for this purpose [the treatment of homosexuality], is expensive. Surely in our post-war world we should see that all medicinal products are for international use and should be free to all sufferers." See Michael Dillon, *Self: A Study in Ethics and Endocrinology* (London: William Heinemann Medical Books, 1946), 53.

290. On Dillon's transformation, see Liz Hodgkinson, *Michael née Laura* (London: Columbus Books, 1989), and Bernice L. Hausman, *Changing Sex: Transsexualism, Technology, and the Idea of Gender* (Durham, N.C.: Duke University Press, 1995). His doctors claimed to be convinced that he was genetically male with hypospadias that was so severe that his scrotum was divided into two and looked like the female labia (see Hausman, *Changing Sex,* 21–22, 43–48). However, Laura's uterus or ovaries were never removed, and Sir Harold Gillies, who constructed Michael's male genitalia, was certainly well aware of his patient's biological sex. Apparently, Gillies was very sympathetic to "nature's mistakes" and performed several pioneering sex-change

operations over his career. As Michael's biographer says, "Gillies probably realized that Michael was anatomically and chromosomally female, but the diagnosis of hypospadias gave him a valid reason for carrying out the operation, should awkward questions be asked." See Hodgkinson, *Michael née Laura*, 55–66, at 66. See also Harold Gillies and Ralph Millard Jr, *The Principles and Art of Plastic Surgery* (Boston: Little, Brown, 1957), 2: 383–84, which has a case history of a "female with male outlook" that, Hausman feels, may well be a slightly altered history of Dillon. See Hausman, *Changing Sex,* 207 n. 4.

291. Until 1970, British nationals could re-register their sex if they could establish that they "belonged more to the opposite sex than to the one in which they had been born." See Hodgkinson, *Michael née Laura*, 63.

292. Hausman, *Changing Sex,* 47.

293. Nelly Oudshoorn, *Beyond the Natural Body: An Archaeology of Sex Hormones* (London: Routledge, 1994), 24–34.

294. For these doubts, see A. C. Crooke, "The Present Clinical Status of the Androgens," *The Practitioner* 169 (1952): 253–59, at 253.

295. Anne C. Carter, Eugene J. Cohen, and Ephraim Shorr, "The Use of Androgens in Women," *Vitamins and Hormones* 5 (1947): 317–91, at 318.

296. See H. Husslein, "Die Androgen-Therapie in der Gynäkologie," in Tassilo Antoine, ed., *Klinische Fortschritte: "Gynäkologie"* (Vienna: Urban and Schwarzenberg, 1954), 339–63, at 339.

297. Husslein, "Die Androgen-Therapie in der Gynäkologie," 339–40.

298. It was also likely that androgens and estrogens had antagonistic functions only above a certain threshold. As long as the level of androgen remained below this threshold, it synergized with estrogen. See Husslein, "Die Androgen-Therapie in der Gynäkologie," 340.

299. Ivo Geikie Cobb, *The Glands of Destiny,* 3d ed., 197.

300. Husslein, "Die Androgen-Therapie in der Gynäkologie," 359.

301. Carter, Cohen, and Shorr, "The Use of Androgens in Women," 339, 360–61, 376; and Husslein, "Die Androgen-Therapie in der Gynäkologie," 351.

302. Carter, Cohen, and Shorr, "The Use of Androgens in Women," 342–43.

303. Ibid., 345.

304. Ibid., 349, 355, 357–58.

305. Ibid., 362.

306. Ibid., 362–63; and Greene, "Androgens," 515.

307. A. E. Rakoff, "Female Hypogonadism; Frigidity," in Max A. Goldzieher and Joseph W. Goldzieher, eds., *Endocrine Treatment in General Practice,* 285–95, at 294–95. For a more recent perspective see J. W. W. Studd, W. P. Colins, and S. Chakravarti, "Estradiol and Testosterone Implants in the Treatment of Psychosexual Problems in Postmenopausal Women," *British Journal of Obstetrics and Gynaecology* 84 (1977): 314–15.

308. Husslein, "Die Androgen-Therapie in der Gynäkologie," 353.

309. Ibid., 351. For recent overviews on the use of androgens in women, see D. Abraham and P. C. Carpenter, "Issues concerning Androgen Replacement Therapy

in Postmenopausal Women," *Mayo Clinic Proceedings* 72 (1997): 1051–55; H. M. Buckler, W. R. Robertson, and F. C. W. Wu, "Which Androgen Replacement Therapy for Women?," *Journal of Clinical Endocrinology and Metabolism* 83 (1998): 3920–24; and Susan R. Davis, "Androgen Treatment in Women," *Medical Journal of Australia* 170 (1999): 545–49.

310. It was gradually determined that the acne and hirsutism were temporary and depended on the duration of treatment and on individual susceptibility. Any deepening of voice, however, was permanent. See Carter, Cohen, and Shorr, "The Use of Androgens in Women," 324–25.

311. Ibid., 364.

312. Max A. Goldzieher and Joseph W. Goldzieher, "Senescence," in Max A. Goldzieher and Joseph W. Goldzieher, eds., *Endocrine Treatment in General Practice*, 23–30, at 29.

313. Ira T. Nathanson, "Endocrine Aspects of Human Cancer," *Recent Progress in Hormone Research* 1 (1947): 261–91, at 261. See also E. C. Dodds, "Hormones in Cancer," *Vitamins and Hormones* 2 (1944): 353–59, at 353.

314. Nathanson, "Endocrine Aspects of Human Cancer," 261.

315. Ira T. Nathanson and Rita M. Kelley, "Hormonal Treatment of Cancer," *New England Journal of Medicine* 246 (24 January 1952): 135–45, 180–89, at 135.

316. This suggestion was controversial and depended to a large extent on how one defined a carcinogen. Dodds, for example, insisted that the term "should be restricted to those compounds which produce cancer when painted on the skin of mice over a prolonged period." See Dodds, "Hormones in Cancer," 354.

317. Ibid., 358.

318. "Briefly," explained Dodds, "it would appear that estrogenic stimulation will increase the incidence of mammary carcinoma in the mouse." But since the synthetic estrogens stilboestrol and hexestrol, which did not bear any structural chemical similarity to the recognized carcinogens, also stimulated the development of mammary carcinoma, the stimulation of malignant growths must be a biological property of all estrogenic substances, and could not have anything to do with the *structural* similarity of the natural estrogens to some chemical carcinogens. See Dodds, "Hormones in Cancer," 358; and H. Auchincloss and C. D. Haagensen, "Cancer of Breast possibly induced by Estrogenic Substance," *JAMA* 114 (1940): 1517–23.

319. Ira T. Nathanson, "Endocrine Aspects of Human Cancer," 268.

320. Ibid., 265.

321. Dodds, "Hormones in Cancer," 359.

322. See F. E. Adair et al., "Use of Estrogens and Androgens in Advanced Mammary Cancer: Clinical and Laboratory Study of 105 Female Patients," *JAMA* 140 (1949): 1193–1200; A. L. Walpole and E. Paterson, "Synthetic Oestrogens in Mammary Cancer," *Lancet* 2 (1949): 783–86; and Edward F. Lewison and Robert G. Chambers, "The Sex Hormones in Advanced Breast Cancer," *New England Journal of Medicine* 246 (3 January 1952): 1–8.

323. The same effect could, of course, be achieved by massive doses of androgens too, but it had been found that the administration of androgens to postmenopausal or castrated women raised the levels of estrogenic substances excreted in urine. That

suggested that "under some circumstances, at least, androgens are involved in increased estrogen production." Nathanson and Kelley, "Hormonal Treatment of Cancer,"142. See also A. A. Loeser, "Male Hormone in Gynaecology and Obstetrics and in Cancer of Female Breast," *Obstetric and Gynecologic Surgery* 3 (1948): 363–81; and I. T. Nathanson, L. L. Engel, B. J. Kennedy, and R. M. Kelley, "Screening of Steroids and Allied Compounds in Neoplastic Disease," in A. White, ed., *Symposium on Steroids in Experimental and Clinical Practice* (Philadelphia: Blakiston, 1951), 375–78.

324. Dodds, "Hormones in Cancer," 357; and Edith Paterson, "The Endocrine Therapy of Cancer of the Breast," *The Practitioner* 165 (1950): 488–96, at 490–91.

325. The Council stressed, however, that "any patient who still menstruates or who has menstruated within a five year period should definitely not receive estrogen therapy, as it accelerates the growth of the carcinoma." See Council on Pharmacy and Chemistry, "Estrogens and Androgens in Mammary Cancer," *JAMA* 135 (13 December 1947): 987–89, at 988; and Edith Paterson, "The Endocrine Therapy of Cancer of the Breast," 493.

326. See G. W. Taylor, "Evaluation of Ovarian Sterilization for Breast Cancer," *Surgery, Gynaecology and Obstetrics* 68 (1939): 452–56; and F. E. Adair, N. Treves, J. H. Farrow, and I. M. Scharnagel, "Clinical Effects of Surgical and X-Ray Castration in Mammary Cancer," *JAMA* 128 (1945): 161–66.

327. A. A. Loeser, "Male Hormone in the Treatment of Cancer of the Breast," *Acta Unio Internationalis contra Cancrum* 4 (1939): 377, quoted in Frank E. Adair, "Testosterone in the Treatment of Breast Cancer," *Medical Clinics of North America* 32 (1948): 18–36.

328. Alfred A. Loeser, "Mammary Carcinoma: Response to Implantation of Male Hormone and Progesterone," *Lancet*, 6 December 1941, 698–700, at 698.

329. In two of these cases, the implant was of progesterone, not testosterone. See ibid., 699.

330. Ibid. Although clinicians such as the New York surgeon Frank Adair often dismissed these as "of no great consequence, considering the seriousness of the disease being treated," they felt more concerned over the enhanced libido. It could, during long-term androgen treatment, become "quite a difficult problem to handle. In those who are unable to sleep it becomes necessary to give fairly heavy sedation and also to apply a local sedative ointment to the clitoric area." Frank E. Adair, "Testosterone in the Treatment of Breast Carcinoma," *Medical Clinics of North America* 32 (1948): 18–36, at 24–25.

331. Alfred A. Loeser, "Mammary Carcinoma," 699.

332. A. Prudente, "Postoperative Prophylaxis of Recurrent Mammary Cancer with Testosterone Propionate," *Surgery, Gynecology, Obstetrics* 80 (1945): 575–92; and James R. Watson and George H. Fetterman, "Testosterone Propionate in the Treatment of Advanced Carcinoma of the Breast," *Surgery, Gynecology and Obstetrics* 88 (1949): 702–10, at 704.

333. Watson and Fetterman, "Testosterone Propionate in the Treatment of Advanced Carcinoma of the Breast," 707. Even this negative report, however, emphasized that the patients had all reported "an improved sense of well being, so marked in several instances that they volunteered the information that they felt better

than they had for years and that their friends remarked on how much better they looked." Weight gain and pain relief were also noted. See ibid., 707–8.

334. Nathanson and Kelley, "Hormonal Treatment of Cancer," 136.

335. Paterson, "The Endocrine Therapy of Cancer of the Breast," 488.

336. Response to hormone treatment—whether androgens or, as we shall see below, estrogens—was not limited to any one set of manifestations. The response was generalized and "the explanation," remarked Edith Paterson, "must be sought in the patient as a whole, or maybe in the tumour as a whole." Paterson, "The Endocrine Therapy of Cancer of the Breast," 494.

337. Frank E. Adair, "Testosterone in the Treatment of Breast Carcinoma," 22. Others questioned whether the action on bone metastases was due to any direct effect on tumor cells or to a nonspecific "stimulation of bone repair at the site of the metastases." Two American investigators argued that "it is not logical to assume that cancer cells would be altered when they reside in bone but not when they occur in soft tissue." James R. Watson and George H. Fetterman, "Testosterone Propionate in the Treatment of Advanced Carcinoma of the Breast," 709.

338. Council on Pharmacy and Chemistry, "Estrogens and Androgens in Mammary Cancer," *JAMA* 135 (13 December 1947): 987–89, at 988. The report was authored by the members of the Council's Subcommittee on Steroids and Cancer, chaired by Ira Nathanson. Frank Adair was also a member of the subcommittee. See ibid., 989.

339. Nathanson and Kelley, "Hormonal Treatment of Cancer," 138. Not everyone agreed with this—Hans Husslein argued that pure testosterone should now be reserved for cases where one *wanted* to raise the libido. In most other indications, the "less-virilizing" androgens such as androstenediol were the agents of choice. See Husslein, "Die Androgen-Therapie in der Gynäkologie," 359–60.

340. See T. Leucutia, "Value of Orchiectomy in Treatment of Carcinoma of Male Breast," *Radiology* 46 (1946): 441–47; and J. B. Herrmann, "Effect of Hormonal Imbalance on Advanced Carcinoma of Male Breast," *Annals of Surgery* 133 (1951): 191–99.

341. See, for examples, L. Halberstaedter and A. Hochman, "Artificial Menopause and Cancer of Breast," *JAMA* 131 (1946): 810–16; G. W. Horsley, "Treatment of Cancer of Breast in Premenopausal Patients with Radical Amputation and Bilateral Oophorectomy," *Annals of Surgery* 125 (1947): 703–17; and R. W. Raven, "Cancer of Breast treated by Oöphorectomy," *BMJ* 1 (1950): 1343–45.

342. The spirit of Steinach did not, of course, hover over every clinician. Some used crossed-sex hormones only to suppress the pituitary gonadotropins, a concept that Steinach never embraced wholeheartedly. See, for instance, W. W. Faloon, L. A. Owens, M. C. Broughton, and L. W. Gorham, "Effect of Testosterone on Pituitary-Adrenal Cortex Mechanism in Patients with Breast Cancer," *Cancer Research* 10 (1950): 215.

343. Edward F. Lewison and Robert G. Chambers, "The Sex Hormones in Advanced Breast Cancer," *New England Journal of Medicine* 246 (3 January 1952): 1–8, at 7.

344. Paterson, "The Endocrine Therapy of Cancer of the Breast," 495.

345. Husslein, "Die Androgen-Therapie in der Gynäkologie," 355.

346. Bernhard Zondek, *Hormone des Ovariums und des Hypophysenvorderlappens: Untersuchungen zur Biologie und Klinik der weiblichen Genitalfunktion,* 2d ed. (Vienna: Springer, 1935), iii–iv.

347. Parkes, "The Rise of Reproductive Endocrinology, 1926–1940," xxviii.

348. Harvey Cushing, *The Pituitary Body and its Disorders: Clinical States produced by Disorders of the Hypophysis Cerebri* (Philadelphia: Lippincott, 1912), 292.

349. See Zondek, *Hormone des Ovariums und des Hypophysenvorderlappens,* iii–iv.

350. For a historical overview of research on the gonadotropins, see Christian Hamburger, "Historical Introduction," in G. E. W. Wolstenholme and Julie Knight, eds., *Gonadotropins: Physicochemical and Immunological Properties* (London: Churchill, 1965), 2–10.

351. See also this British report on the clinical use of gonadotropic extracts obtained from pregnancy urine: R.W. Johnstone, B. P. Wiesner, and P. G. Marshall, "The Therapeutic Application of Gonadotropic Hormones," *Lancet,* 3 September 1932, 508–11. For Zondek's own discussion of the evidence and insistence that although the evidence was less than definitive as yet, it was strongly likely that there were two separate gonadotropic hormones, see Zondek, *Hormone des Ovariums und des Hypophysenvorderlappens,* 203–16.

352. Novak, "Anterior Pituitary and Anterior Pituitary-Like Substances," 999. Partly because of similar concerns, a German gynecologist had attempted to treat cases of amenorrhea and uterine hypoplasia with intramuscular injections of the blood of pregnant women, which was naturally rich in gonadotropins and any unknown substances that might be necessary to potentiate their action. As far as I know, this remained an idiosyncratic procedure. See Paul Esch, "Hormontherapie durch intramuskuläre Injektionen von Schwangerenblut bei Menstruationsstörungen," *CfG* 54 (1930): 19–26.

353. One of the conditions where gonadotropins were found very helpful was in functional menorrhagia—this was defined as excessive menstrual bleeding at any age not due to any detectable structural abnormality such as fibroids, often associated with absence of corpora lutea, and therefore, hypothetically due to deficiency of the luteinizing principle of the pituitary. An authoritative American textbook of gynecological endocrinology declared that gonadotropin treatment was "almost a specific in the treatment of functional uterine bleeding." See Mazer and Goldstein, *Clinical Endocrinology of the Female,* 267, 270.

354. See Pratt, "Endocrine Disorders in Sex Function in Man," in Edgar Allen, ed., *Sex and Internal Secretions* (1932), 909.

355. Emil Novak, "Anterior Pituitary and Anterior Pituitary-Like Substances," *JAMA* 104 (1935): 998–1002, at 998. Even the growth hormone was not yet available commercially in its pure form, although there were three relatively crude pituitary products that contained the growth-promoting factor (ibid., 998). As for the gonadotropin(s), Zondek's claim that there were two "prolans," one responsible for the maturation of follicles and the other for luteinization, had been questioned later by "many excellent investigators" who argued that the two activities represented two different effects or two distinct phases in the activity of the same hormone (ibid.,

999). Even in 1958, Carl Gemzell and his collaborators reported that many investigators still argued that the two gonadotropins were not found in humans. See Carl A. Gemzell, Egon Diczfalusy, and Gunnar Tillinger, "Clinical Effect of Human Pituitary Follicle-Stimulating Hormone (FSH)," *Journal of Clinical Endocrinology* 18 (1958): 1333–48, at 1344.

356. See Novak, "Anterior Pituitary and Anterior Pituitary-Like Substances," 1001. Primary dysmenorrhea, Novak declared, was often psychogenic, but since the luteal hormone was supposed to inhibit uterine contractions, it was not illogical to use the anterior-pituitary-like gonadotropic substance to stimulate luteinization in the ovary in *some* cases of dysmenorrhea, where there was a possible endocrine cause. "At times the results seem brilliant, in other cases only improvement without disappearance of the pain is reported by the patient, and in still others the results are disappointing" (ibid.). The endocrine theory of dysmenorrhea was far from uncontested. An influential hypothesis held that it was due to "the increased irritability of the autonomic nerve endings in the uterus." See Mazer and Goldstein, *Clinical Endocrinology of the Female,* 247. In 1932, a team of researchers in Edinburgh argued that dysmenorrhea was in all probability caused by uterine contractions during menstrual bleeding and that these contractions could be reduced by the secretion of the corpus luteum. In the absence of reliable, standardized luteal hormones, one could either use more-or-less crude extracts of the corpus luteum or use gonadotropic extracts—which, of course, were not necessarily much more pure as yet—to stimulate the ovary's own corpora lutea. See R. W. Johnstone, B. P. Wiesner, and P. G. Marshall, "The Therapeutic Application of Gonadotropic Hormones," *Lancet,* 3 September 1932, 508–11, at 510. On the use of gonadotropins in preventing habitual abortions, see ibid., 511. Low-dose irradiation of the pituitary was still used to stimulate secretion. See D. G. Drips and F. A. Ford, "Irradiation of the Ovaries and Hypophysis in Disturbances of Menstruation," *JAMA* 91 (1928): 1358. The technique was applauded as giving "phenomenal results," albeit only in a few cases. See Mazer and Goldstein, *Clinical Endocrinology of the Female,* 24.

357. "Sex Hormone Therapy," *Lancet,* 3 September 1932, 525.

358. C. F. Fluhmann, "The Interrelationship of the Anterior Hypophysis and the Ovaries," *American Journal of Obstetrics and Gynecology* 26 (1933): 764–75, at 772.

359. On testicular descent, see Samuel Cohn, "Anterior Pituitary-Like Principle in the Treatment of Maldescent of the Testicle," *JAMA* 103 (1934): 103–5; and S. B. D. Aberle and Ralph H. Jenkins, "Undescended Testes in Man and Rhesus Monkeys treated with the Anterior Pituitary-Like Principle from the Urine of Pregnancy," *JAMA* 103 (1934): 314–18. "In view of the frequent failure of surgery to correct this condition," commented Emil Novak, "the endocrine treatment...would seem to be indicated in cases of undescended testis before operation is resorted to" (Novak, "Anterior Pituitary and Anterior Pituitary-Like Substances," 1001). On use of pituitary gonadotropin-like substance in certain cases of baldness, see Bengt Norman Bengtson, "Pituitary Therapy of Alopecia," *JAMA* 97 (1931): 1355–58. On psoriasis and menopausal eczema, see Bernhard Zondek, *Hormone des Ovariums und des Hypophysenvorderlappens,* 2d ed., 533.

360. Novak, "Anterior Pituitary and Anterior Pituitary-Like Substances," 999. By the early 1930s, the debate over the origin of Zondek's prolan was intense, and, at least

in America, there was strong consensus that the gonadotropic hormone in the urine of pregnant women originated from the placenta, not the pituitary gland. See E. T. Engle, "Effects of Extracts of the Anterior Pituitary and Similar Active Principles of Blood and Urine," in Edgar Allen, ed., *Sex and Internal Secretions* (1932), 765–804, at 792–93. For Zondek's response to the controversies about the differences between pituitary gonadotropins and urinary prolan, see B. Zondek, *Hormone des Ovariums und des Hypophysenvorderlappens*, 254–64.

361. M. Reiss, "The Role of the Sex Hormones in Psychiatry," *Journal of Mental Science* 86 (1940): 767–89, at 784.

362. William R. Brosius and Robert L. Schaffer, "Spermatogenesis following Therapy with the Gonad Stimulating Extract from the Urine of Pregnancy," *JAMA* 101 (1933): 1227.

363. Novak, "Anterior Pituitary and Anterior Pituitary-Like Substances," 1001.

364. Mazer and Goldstein, *Clinical Endocrinology of the Female*, 333. This, of course, was also why gonadotropins were expected to be of help in preventing habitual abortions.

365. Naomi Pfeffer, "Pioneers of Infertility Treatment," in Lawrence Conrad and Anne Hardy, eds., *Women and Modern Medicine* (Amsterdam: Rodopi, 2001), 245–61, at 250.

366. Human Chorionic Gonadotropin (HCG) obtained from the urine of pregnant women was available, but it was prohibitive in price and, we now know, was almost entirely luteinizing (i.e., its effects were identical to Zondek's Prolan B). In the absence of the other pituitary gonadotropin (Zondek's Prolan A, now called the Follicle Stimulating Hormone), HCG was not very effective. Although it would take us far out of our orbit to discuss the later history of gonadotropic treatment and the new kind of fertility treatment emerging in the 1950s, it is important to record that advances in gonadotropin research were most marked in Sweden during the mid-twentieth century because of that country's relaxed attitude toward the harvesting of pituitary glands from corpses. See the excellent discussion in Pfeffer, "Pioneers of Infertility Treatment"; and Carl A. Gemzell, Egon Diczfalusy, and Gunnar Tillinger, "Clinical Effect of Human Pituitary Follicle-Stimulating Hormone (FSH)," 1333, 1345.

Epilogue

1. Ernst Scharrer and Berta Scharrer, *Neuroendocrinology* (New York: Columbia University Press, 1963), viii.

2. Ibid., 2.

3. Walter Langdon-Brown, *The Integration of the Endocrine System* (Cambridge: Cambridge University Press, 1935), 40–41.

4. A. S. Parkes, "The Rise of Reproductive Endocrinology, 1926–1940," *Journal of Endocrinology* 34 (1966): ix–xxxii, at xxvii.

5. The term "hypothalamus" and "hypothalamic nuclei" are used rather loosely in this chapter. The nomenclature and classification of the different hypothalamic nuclei were complex, and the term hypothalamus had been incorporated into the *Basler Nomina Anatomica* only in 1895. See Rudolf Pappenberger, "Abhängigkeit der gonadalen Funktion vom zentralen Nervensystem: Klinische Beobachtungen und

Tierexperimente zwischen 1850 und 1912," Inaugural dissertation, Friedrich-Alexander-Universität, Erlangen-Nürnberg, 1985, 156–57.

6. Herbert M. Evans, "The Function of the Anterior Hypophysis," *Harvey Lectures* 19 (1923–24): 212–35, at 212.

7. Louis Berman, *The Glands Regulating Personality: A Study of the Glands of Internal Secretion in relation to the Types of Human Nature* (New York: Macmillan, 1922), 193.

8. Parkes, "The Rise of Reproductive Endocrinology, 1926–1940," xxix.

9. For an excellent survey of these, see Pappenberger, "Abhängigkeit der gonadalen Funktion vom zentralen Nervensystem."

10. Scharrer and Scharrer, *Neuroendocrinology,* 5.

11. "The hope that endocrine diseases might be heralded by discernible mental changes or vice versa, that developing neuro- or psychopathic states might provide early warnings of impending psychological crises in the form of measurable changes in hormone output, has not been fulfilled." Scharrer and Scharrer, *Neuroendocrinology,* 5. They emphasized, however, that "subclinical symptoms such as premenstrual tension, irritability, melancholia, etc. often respond to very small doses of hormones. Presumably the normal hormone ratio has been disturbed and can be restored" (ibid.).

12. Pappenberger, "Abhängigkeit der gonadalen Funktion vom zentralen Nervensystem," 33–39.

13. Alfred Götzl and Jakob Erdheim, "Zur Kasuistik der trophischen Störungen bei Hirntumoren," *Zeitschrift für Heilkunde* 26 (1905): 372–401, esp. 399; and Pappenberger, "Abhängigkeit der gonadalen Funktion vom zentralen Nervensystem," 70.

14. Bernhard Aschner, "Zur Physiologie des Zwischenhirns," *WkW* 25 (1912): 1042–43; and Pappenberger, "Abhängigkeit der gonadalen Funktion vom zentralen Nervensystem," 145.

15. Quoted by Langdon-Brown, *The Integration of the Endocrine System,* 16.

16. "The anatomists of the past, looking at the brain enclosed in bone, and joined by a narrow stalk to this small pituitary body also thus enclosed, like a brain in miniature, were struck with the idea of a little shrunken brain, which, as it were, responded to, or repeated, the actions of the big brain above. Modern research has shown that there is more in this idea than was supposed in the nineteenth century." Langdon-Brown, *The Integration of the Endocrine System,* 15.

17. Samson Wright, *Applied Physiology,* 6th ed. (London: Oxford University Press, 1936), 206–7.

18. Langdon-Brown, *The Integration of the Endocrine System,* 40–41.

19. H. Cushing, "Disorders of the Pituitary Gland: Retrospective and Prophetic," *JAMA* 76 (18 June 1921): 1721–26, at 1724.

20. The identities of these substances was clarified only in the 1950s, and most endocrinologists of the twenty-first century would probably agree that they still have much to learn about the details. On the history of the nomenclature of the hypothalamic secretions, see G. W. Harris, "Humours and Hormones," *Journal of Endocrinology* 53 (1972): i–xxiii, at xi. On continuing uncertainties, see V. C. Medvei, *History of Clinical Endocrinology* (Carnforth, Lancs.: Parthenon, 1993), 356–59.

21. On Harris, see Marthe Vogt, "Geoffrey Wingfield Harris 1913–1971," *Biographical Memoirs of Fellows of the Royal Society* 18 (1972): 309–29.

22. "During the first twenty years of my life," recalled Harris at its end, "I lived about half a mile from the building in which Sir Henry Dale discovered the oxytocic activity of posterior pituitary extracts." G. W. Harris, "Humours and Hormones," ii. Harris's MD thesis of 1944 was also on the posterior pituitary. See Marthe Vogt, "Geoffrey Wingfield Harris 1913–1971," 311.

23. G. W. Harris, "Humours and Hormones," iii.

24. Vogt, "Geoffrey Wingfield Harris 1913–1971," 311–12.

25. F. H. A. Marshall and E. B. Verney, "The Occurrence of Ovulation and Pseudopregnancy in the Rabbit as a Result of Central Nervous Stimulation," *Journal of Physiology* 86 (1936): 327–36; G. W. Harris, "The Induction of Pseudopregnancy in the Rat by Electrical Stimulation through the Head," *Journal of Physiology* 88 (1936): 361–67; and G. W. Harris, "The Induction of Ovulation in the Rabbit by Electrical Stimulation of the Hypothalamico-Hypophysial Mechanism," *Proceedings of the Royal Society of London (B)* 122 (1937): 374–94, at 375.

26. G. W. Harris, "Humours and Hormones," v–vi.

27. Ibid., vii–ix.

28. The electron microscope could and did—but that was only in the 1960s. See ibid., x.

29. Ibid.

30. Douglas Hubble, "The Endocrine Orchestra," *BMJ*, 25 February 1961, 523–28, at 524–25. Similar questions were applicable to the secretion of the other tropic hormones by the pituitary.

31. G. W. Harris, "Humours and Hormones," x.

32. R. Guillemin and B. Rosenberg, "Humoral Hypothalamic Control of Anterior Pituitary: A Study with Combined Tissue Cultures," *Endocrinology* 57 (1955): 599–607; and M. Saffran, A. V. Schally, and B. G. Benfey, "Stimulation of the Release of Corticotrophin from the Adenohypophysis by a Neurohypophysial Factor," *Endocrinology* 57 (1955): 439–44.

33. Medvei, *History of Clinical Endocrinology*, 351.

34. Steven W. J. Lamberts, Annewieke W. van den Beld and Aart-Jan van der Lely, "The Endocrinology of Aging," *Science* 278 (1997): 419–24, at 422–23.

35. Medvei, *History of Clinical Endocrinology*, 356–57.

Index

Achard-Thiers syndrome, defined, 291n163
acromegaly, 278n50
Acton, William, 310n181
Adair, Frank, 198, 323n330
adiposogenital syndrome, 126, 208–9; Fröhlich on, 279n54
adrenal gland: and ACTH, 144; adrenaline secretion, 141; ambisexuality of, 143; Baur on, 143–44; Brown-Séquard on, 36; cortex of, 141; and courage, 72–73; embryology of, 144; estrogenic secretion, 144; extracts of, 69, 144; and feminization, 143; and gonadotropic hormones, 144; "infant Hercules" syndrome, 142–43; and homosexuality, 257n86; Lipschütz on, 291n166; as "male gland," 141–42, 143; and masculinization, 118, 141–45, 192, 291n157, 291n159, 292n180; and neurasthenia, 272n264; progesterone secretion, 144; psychological symptoms of hyperactivity, 143; as "third gonad," 139, 144
Adrenocorticotropic Hormone (ACTH), 144, 210
adrenogenital syndrome, 145, 291n163, 292n180
African-Americans, and homosexuality, 318n269

aging: and brain, 213; as deficiency disorder, 101, 105–6, 109, 269n216; endocrine analogies of, 257n93; Mechnikov on, 101; theories and debates on, 100–105; in rats, 258n97, 258n99; and sexual desire, 177; and testicular histology, 258n97; and thyroid, 257n93; van Swieten on, 101
Albright, Fuller, 168
Allbutt, Thomas Clifford, 50
Allen, Edgar, 156; "An Ovarian Hormone" (1923), 297n19. *See also* Allen-Doisy test; Doisy, Edward A.
Allen-Doisy test, 122, 129, 157; defined, 138, 156; non-specificity of, 288n127
ambisexuality, 4; adrenal, 143; hormonal, 5
amenorrhea: causes of, 15, 126, 160, 162, 208, 298n34; hormone therapy of, 159–63, 201–2, 298n37, 299nn41–42, 325n352; primary and secondary, 160
Amann, Joseph, 221n39
Ancel, Paul: experiments on testicles, 56. *See also* Bouin, Pol André
androgens. *See* sex hormones: male
andrology, 29, 225n93
andropause. *See* climacteric, male
angina pectoris, treatment with male sex hormones, 175

food, as source of female sex hormones in males, 139, 289n135
Foges, Arthur, 35
Follicle Stimulating Hormone (FSH). *See* gonadotropic hormones
Foucault, Michel: *The History of Sexuality* (1978), 249n40; on homosexuality, 75–76
Fraenkel, Ludwig: experiments by, 46–49, 233n71; on "functional" disorders, 19; on female sex hormone, 300n60; "Die Funktion des Corpus luteum" (1903) 232n65; Halban on, 232n68; and Hegar, 232n64; Lutein, 47–48, 167, 233n72; on ovary as "luxury organ," 300n58; Schauta on, 232n68
Frank, Robert T.: on Allen-Doisy test, 157; on amenorrhea, 299n45; on estrogens, 174; *The Female Sex Hormone* (1929), 217n22; on homosexuality, 187; on hormone therapy and costs, 161, 300n58; on menopause, 304n97
Frankl-Hochwart, Lothar von: on pineal tumors, 146–47
freemartin, 119–20, 275n9
Freud, Sigmund: criticism of, 72; on Steinach and "Steinach Operation," 80–81, 87, 255n76, 260n123
frigidity, treatment with male sex hormones, 194
Fröhlich, Alfred: on adiposogenital syndrome, 208, 279n54
"functional" diseases, defined, 220n19
Funk, Casimir: and androgens, 175, 308n162; "The Male Hormone" (with B. Harrow, 1928–29), 308n164
Funk, Thilo: "Uterine Fibromyome und Blutungen" (1986), 219n17

Geller, Friedrich, 163; "Die Hormontherapie in der Gynäkologie" (1934), 298n36
gender, 2; Berman on, 113; and biology, 114; chemical theory of, 113; deconstruction and stabilization of, 9, 64–65, 67, 110, 113–15, 151, 274n280
gigantism, 125–26; in rats, 278n50

Gillies, Sir Harold: sexual reconstructive surgery, 192; treatment of Dillon, 320n290
Gilman, Sander L.: *Making the Body Beautiful* (1999), 265n168
Glaevecke, Ludwig: critique of Hegar, 222n50
glands, endocrine, 1, 6, 117–18, 143, 157, 232n59, 245n178; defined, 215n3
glands, exocrine: defined, 215n3
Glass, James: and ovarian transplantation, 230n41
Glynn, Ernest: "The Adrenal Cortex" (1912), 291n162; on masculinization, 142
Goldschmidt, Richard: "Die biologischen Grundlagen der konträren Sexualität" (1916–18), 245n179; on determination of sex, 64–65; on homosexuality, 78–79, 253n64; on Steinach, 65
Goldzieher, Max: *The Adrenal Glands* (1944), 291n160; on libido and male sex hormones, 194; on male climacteric, 185; on endocrine therapy, 70, 307n143
Goltz, Friedrich: career of, 227n13; experiments by, 35–36; Weininger on, 228n15
gonadocentrism, 119, 260n118
gonadotropic hormones: in abortion, 327n364; in amenorrhea, 201–2, 299n45; in baldness, 202; in cancer, 283n88, 284n92; in male climacteric, 184; clinical uses, 200–204, 325nn351–53; dangers of, 202; diagnostic uses, 202; in dysmenorrhea, 326n356; Follicle Stimulating Hormone (FSH), 201, 211–12, 302n78; 327n366; and gonadal hormones, 129, 132–34, 160, 183–84, 188, 193–94, 196, 201, 283n85, 286n109, 287n115, 314n237; and Growth Hormone, 282n81; in homosexuality, 188–90; Human Chorionic Gonadotropin (HCG), 327n366; hypothalamic control of, 210–12; in infertility, 203;

Marshall, Francis Hugh Adam: *Physiology of Reproduction* (2d ed., 1922), 244n170; on nervous system, 210; on Steinach, 64, 157

masculinity: and biology, 29; and female sex hormones, 137–38; and sexual orientation, 118, 190; quest for, 114; restitution of, 113

Mateer, Florence: *Glands and Efficient Behavior* (1935), 247n6

Mayreder, Rosa, 114

McCrea, Frances: "The Politics of Menopause" (1983), 303n93

McGee, Lemuel: develops testicular extract, 158, 174–75. *See also* Koch, Fred

mechanics, developmental: defined, 238n132. *See also* Roux, Wilhelm

Mechnikov, Ilya Ilyich: on aging, 101; *The Prolongation of Life* (1910), 268n211

Medvei, Victor Cornelius, 8–9; *History of Clinical Endocrinology* (1993), 216n3; on hypothalamus, 213; on pineal, 293n198

melancholia, involutional: conceptual uncertainties, 307n144; electroconvulsive treatment in, 172; and heredity, 171–72, 307n140; in men, 182–83, 313n222; Molony on, 313n222; Skae on, 170, 313n222; suicide in, 313n222; symptoms and pathophysiology of, 170–72; therapy of, 171–73, 306n131, 306n138; Werner on, 171, 306n131, 314n231

Mendel, Kurt: on male climacteric, 178–83, 311n193, 311nn195–96, 311n199; on "Steinach Operation," 180–81, 312n213

Mengert, William: on sex chromosomes, 66. *See also* Witschi, Emil

menopause: Aschheim-Zondek reaction in, 131; Bishop on, 304n101; conceptual and cultural issues, 166, 170; depression in, 170–73, 307n143; diathermy in, 264n159; Hegar on,

222n45; hormonal fluctuations in, 131, 304n96; endocrine therapy of, 166–70, 295n4, 301n61, 303n94, 304n97, 307n143, 312n200; and male climacteric, 310n185, 311n199, 312n200, 314n223; male sex hormones in, 193, 307n143; medicalization of, 303n94; as natural, 167, 170, 303n94; neurohumoral factors in, 212–13; after oophorectomy, 230n43; premature, 230n43; psychological features, 167, 168, 178; sedative treatment of, 304n97; as social threat, 169; stages of, 131, 166–67; suggestibility in, 304n97; symptoms of, 163, 167, 178, 193; x-ray treatment of, 296n13, 303n94. *See also* hormone replacement therapy, long-term (HRT)

menopause, male. *See* climacteric, male

menorrhagia: defined, 325n353; and gonadotropic hormones, 325n353

menstruation: theories of, 13–14, 16, 220n20; hormonal basis of, 156, 159–60; analogous bleeding in male homosexuals, 252n56; re-establishment of, 155

Meyer, Robert: on masculinization, 140

Meyer-Bahlburg, Heino F. L.: "Sex Hormones and Female Homosexuality" (1979), 320n286; "Sex Hormones and Male Homosexuality" (1977), 319n282

Mikulicz-Radecki, Friedrich von: on endocrine approaches in gynecology, 161, 300n49; "Die praktische Bedeutung der Hormonbehandlung" (1934), 294n212

Moll, Albert, 252n55

mongolism, 246n4

"monkey gland" grafts, 85, 95; cost of, 267n187; histology of, 266n180; indications and outcome, 98–99; in male climacteric, 181; procedure of, 267n184; and Steinach, 95, 97–98, 266n174; Thorek on, 97, 99; Walker on, 181; Williams on, 104; S. Wright on, 103. *See also* Voronoff, Serge

symptoms, 22, 223n56, 224n72; in men, 223n55; oophorectomy in, 22–24, 48, 224n68; ovarian abnormalities in, 223n63

osteoporosis, senile, 169; hormonal therapy of, 168, 174, 193

Oudshoorn, Nelly, 5, 8–9, 12; *Beyond the Natural Body*, 217n18; on biochemistry, 137, 204; on "crossed-sex" hormones, 288n132, 290n143; on gender, 151; "On the Making of Sex Hormones" (1990), 216n11; "On Measuring Sex Hormones" (1990), 216n11; on menopause, 303n94

ovariotomy, 219nn16–17, 220n18; "normal," 219n16. *See also* oophorectomy

ovary: and cancer, 26; corpus luteum, 46–49, 164–65; extracts and hormones, 40–41, 154–56, 165, 173, 229nn35–36, 282n80, 300n58, 312n200; and "female" disorders, 15, 19; and femininity, 14; follicular hormone, 10, 173–74, 282n84, 288n127; histology of, 45, 48, 262n141, 282n84; irradiation of, 80, 91, 92, 140–41, 169, 241n158, 255n75, 262n140, 263n144, 296n13; as "luxury organ," 300n58; and menstruation, 220n20; as neural node, 3, 18–19, 21; and pituitary, 126, 296n10; regeneration of, 89, 123, 128, 134, 265n164; removal of, 15–16; secretion of male hormone by, 4, 10, 118, 139–41, 290n155; and sexual development, 42–43, 59; Steinach's experiments, 60–63, 89, 140–41; analogy with thyroid, 40, 229n35; transplantation of, 12, 39, 41–45, 60–63, 89, 121, 165, 230n41; 231n56, 231n58, 241n151, 262n135, 285n96, 290n153; tumors of, 140. *See also* oophorectomy

ovotestis, 63

ovulation, 46; and corpus luteum, 164–65; and prolan, 130; Virchow's theory, 13

oxytocin, 210

Papanicolaou, G. N.: "The Existence of a Typical Oestrous Cycle in the Guinea Pig" (with C. R. Stockard, 1917), 297n20

Pappenberger, Rudolf, "Abhängigkeit der gonadalen Funktion vom zentralen Nervensystem" (1985), 327n5

Parkes, Alan: on adrenal, 143; "Ambisexual Activity of the Gonads" (1938), 290n155; on gonadotropin therapy, 200; on identification of hormones, 158–59; on ovary, 141, 290n155; on nervous and endocrine functions, 206–7; "The Rise of Reproductive Endocrinology" (1966), 297n29; on testosterone, 175

Paterson, Edith: on hormone therapy of breast cancer, 196, 198, 323n324, 324n336

Pearl, Raymond, 270n222

periodicity, 311n188; in men, 178; and nerves, 13

personalities, endocrine. *See* temperaments, endocrine

Pfeffer, Naomi: on gonadotropic hormones, 203; on infertility, 163, 301n65, 301n71

pharmaceutical industry, 7–8, 154, 165, 169–70, 175, 281n69

phylogeny, human: and sexual development, 77, 245n176

pill, contraceptive, 164, 166; in Hungary, 303n91

pineal: as "anti-puberty gland," 146–48; experiments on, 148; Frankl-Hochwart on, 146–47; and pituitary, 147; and sexual development, 148, 293n198

pituitary: and adiposogenital syndrome, 126; anatomy of, 124; Aschner on, 279nn52–53; Berman on, 125, 280n60; and courage, 71–72, 73, 125; Cushing on, 124–26, 279n55; deficiency of, 176–77; defined, 277n35; diathermy of, 92, 155, 169, 264n159; extracts of, 155, 164; governance of, 209; and growth, 278n50, 279n52; Growth Hormone,

transsexuality, 192
Turing, Alan, 320n284

Ulrichs, Karl Heinrich: on male
homosexuality, 77, 251n53, 252n57
United States of America: attitudes toward
homosexuality in, 318n269; concern
with energy in, 272n264; ideals of
youthfulness in, 109; interest in glands
in, 70, 97, 109
urine, use in gynecological therapy,
298n38
uterus: bleeding of, 193, 201, 220n28,
325n353; cancer of, 195; and corpus
luteum, 46–47, 202, 232n68, 282n84;
and dysmenorrhea, 326n356;
endometriosis, 194; and femininity, 14,
218n11; metropathia hemorrhagica,
201

Vaerting, Mathias: on male climacteric,
311n198
van Swieten, Gerard: on aging, 101
vas deferens: ligation of, 56, 60, 83, 88,
257n95, 258n100; vasectomy,
257nn95–96, 258n100, 258n102,
258n105. *See also* "Steinach
Operation"
vasoligature. *See* vas deferens: ligation of
vasopressin, 210
Velle, Weiert: "Urinary Oestrogens in the
Male" (1966), 287n123; on B. Zondek,
288n133
Vienna, debates on gender and civilization
in, 114
Viereck, George Sylvester: on rejuvenation
in America, 109; on eugenics and
evolution, 108, 272n258; on Freud,
87; *Glimpses of the Great* (1930),
260n122; political beliefs of, 272n258;
on Progynon, 174; *Rejuvenation: How
Steinach Makes People Young* (as
George F. Corners, 1923), 265n165;
on Steinach, 81, 100, 107; on
rejuvenation of women, 93,
265n168
Vincent, Swale: on organotherapy, 73–75,
248n28; *Internal Secretion and the*

Ductless Glands (1912), 294n211; on
thymus, 150
Virchow, Rudolf: on femininity, 14; Hegar
on, 20; on humoralism, 218n7; on
ovarian functions, 12–15; on pathology,
251n49; "Der puerperale Zustand"
(1856), 218n4; on Siamese twins, 13
Vivarium. *See* Institute for Experimental
Biology (Vienna)
von Baer, Karl Ernst, 14
Voronoff, Serge: career, 95, 99, 266n177;
and eugenics, 272n256; on eunuchs,
96; fees of, 267n187; *Life: A Study of the
Means of Restoring Vital Energy* (1920),
266n179; and Lydston, 267n189;
"monkey-gland" operations, 96, 98,
267n184; murder allegation against,
266n177; on rejuvenation of women,
97; and Steinach, 97–98, 267n195;
Thorek on, 97, 99; S. Wright on, 103

Wagner, Rudolf, 34
Waldeyer, Wilhelm, 244n176
Walker, Kenneth: on adrenal, 118; on
castration, 222n45; *The Circle of Life*
(1942), 268n202; on male climacteric,
181, 313n216; *Male Disorders of Sex*
(1930), 268n202; on Napoleon,
278n45; on pituitary, 125; on
rejuvenation, 268n202; on sexual
difference, 118–19; on endocrine
temperaments, 278n44; on testicle
transplants, 99, 267n184; on
vasectomy, 259n105
Warthin, Alfred Scott: on involution,
269n216; on rejuvenation as
re-erotization, 103
Watkins, Elizabeth Siegel, 164, 169;
"Dispensing with Aging" (2001),
305n116
Watson, John B.: on the "Steinach
Operation," 105. *See also* Lashley, Karl
Weindling, Paul: *Health, Race and German
Politics* (1989), 271n255; on
regenerationist biology, 108
Weininger, Otto: on effeminacy, 109, 145;
on Goltz, 228n15; on homosexuality,
253n59